Science and Empire in the Atlantic World

The history of science beyond Europe has traditionally been understood through heroic narratives of discovery and exploration. But in the early modern Atlantic world, it was commercial travel, back and forth across the ocean to the "New World," that made new knowledge of all kinds.

Science and Empire in the Atlantic World is the first book to examine the making of scientific knowledge in the early modern Americas from a comparative and international perspective. Twelve essays from leading scholars range from the science of navigation in Seville to the creation of medical knowledge in Brazil, from experiments with electricity in British America to the practice of Mesmerism in Haiti. Connecting Atlantic history with the history of science, the chapters explore how knowledge and the colonial order were made together, through complex interactions between metropolitan travelers, Creole settlers, Amerindians, and African slaves.

Re-orienting our view of knowledge's movement along the networks between center and periphery, *Science and Empire in the Atlantic World* shows just how challenging it was to make knowledge – and impose control – at a distance.

James Delbourgo is Assistant Professor of History and Chair of History and Philosophy of Science at McGill University. He is the author of *A Most Amazing Scene of Wonders: Electricity and Enlightenment in Early America*.

Nicholas Dew is Assistant Professor of History at McGill University, where he teaches early modern European history and history of science. He is the author of *Orientalism in Louis XIV's France*.

Science and Empire in the Atlantic World

EDITED BY
JAMES DELBOURGO AND
NICHOLAS DEW

Routledge
Taylor & Francis Group
NEW YORK AND LONDON

First published 2008
by Routledge
270 Madison Avenue
New York, NY 10016

Simultaneously published in the UK
by Routledge
2 Park Square, Milton Park
Abingdon, Oxon OX14 4RN

Routledge is an imprint of the Taylor & Francis Group, an informa business

© 2008 Taylor & Francis

Typeset in Minion by
HWA Text and Data Management, Tunbridge Wells
Printed and bound in the United States of America on acid-free paper by
Edwards Brothers, Inc, Lillington, NC.

All rights reserved. No part of this book may be reprinted or reproduced or utilised in any form or by any electronic, mechanical, or other means, now known or hereafter invented, including photocopying and recording, or in any information storage or retrieval system, without permission in writing from the publishers.

Trademark Notice: Product or corporate names may be trademarks or registered trademarks, and are used only for identification and explanation without intent to infringe.

Library of Congress Cataloging in Publication Data
A catalog record has been requested for this book

ISBN10: 0–415–96126–2 (hbk)
ISBN10: 0–415–96127–0 (pbk)
ISBN10: 0–203–93384–2 (ebk)

ISBN13: 978–0–415–96126–4 (hbk)
ISBN13: 978–0–415–96127–1 (pbk)
ISBN13: 978–0–203–93384–8 (ebk)

All sign of life had disappeared from the sea. In front of the ship, black ripplings of dolphins, more solid and rhythmical than the breaking of the foam against the prow, had gracefully preceded the white tips of the retreating waves; now no jet from a blower-dolphin cut across the horizon; nor was the sea any longer intensely blue and peopled by fleets of nautili, with their delicate mauve-and-pink membranes outstretched as sails. And, when we got to the far side of the oceanic depths, would all the marvels seen by the old navigators still be there to greet us?

Claude Lévi-Strauss, *Tristes Tropiques*

Contents

List of illustrations — xi
Acknowledgments — xiii

Introduction: The Far Side of the Ocean — 1
    JAMES DELBOURGO AND NICHOLAS DEW

Section I
Networks of Circulation — 29

Chapter 1   Controlling Knowledge: Navigation, Cartography, and Secrecy in the Early Modern Spanish Atlantic — 31
    ALISON SANDMAN

Chapter 2   *Vers la ligne*: Circulating Measurements Around the French Atlantic — 53
    NICHOLAS DEW

Chapter 3   Knowing the Ocean: Benjamin Franklin and the Circulation of Atlantic Knowledge — 73
    JOYCE E. CHAPLIN

Section II
Writing the American Book of Nature — 97

Chapter 4 A New World of Secrets: Occult Philosophy and
Local Knowledge in the Sixteenth-Century Atlantic — 99
RALPH BAUER

Chapter 5 Tropical Empiricism: Making Medical Knowledge in
Colonial Brazil — 127
JÚNIA FERREIRA FURTADO

Chapter 6 American Climate and the Civilization of Nature — 153
JAN GOLINSKI

Section III
Itineraries of Collection — 175

Chapter 7 Empiricism in the Spanish Atlantic World — 177
ANTONIO BARRERA-OSORIO

Chapter 8 Fruitless Botany: Joseph de Jussieu's South
American Odyssey — 203
NEIL SAFIER

Chapter 9 Atlantic Competitions: Botany in the
Eighteenth-Century Spanish Empire — 225
DANIELA BLEICHMAR

Section IV
Contested Powers — 253

Chapter 10 The Electric Machine in the American Garden — 255
JAMES DELBOURGO

Chapter 11 Diasporic African Sources of Enlightenment
Knowledge — 281
SUSAN SCOTT PARRISH

Chapter 12	Mesmerism in Saint Domingue: Occult knowledge and Vodou on the Eve of the Haitian Revolution FRANÇOIS REGOURD	311
Afterword:	Science, Global Capitalism, and the State MARGARET C. JACOB	333
Contributors		345
Index		349

Illustrations

Figures

	The Pillars of Hercules	3
3.1	Samuel Sturmy, *The Mariners Magazine*	76
3.2	William Dampier, "A View of the General Coasting & Trade Winds in the Atlantick & Pacific Oceans"	80
3.3	Benjamin Franklin and Timothy Folger, chart of the Gulf Stream	88
3.4	Benjamin Franklin "A Chart of the Gulf Stream"	91
5.1	*The Natural History of Brazil*	134
5.2	Ex-voto de Bom Jesus de Matosinhos	139
5.3	The Dutch invasion of Brazil	141
5.4	Luís Gomes Ferreira	143
7.1	Martín Cortés, *Breve Compendio de la Sphera*	185
7.2	Juan López de Velasco, *Descripción y división de las Yndias*	187
10.1	The Electric Machine	256
10.2	Enlightenment on board ship	264
10.3	Political Automatism	268
10.4	Gymnotus in the Garden	273
11.1	The nativity of New Science	284
11.2	The celebrated Graman Quacy	292

Table

9.1	Botanical expeditions conducted in the Spanish Empire during the second half of the eighteenth century	227

Acknowledgments

This book has its origins in a series of conversations that began in the summer of 2002 and crystallized in a session at the annual meeting of the History of Science Society held in Cambridge, Massachusetts, in 2003. In February 2005, with the generous support of Margaret C. Jacob, we organized "Atlantic Knowledges: The Sciences and the Early Modern Atlantic World," a two-day workshop sponsored by the UCLA Center for Seventeenth and Eighteenth Century Studies and held at the William Andrews Clark Memorial Library in Los Angeles. We gratefully acknowledge the central role Peg has played in helping to move this project forward from the beginning. Thanks also go to Peter Reill, Director of the Center, and his staff, especially Anna Huang and Candis Snoddy, for their professionalism and courtesy in handling the organization of our meeting. It has been a great pleasure to work with all of our contributors: from the very beginning of the project they have been immensely supportive, and we thank them wholeheartedly for their hard work, their advice, their promptness, and their patience. We also gratefully acknowledge the contributions of the other participants at UCLA, especially Deborah Harkness, María-Elena Martínez, Anthony Pagden, Ben Schmidt, Pamela Smith, Sanjay Subrahmanyam, and Mary Terrall. For advice and assistance along the way, we thank John Hall, David Armitage, Simon Schaffer, Daniela Bleichmar, Pete Beatty, Brendan O'Neill, Chris Lyons, Laura Kopp, and Neil Safier. Jorge Cañizares-Esguerra participated in the meeting at UCLA and provided timely support for the project as it moved forward at Routledge, as did a second anonymous reviewer. For advice on the introduction, we thank Elizabeth Elbourne, Daviken Studnicki-Gizbert, and Kristen Keerma, who also helped to proofread the entire manuscript. Richard Drayton provided an incisive and stimulating critical reading of the individual essays and the introduction, as well as commentary on the project as a whole. Finally, we extend our warm thanks to Kimberly Guinta for her vision and energy in helping us to bring this project to fruition at Routledge.

Introduction

The Far Side of the Ocean

JAMES DELBOURGO AND NICHOLAS DEW

The entire body of the sciences can be thought of as being like the ocean, which is everywhere continuous and without division, even though men have conceived parts for it, and give them names according to their convenience. And just as there are seas that are unknown, or which have only been navigated by a few vessels cast there by fortune, one can say that there are sciences of which we only have knowledge through chance encounters and without design.

—Leibniz, *De l'horizon de la doctrine humaine* (1693)

The vast ocean cannot be possessed.

—Cornelius van Bynkershoek, *De dominio maris* (1702)

What the *New Atlantis* Conceals

How did the sciences shape the Atlantic world, and how did the Atlantic shape the sciences? Francis Bacon's *New Atlantis* (1627) suggests how at least one influential commentator envisioned this relationship as unfolding. In Bacon's fable, a crew of sailors loses their way after setting out from Peru in search of a passage to China and Japan via the South Seas and comes across the mysterious, hitherto uncharted island society of Bensalem. Initially wary of their situation, the sailors are reassured by the sight of a Christian cross, "a presage of good," and by the kindness of the islands' inhabitants, who turn out to be fluent in Spanish, Latin, Greek, and Hebrew. To their surprise, the sailors learn that the islanders possess an advanced technological command over nature. Strikingly, Bensalem is a society that is knowing, yet unknown: it dispatches "Merchants of Light" to gather information for study by "Interpreters of Nature" while zealously concealing its existence from the outside world. The key to this epistemological paradox is Salomon's House—a potent yet secretive institution for "enlarging the bounds of Human Empire," for "finding out the true nature of all things, whereby God might have the more glory in the workmanship of them, and men the more fruit in the use of them." The range of technologies its members claim to possess is astonishing: "Chambers of Health" for curing diseases; medicine shops with a far greater variety of drugs than in Europe; "perspective-houses" and "sound-houses" for experimental demonstrations that divert the senses; and "engine-houses" containing machines that imitate the "motions of living creatures, by images of men, beasts, birds, fishes, and serpents." There is even a gallery to commemorate "principal inventors." The first mentioned? "Columbus, that discovered the West Indies."[1]

The moral of Bacon's parable was that venturing beyond the Pillars of Hercules, the limits of the known world—famously represented in the frontispiece to his *Instauratio Magna* (1620) (shown opposite)—promised untold power through dominion over nature.

Several key components of early modern colonial knowledge are in play here: the cultivation of exotic foodstuffs and drugs; the exploration of strange natural landscapes; the enabling function of Christianity; the utility of commercial travel (despite its contingencies: Bacon's sailors *lost* their way); and the organization of useful knowledge from distant peripheries in metropolitan institutions. Rather than simply follow Bacon's well-known text, however, we can read the *New Atlantis* for what it *conceals*. Three aspects of the relationship between knowledge and colonialism in Bensalem are particularly important for our purposes. First, Bensalem is a political ambiguity rather than a transparent utopia. Thanks to its wizardry, Salomon's House affords the island peace and prosperity, but the true political objectives of its members, who represent the state and true Christian faith, are shrouded in secrecy. Second, the narrators appear to be *Spanish* sailors, not English—a clue to the geopolitical contest in the early modern world that Bacon's account otherwise glosses over. This prophecy

The Pillars of Hercules. This icon of the limits of the known world had long been employed by the Spanish under the Habsburg emperor, Charles V, before Bacon borrowed it to emblazon his call for the refoundation of natural knowledge in England. Title page to Francis Bacon, *Great Instauration* (London, 1620). Reproduction courtesy of Osler Library of the History of Medicine, McGill University, Montreal, Quebec, Canada.

of knowledge and power was an English statesman's (and Virginia Company investor's) lament over English weakness in the face of Iberian power, the result of more than a century of American colonization by rival states. Dreams of Atlantic knowledge and power were Iberian long before they were English: even Bacon's use of the Pillar of Hercules, with the legend "plus ultra" ("further still"), was borrowed from iconography long associated with the Habsburg emperor Charles V.[2] And third, Bensalem is not inhabited by indigenous peoples or African slaves. In reality, violence, slavery, disease, and evangelization forcefully reshaped the Atlantic world shared by Europeans, Amerindians, and Africans. However, in Bensalem no coercion is evident. Bacon's utopia is both technological and social: knowledge is power, produced magically without conflict.

What actually happened when Europeans crossed the Atlantic, seeking knowledge as power, and how did the pursuit of "science" and "empire" intersect? One classic answer has been to jump forward to the late eighteenth-century career of that seemingly paradigmatic Atlantic traveler, Alexander von Humboldt. Together with his companion, Aimé Bonpland, Humboldt obtained an official passport that made him one of the first foreign travelers in many years to move freely through Spanish America, during the period 1799–1804. The result was a voluminous documentation of many aspects of the natural and social environments he encountered and, as a result, a richly varied intellectual and political legacy. In this view, Humboldt was *the* prototypical modern scientific traveler. What interested him were measurements rather than wonders. He brought into the field an armory of instruments for precisely recording American atmospheric conditions, as part of a larger project in "global physics" to explain and predict biogeographical diversity. Yet Humboldt was also a Romantic natural philosopher whose writings, influenced equally by German *Naturphilosophie* and Creole-American views of the natural richness of the Western hemisphere, celebrated the aesthetic grandeur of American nature—celebrations that were to play a prominent role in the articulation of revolutionary nationalist ideologies in nineteenth-century Latin America.[3]

The "legendary" Humboldt was far from typical of early modern scientific travelers, however. Even his was an itinerary of contingency in that his departure for South America was partly determined by his failure to join Napoleon's expedition to Egypt. Atlantic voyages that yielded natural knowledge were, indeed, more often itineraries of contingency than precision. Where such Pacific explorers as Malaspina, Bougainville, and Cook enjoyed state support for official expeditions, Atlantic travelers—such figures as Sebastian Cabot, Joseph de Jussieu, or Edward

Bancroft—were creatures of the commercial networks that connected Europe, Africa, and the Americas. Finding themselves in unintended destinations, losing their way, and being assailed (sometimes fatally) by the natural and social dangers of the colonial world were just some of the delights that lay in store for them. Rarely if ever did such travelers possess the high standing and connections Humboldt enjoyed. Rather than using their travels to construct novel theories of nature as a whole, most Atlantic travelers acted on behalf of metropolitan patrons who were their social and institutional superiors. Moreover, if their itineraries were not always precise, neither were their methods for recording New World natures. Travelers before Humboldt were not always equipped with precision instruments; they often relied instead on verbal and visual descriptions to make knowledge.[4] Most important, however, the history of science in the Atlantic world cannot be understood simply as a history of scientific travel from center to periphery and back again, because many who made knowledge in this world never made any such journey. They were members of the sizeable settler communities that formed throughout the Americas, composed of Creolized, mestizo, and enslaved peoples, whose knowledge often drew heavily on local Amerindian and African labor and expertise. Large-scale settlement, demographic expansion, and Creolization made the Americas distinctive from early modern European incursions into Africa, South Asia, and the Pacific, where Europeans achieved little or no interior penetration, maintaining coastal military and commercial outposts at best. Making scientific knowledge in the Atlantic world was not simply a question of metropolitan travel, therefore, or even purely of commercial networks but, increasingly as time passed, one that involved the politics of empire in communications with settler populations.

The history of science beyond Europe has often been made synonymous with heroic narratives of discovery, so that the knowledge made by quotidian crossings and re-crossings of the early modern Atlantic has remained almost invisible next to celebrated expeditions, such as those of Cook, Bougainville or, indeed, Humboldt. Owing in part to the seductive power of the heroic narrative, we ironically know much more about the knowledge made on those few specific journeys than on the thousands of commercial voyages less glamorously made around the Atlantic. Turning away from such heroic narrative structures, *Science and Empire in the Atlantic World* assembles a different cast of characters to shed light on the making of natural knowledge in the colonial Americas: common mariners and navigators, magicians and artisans, errant astronomers and botanists, civil servants and physicians, Amerindian guides, and African slaves. The twelve essays in this book explore the period 1500–1800, ranging from Portuguese Brazil and New Spain to the West Indies and mainland North America, examining a diverse range of practices from cartography and botany to natural philosophy, medicine, collecting, and Mesmerism. Taken together, these studies work against traditional national narratives of center-periphery relationships, allowing us to grasp the commonalities and contrasts among European projects to turn the Americas into a source of knowledge and power and how Creole populations in the Americas responded to and engaged with such projects. Bringing to life a world of criss-crossing networks, heterogeneous practices, and multiple itineraries, they build a complex picture of the relationship between science and empire in the age before the modern nation-state and professionalized scientific disciplines, showing just how difficult it was to make knowledge—and impose control—at a distance.[5]

Early Modern Science and Empire: A Hitchhiker's Guide

The notion of early modern "science" requires careful scrutiny. In the *New Atlantis*, one of the galleries in Salomon's House contains statues of "all [the] principal inventors," in which Columbus is ranked alongside the inventors of ships, gunpowder, music, writing, and printing. The notion that the discovery of America was one of the great inventions of human (European) history has a venerable pedigree. The New World's discovery rapidly became part of a rhetorical armory with which Europe's "modernity" came to be asserted and debated in the long-running seventeenth- and eighteenth-century *querelle des anciens et des modernes*.[6] The whiggish histories of scientific and technological progress of the twentieth century, however, were usually chauvinistic declarations of technological determinism or naively trumpeted the rise of allegedly transparent empirical and experimental methods, overlooking the complex processes of mediation by which "modern" knowledge inevitably continued to be made. Recently, scholars have moved from decontextualized histories of scientific thought constructed on the texts of canonic figures such as Descartes and Newton toward a social history of knowledge-making as a form of local practical labor that required collectives of practitioners (usually invisible in older accounts), financial resources, social discipline, and the mobilization of public assent through cultures of print and performance. Long-standing reliance on nineteenth-century conceptions of what should count for science—Newton's mathematics, for example, but not his alchemy—have been replaced by a rigorous historicism that recognizes the conceptions of natural knowledge mobilized by historical actors themselves. Recent history of science can thus serve Atlantic history extremely well, by offering an acute understanding of how networks of practical expertise functioned across often remarkable distances.⁷

Until recently, the connections between early modern European science and Europe's increasing engagement with the rest of the world have been overlooked by historians. But as Steven Harris has pointed out, if we charted the journeys that the objects in a Baroque cabinet of curiosities had to make, from their point of origin to their resting place in the cabinet, the resulting map would be virtually indistinguishable from the trade routes of European commerce. Natural philosophy had always made universal claims, but by the eighteenth century, the practical basis for these claims was increasingly global, fostering new confidence in the universal validity of such knowledge. Newtonian natural philosophy now insisted on the uniform operation of gravitational force, according to laws that rendered all bodies in motion, everywhere, susceptible to mathematical calculation. Linnaean botany offered a logical system for classifying plants according to their sexual characteristics and a clear method of relatively simple binomial nomenclature, consciously designed (unlike earlier Baroque classification systems) to incorporate the world's flora into a single taxonomic scheme. Such projects were responses to global commercial networks in two important ways: they used local knowledge, such as astronomical observations, to build universal understanding of such phenomena as gravitation, and they aimed to come to terms with the variety of the world's natural productions by constructing for them a single coherent catalogue.⁸

A reflexive questioning of categories also helps to clarify what we mean when we talk about the "Atlantic world." Because oceans, too, are socially constructed, with meanings that vary across space and time, any definition of the "Atlantic" will be local and period-bound. As Joyce Chaplin shows in her essay in this book, the very term *Atlantic Ocean* was not commonly used for the entire Atlantic basin until the mid-eighteenth century. Prior to then, Europeans and colonizers spoke of "the Ethiopic sea" (for the area south of the Gulf of Guinea), the "Atlantic Sea," the "Mar del Norte" (or Northern Sea, for the North Atlantic) or, most Eurocentrically, the "Western Ocean" (for the waters immediately west of Europe).⁹ This "Atlantic" was also far from being a closed system. South American bullion flowed eastward through Europe to finance the spice trade; European navigators (most famously Columbus) were looking for passages to the East rather than any "New World" in the West; and all the European powers were no less preoccupied with advancing their interests in Asia and the South Seas, as well as Africa and the Americas. Nevertheless, as a number of scholars have recently argued, the Atlantic constitutes a "logical unit of analysis" because of its specific history of interactions between Amerindians, Africans, and Europeans through varieties of triangular trade—above all the slave trade— that transformed the culture and economy of each continent. Nowhere else in the early modern world was there such a seismic displacement and recombination of different peoples in new settler societies. In his magisterial history of the Mediterranean world in the age of Philip II, Fernand Braudel noted that "in the sixteenth century, the [Atlantic] ocean did not yet have a fully independent existence. Man was only just beginning to take its measure and to construct an identity for it with what could be found in Europe, as Robinson Crusoe built his cabin from what he could salvage from his ship." As a matter of historical fact, however, Braudel could not have been more wrong: the Atlantic was not the autonomous creation of heroic European men but a transformative mixing of men and women from Africa, the Americas, and Europe, the results of whose coming together no one group could predict or fashion.[10]

Suggestively, early modern languages of epistemological and political dominion often intertwined. The English experimenter Robert Hooke, in his *Micrographia* (1665), wrote, for example, of the microscope's potential for enlarging "the empire of the senses," and of its "new visible world discovered."[11] However, what form of "empire" were Europeans aiming to construct across the Atlantic? While territorial charts puffed European chests with their inspiring cartographic neatness, the Atlantic could not be peacefully divided into discrete zones of influence. Though the Atlantic constitutes a coherent unit of historical analysis, its history is necessarily one of competition and conflict. The two most important connotations of "empire" in the early modern period were of undivided sovereignty (England was declared "an empire of itself" when it separated from Rome in the 1530s) and, most fundamentally, the power to command and coerce. In its classical genealogy, of which early modern statesmen and scholars would have been well aware, empire specifically connoted territorial dominion and the incorporation of conquered peoples into the culture of the *civitas*—the imperial center, exemplified by ancient Rome.[12] Like "science," therefore, "empire" also raises, in its own way, the question of universalism. The political rhetoric of the Spanish monarchy (and at certain points also of the French) included the claim to a universal Catholic monarchy. The Christian churches, with their "universalist" ideology and a series of long-range missionary networks, had a complex relationship with both "science" and "empire" across the period. Indeed, as the travels of Jesuit missionaries vividly show, natural knowledge could be produced across vast distances along networks defined by ecclesiastical community rather than national affiliation, crossing political lines, and stretching east to China and west to Spanish America and New France. The English and Dutch, meanwhile, in self-conscious Protestant reaction to Catholic success in America, cast themselves as "planters" and maritime commercial traders, articulating ostensibly more limited claims to New World dominions based on agricultural property in land more than divine evangelical right over "heathen" peoples, though they proved no less flagrant in their violent displacement of Amerindians and enslavement of Africans.[13]

Empire's conceptual history is important to understanding its particular early modern manifestations, but—like science—empire is inevitably a social project as well, an aspiration to dominion that must ultimately be assessed through the social history of its networks and agents and of those human actors and material conditions that defy it. Would-be empire builders drew lines on maps of water, not just land, such as the line established by the Treaty of Tordesillas in 1494 to distinguish Spanish from Portuguese dominions. However, the very possibility of staking national claims on the seas beyond the first meridian (usually marked at the Canaries) and south of the Tropic of Cancer was intensely debated and remained unresolved, as Dutch jurists such as Hugo Grotius took the lead in arguing strenuously in favor of *mare liberum* to keep the Atlantic free for competitive commercial shipping. On the high seas, the disciplinary arm of the maritime state was increasingly weak—no peace beyond the line, as the saying went—and for much of the early modern era, privateers and pirates of all nations flourished. There were, of course, culturally discrete zones of maritime and territorial influence carved out by the different European nations, shaped by complex and shifting indigenous alliances (such as the "middle ground" of the Great Lakes region), specific histories of cultural mixing (Spanish-American *mestizaje* took place on a scale unknown in British America), and dramatically varying dependencies on African slave labor (New France and Brazil were, in this respect, radically different American societies). However, the empires of cartographic discretion and mercantilist discipline drawn up on maps and in navigation acts in Europe often clashed with the local necessity of negotiations and alliances that allowed practices such as smuggling and piracy to flourish, fostered the production of complex mestizo communities, frustrated attempts to command political and economic obedience from settlers, and persistently provoked resistance, if not rebellion, from peoples as distinct as the Pueblo Indians in New Mexico, the slaves of Saint Domingue, and the slave owners of British America. Though colonization's abuses were brutally real, visions of undivided sovereignty and the power to command proved more imagined than actual in the Atlantic world, often disintegrating precisely when asserted. As the Dutch jurist Cornelius van

Bynkershoek, among many others, argued, the vast ocean could not be possessed.[14]

New Journeys between Center and Periphery

This section briefly outlines some of most important recent developments in approaches to the history of science and empire in the early modern world. Unquestionably, the fundamental project that has emerged in recent years is a profound rethinking of traditional metropolitan narratives of center-and-periphery, involving the recovery of a variety of "peripheral" actors' agencies, and the reimagination of knowledge production from their perspectives. The straight lines of communication that scholars of scientific "diffusion" once traced from center out to periphery, through which European science was seen as the engine of "modernization" and "development" for the entire world, are now being replaced by an intricate latticework of intersecting itineraries and competing agencies. In the diffusionist vision, colonial settlement played a redemptive, progressive historical role in transmitting European scientific traditions to new societies that would eventually find their path to modernity precisely through the reproduction of European scientific culture. This model was teleological and deceptively consensual. Diffusion implied inevitability: the identity and value of modern science were self-evident and did not require explanation. Its circulation and reproduction could be taken for granted, historically and morally.[15]

Where, though, is scientific knowledge actually made? Because science is a collective process composed of many local activities, recent studies have usefully employed networks as categories of analysis for examining the movement and translation of natural phenomena around the world into recognizable and useful signs. The concept of "metrology," according to which local specimens and phenomena in the world "outside" are translated into a system of stable and mobile signs for interpretation and manipulation "inside" what Bruno Latour has called "centers of calculation," offers a valuable model for understanding how long-distance relationships make scientific knowledge. A key example of "science in the making" deployed by Latour in his highly influential *Science in Action* (1987) was provided by an eighteenth-century expedition involving the accumulation of non-European expertise: Jean-François de La Pérouse's translation of indigenous navigational knowledge from the Pacific into a map of the world to be drawn in France. Science, according to this account, can usefully be regarded as a function of the movement of translated particulars along networks connecting peripheral actors to metropolitan ones. Understanding the production of knowledge thus requires, above all, understanding movement: of people, things, "languages," and techniques.[16]

The fruitful example of La Pérouse may, however, be misleading and limiting, if it is taken to reinstate the adequacy of the classic diffusionist itinerary of center-periphery-center voyages to describe the process of making knowledge across vast and complex spaces, such as the Atlantic world.[17] Agents of the state are clearly important in distinct phases of Atlantic history. The institutional and legal structures of Spain, Portugal, and France and their colonial settlements were clearly more centralizing than their Dutch and British counterparts, for example (although British merchants proved very effective at persuading Parliament to deploy the resources of the Royal Navy against the pirates who threatened their profits). The *Casa de la Contratación* existed in Seville for two and a half centuries before Joseph Banks began to place economic botany on the footing of a coordinated national policy in Britain, a strategy inspired by the success of Britain's imperial rivals, France. The point is not now to banish agents of the state from new narratives but to understand them as one kind of actor among many and by no means necessarily the dominant kind.[18] Understanding how the networks of commercial trade behaved and the challenges in exploiting them for the pursuit of knowledge helps to correct the overdrawn historical image of states efficiently penetrating and exploiting their distant peripheries. As Nicholas Dew's essay shows, in the late seventeenth century, even the astronomers who set out from "absolutist" France relied on commercial shipping, which landed them in destinations not of their own choosing—the merchants decided where their ships went, not the state or its academies. Merchants not only moved people who made knowledge; they made and sold knowledge themselves, above all botanical knowledge of lucrative exotic commodities. What needs exploring, therefore, is not so much how European centers managed peripheral accumulation, as how the production of knowledge resulted from specific and sometimes quite temporary intersections with the quotidian and autonomous networks of commerce.[19]

Sociologists and historians of science have drawn important inspiration from anthropology, imagining themselves as "playing the stranger" in such settings as the modern laboratory, to explain the emergence of seemingly self-evident practices, such as experiment.[20] History of science can return the compliment. Where European networks end, the agency of others becomes especially visible and puts European universal claims into perspective. As Simon Schaffer has shown for seventeenth-century Guinea, English traders who lacked access to the African interior were forced to accept the value of gold as specified by Akan traders' measurement systems rather than the values they sought to impose through the use of their own assaying techniques. Conflicts and controversies over establishing "true" value across cultures can help to display the dynamics of colonial contests, expose European knowledge's dependencies and limits, and situate it as part of a larger world-historical interaction of competing *local* knowledge systems.[21] The "power of the weak" lies not only in the laboring capacity of disenfranchised peoples, broadly understood, but in specific knowledges held in non-European communities that colonizers were desperate to gain access to—such as the expertise of rice growers (many of them women) who were forcibly transported as slaves from the "Rice Coast" of West Africa to the plantations of such American colonies as Carolina.[22] Gradually but surely, scholars are moving away from histories focused on seemingly isolated metropolitan knowers, such as the paradigmatic gentlemanly natural philosopher, and preoccupations with alterity and the textual erasure of indigenous presence, toward a social history of the interconnections between the radically different peoples that made and circulated early modern knowledge.[23]

Economic botany helped to identify and organize New World plants for commercial cultivation; cartography and hydrography charted land and water for strategic access; detailed ethnographies provided cultural and military intelligence with respect to indigenous cultures. However, what counted for "useful knowledge," and what wider cultural purposes did such knowledge also serve? Narratives of science and empire as the history of practical utilitarianism should not obscure the difficulties of identifying what was useful or the persistent symbolic importance of natural knowledge. Early modern Europeans marveled at the strange "things" yielded by long-distance travel without always understanding their significance. Bernardino de Sahagún wrote a *General History of the Things of New Spain*, whereas Nicolás Monardes compiled information about the "things which are found in our West Indies useful to medicine." The use of "things" ("cosas") signals a telling uncertainty about exotic specimens and artifacts: their utility, indeed, their identity, was never automatically clear.[24]

The quest for practical utility, moreover, by no means precluded attempts to invest American nature with wondrous symbolism. Theological frameworks—especially Christian visions of "improvement" as redemption for the Fall—were powerful ideological discourses framing the accumulation and use of American natural knowledge. To be reminded of this, one need only recall that Linnaeus considered his taxonomic system, widely adopted around the Atlantic world, as an Adamic act of divine renaming that would facilitate human dominion. In his utilitarian natural history of Barbados first published in 1657, designed to encourage sugar cultivation, Richard Ligon saw in the form of a banana tree the "picture of Christ upon the Crosse," which he interpreted as a divine rebuke to masters for not converting their slaves. Practical activities such as botany produced economic, natural, and cosmological orders simultaneously. The Society of Jesus is perhaps the best-known example, but numerous other missionary organizations (Catholic and later Protestant) were engaged in the early modern period in the production of natural knowledge, from astronomy to botany. The Jesuits would have been the first to admit that their natural-scientific pursuits were at once the by-product and the handmaiden of their spiritual concerns. The notion that Europeans steadily turned American nature from a wonder into a curiosity, subjecting it to utilitarian taxonomic and economic discipline, should thus be tempered with the recognition that wonder and utility could, and did, coexist, and that utility could be cultural as well as material. The multiple and often uncertain meanings invested in strange natural things should not be bracketed off as epistemological noise obscuring the "proper" functioning of scientific networks but explored in their own right as fundamental to the cultural history of exotic knowledge. In a variety of ways, early moderns simultaneously made use of wonders and wondered at useful things.[25]

The question of who could speak publicly about nature in an expansionist early modern world was, moreover, one of geopolitics, not just the domestic politics of European society. When it came to long-distance reportage, disciplining and recording experience were vital. On what basis could remote reports of distant lands and waters be trusted, and *whose* reports? The depersonalization of witnessing was an effective means to generate trust, through two principal methods: collectivizing the interpretation of particulars through shared languages and signs and using instruments to record natural facts. With their corporate structure, which bound traveling missionaries into a network of correspondents sending reports systematically back to Rome, the Jesuits were one of the best organized collectives for the long-distance circulation of early modern knowledge. State-sponsored scientific institutions aimed to emulate this kind of discipline. Robert Boyle, for example, distributed carefully worded questionnaires to procure more accurate travel accounts for the Royal Society in London. Such methods proved haphazard, however. Scientific instruments for making and recording precise numerical measurements held out the possibility of a more reliable accounting of American natural conditions. However, human error, exotic climates, and dangerous environments meant that the deployment of such devices by no means guaranteed solutions to the problem of remote witnessing.[26]

Because the Atlantic increasingly consisted of expanding settler colonies, issues of trust and discipline in the production of knowledge extended critically to the permanent Creole populations of the Americas, not just travelers. Although the roots of an Atlantic community are often (Anglo-centrically) traced to the Grand Alliance of the Second World War, a less glorious notion of such community had already emerged in pseudoscientific notions of "Anglo-Saxonism" in the late nineteenth century. In this vision of racial confraternity, the lines of imagined "blood ties" ran from East to West across the North Atlantic, uniting European Protestants and their American descendants against "Southerners," Africans, and Jews.[27] This horizontally imagined community was in stark contrast to early modern European emphases on hemispheric difference that downplayed any putative transatlantic unity. American climates fostered humoral "degeneracy" and degraded intellectual capacity, justifying the political subordination of American Creoles. This was not a modern concept of radical racial difference stipulating the permanent inferiority of non-Europeans based on innate physical difference. However, notionally at least, political and epistemological hierarchies were naturalized in terms of the effect of specific climates on human capacities.[28]

As Jorge Cañizares-Esguerra has demonstrated, the dispute of the New World that ensued was not limited to the well-known late-eighteenth-century argument between Thomas Jefferson and the Comte de Buffon but flowed mainly from the pens of numerous Creole Spanish-Americans who wrote back against empire throughout the colonial period. In so doing, Creoles made a powerful claim to the authority to write their own natural and social histories, defying European disparagement. Though modern notions of race as an explanatory force in history would not be articulated before the nineteenth century, it is nonetheless significant that the practice and achievements of science were increasingly invoked to distinguish the capacities of the world's different peoples—a globalizing perspective opened up precisely by the colonization of theaters such as the Atlantic. "Let the flat-faced African with his black complexion and woolly hair, give place to the European, whose regular features are set off by the whiteness of his complexion and beauty of his head of hair," wrote the Swiss naturalist Charles Bonnet in 1766. "To the filthiness of a Hottentot, oppose the neatness of a Dutchman. From the cruel Anthropophagite pass swiftly to the humane Frenchman. Place the stupid Huron opposite the profound Englishman. Ascend from the Scotch peasant to the great NEWTON."[29] Bonnet shows how inhabitants of the European periphery, no less than the New World, could palpably embody savagery from the metropolitan perspective, but more telling for our purposes is his progression from the base physicality of the African to the pinnacle of Newton's scientific genius. The achievements of science were acquiring a unique ability to confer superiority on Europeans (in their own eyes) as they encountered other peoples. Not for nothing did Philippe-Jacques de Loutherbourg place an octant in Cook's hand in his 1794 depiction of his apotheosis over Hawai'i. In the Enlightenment, science increasingly became a ceremony of benign imperial possession, enacted in Europe and abroad.[30]

Assuming the inevitability of revolutionary resistance to such chauvinism—which applied to Creole Americans as well—would, however, oversimplify our understanding of the politics of Atlantic knowledge. Creoles resented their subordinate status to differing degrees depending on social status, place, and time. Despite their formal subordination, many were happy to seek distinction and connections by participating in metropolitan knowledge projects through activities from astronomical observations to botanical collecting. The American correspondents of Europeans were not servile drones but shrewd self-fashioners who sought to turn European recognition to local advantage in cultivating their status as cosmopolitan knowledge makers in the provinces and serving provincial agendas as well as metropolitan ones. The political order of Atlantic knowledge was not, therefore, a revolution perpetually in the making but a series of negotiations that, albeit tensely, knit imperial and local purposes together for a considerable time.[31]

Atlantic Knowledges

The essays that follow are grouped into four sections entitled "Networks of Circulation"; "Writing the American Book of Nature"; "Itineraries of Collection"; and "Contested Powers." These groupings are not intended as strict divisions; there are many themes that range across them, connecting them in different ways. However, they do serve two useful purposes: to understand the process of making Atlantic knowledge from a trans-national perspective, rather than in national isolation, and to clarify certain themes that seem particularly fruitful for understanding the challenges inherent in making knowledge across such a vast geographical space.

"Networks of Circulation" explores the mechanics of circulating knowledge through a variety of agents in networks spanning the Atlantic. It is tempting to assume a certain teleology at work in the history of circulation, from secrecy to dissemination. Circulation and communication are not historical givens, however. Indeed, as Alison Sandman shows, the circulation of knowledge could be regarded as a veritable evil to be guarded against in early modern competitions for global power. She explores a sixteenth-century Spanish quarrel over *what* exactly counted for useful knowledge of the Atlantic Ocean. Was it the experiences of pilots that offered local practical advice to seafarers, or the "universal" mapping of longitude and latitude lines sought by cosmographers at the *Casa de la Contratación* in Seville? Sandman shows that struggles over defining useful maritime knowledge were struggles between different knowledge communities. Despite the Seville cosmographers' ambitions, it proved extremely difficult to keep control over long-distance agents and to keep their knowledge secret.

Nicholas Dew examines the journey of Jean Richer from the Académie des Sciences in Paris to Cayenne in French Guiana in the late seventeenth century. Richer's trip, Dew demonstrates, was to become controversial because his experiments in Cayenne with a pendulum caused trouble both for contemporary theories of weight and for projects in universal metrology. Richer's tropical measurements posed a problem for metropolitan savants who wished to collate local observations from around the world into a network of standardized measures and yet had trouble knowing how to trust tropical measurements. Could there be such a thing as universal knowledge when climatic conditions in the French colonies might alter the behavior of instruments? By exploring the contingencies of journeys dependent on the preexisting networks of the triangular trade, Dew shows that Atlantic geography mattered as much at the macro-level of transport and communication as at the micro-level of experimental practice. Joyce Chaplin shifts our attention away from the official emissaries of European states to see how British and North American sailors produced vernacular knowledge of the ocean. As Chaplin makes clear, conceptions of the "Atlantic Ocean" changed significantly during the early modern period. They appear to have originated with non-elite mariners, gradually becoming less Eurocentric over time thanks to the changing political situation of those who produced them. She argues that, unlike Iberian or French networks, Anglophone knowledge networks were less oriented towards serving imperial purposes. By contrast, Benjamin Franklin's publication of charts of the Atlantic Gulf Stream demonstrate how the conversion of vernacular maritime knowledge could be appropriated and refashioned by middling Creole Americans in imperial, revolutionary and finally republican contexts.

What did American nature mean? For Christians, God's word was written in the Book of Scripture and the Book of Nature equally, to be read and interpreted as a system of signs revealing divine will. Nature, however, was not a text whose meanings simply unfurled themselves to attentive readers. Though modestly claiming merely to read the natural world, observers in reality *wrote* meanings into nature, presenting them as revealed rather than made. "Writing the American Book of Nature" explores the process of making American nature speak. The Baconian injunction to plant colonies and reap empirical knowledge ostensibly makes the New World synonymous with the "New Science." However, Ralph Bauer shows that this was not automatically the case. The issue of secrecy, first explored by Sandman, resurfaces in Bauer's examination of the relationship between Baroque epistemology and American colonization in Walter Ralegh's account of his quest for El Dorado, *The Discoverie of the Large, Rich, and Bewtiful Empyre of Guiana* (1596). Bauer argues that there was indeed a New World before the New Science, one that functioned as a mystical system of secret signs and signatures, waiting to be unlocked by occult philosophers. And because sixteenth-century Europeans embraced occult knowledges, such as alchemy and the doctrine of signatures, they enjoyed some affinity with Amerindian traditions of natural knowledge, leading colonizers to praise and seek out the useful "conjuring" abilities of Native Americans in ways that would become increasingly rare later on.

Júnia Ferreira Furtado examines the interactions between Portuguese surgeon-barbers, Amerindians, and slaves in Minas Gerais, Brazil. She holds that these interactions produced a local "tropical" medicine that found its way into print for a transatlantic Lusophone audience. Colonial Brazilian medical knowledge was highly dependent on the contributions of Amerindians and Creolized Africans. The empirical gathering and testing of plants and drugs necessitated social collaborations and practical exchanges across cultures, even as Europeans tended to reject the interpretative frameworks in which such knowledges were originally embedded. Portuguese colonizers, in other words, collected practical knowledge from the varied populations of colonial Brazil while writing Brazilian nature up as a paradisiac landscape made useful through European ingenuity. Jan Golinski examines the connections that eighteenth-century British Americans, then citizens of the independent United States, forged between climate and identity. This relationship had been defined by European commentators, who disparaged American societies by depicting American nature as puny and miasmatic, deadly and chaotic. Though these kinds of attacks had long been rebutted by Spanish-American Creoles, it was not until the later eighteenth century that North Americans mounted similar patriotic defenses. Golinski shows how the former colonists rewrote the book of American nature in this era, arguing that they themselves had "improved" the American climate through planting, clearing, and agriculture, avoiding the fate of the hapless natives and thus positioning the independent United States as a civilized republic fit by nature for self-government and the occupation of the continent.

"Itineraries of Collection" foregrounds the multiple itineraries of Atlantic travel and the unpredictability of the uses to which the collection of knowledge and specimens was put. The indeterminacy of exotic things, uncertainties about value, the contingencies of communication, and the difficulty of turning New World knowledge to account are here on full view. Antonio Barrera-Osorio connects the development of empirical practices in early modern Spain to the importation of American specimens and technologies. As he makes clear, the Spanish were the first to create institutions (the *Casa de la Contratación*, the Council of the Indies, and the vice-royal courts) for systematically collecting and testing everything from American dyes, such as pastel and indigo, to hydrographic knowledge of Atlantic waterways. Understanding, organizing, and using New World nature in Spain prompted new relationships between artisans, royal officials, and scholars, and fostered a culture of experiment linking private and state interests in the economic service of the metropolis.

Economic botany, as is well known, was a highly fruitful long-distance enterprise for the identification and commodification of agricultural specimens. Neil Safier demonstrates how such ventures could, however, easily result in "thwarted knowledge" and failure. "Live first, then philosophize," Joseph de Jussieu heroically declared, placing experience on a well-worn early modern rhetorical pedestal, over and above mere book learning. However, his career vividly displays just how challenging early modern long-distance travel and communication often proved. The lesser-known member of an illustrious French botanical family, Jussieu enjoyed the freedom to collect a variety of knowledges, botanical and otherwise, over a thirty-year period from his base in the Andes but, as Safier argues, the price of intellectual freedom appears to have been metropolitan irrelevance. Geopolitical contests, the challenges of the physical environment in South America, the vagaries of Atlantic transportation, and Jussieu's own apparent psychological distress prevented him from sending back valuable specimens such as cinnamon, thwarting his ability to make coherent scientific contributions and a respectable name for himself back in Paris. Daniela Bleichmar examines the Spanish turn from mineralogical ventures to economic botany as part of the "Bourbon reforms" of the eighteenth century. Royal institutions based in Spain took unprecedented interest in this era in such commodities as cinnamon, tea, pepper, and the medicinal properties of the cinchona tree. Though the domestication of such specimens did not always succeed, the mobilization of the official bureaucracy as a network for collection brought American things into the display cabinets of Madrid. However, Bleichmar argues, in a useful counterpoint to Barrera-Osorio's article, this centripetal movement of goods and artifacts is only half of the story. She demonstrates how Creole naturalists working in Lima, Mexico City, and Santa Fé de Bogotá came to identify with each other as a regional group whose botanical work should serve local Spanish-American interests as much as metropolitan ones.

Mastery of natural powers could become an explicit, indeed explosive political question. American Revolutionaries, for example, insisted that the capacity of Franklin's lightning rod to control nature was evidence of the ability of British-Americans to control their own political destiny. "Contested Powers" explores the pursuit of natural powers as a series of struggles for political power around the Atlantic. According to center-and-periphery models, power flows back to the center of long-distance networks, thanks to the obedience of peripheral agents. However, such networks harbored resistance, disobedience and, in the later eighteenth century, outright rebellion, showing how political conflicts, conversely, become contests over who will control nature itself. James Delbourgo investigates the changing meanings of the electrical machines imported and constructed in British America beginning in the 1740s. Moving beyond the lone figure of Franklin, he identifies two distinct phases in this history. In the colonial period, the circulation of generators was driven by a market for provincial enlightenment, in which colonial Americans eagerly participated in experimental philosophy as a form of improvement in the context of empire. In the American Revolution, however, electrical machines were suddenly reimagined as engines of republican transformation: Franklinist knowledge of active powers now betokened new political powers for Americans demanding self-government. Linking this discussion to an examination of experiments conducted on electric eels by travelers in Dutch Guiana and New Granada, Delbourgo argues that American electricity involved surprising powers that defied complete reduction to predictable effects via mechanical discipline.

The role of Africa and Africans in diaspora—fundamental to the entire Atlantic economy—in making natural knowledge is undoubtedly the least well understood aspect of the history of science in the Atlantic world. Recent work offers substantial hope that this is changing, however.[32] The last two essays make much needed explorations of this theme. First, Susan Scott Parrish illuminates the social power at stake in botanical knowledges controlled by African slaves. The perennial question of whether American flora would nourish, heal, or kill was an especially agonizing issue for colonizers who feared slave poisons as a technology of resistance. As Parrish demonstrates, Europeans aimed to mobilize the physical resources and vernacular knowledge of slaves who labored in more intensive proximity to colonial landscapes than they did. Slaves were thus deployed as auxiliaries in such activities as natural-historical collecting; in one remarkable instance, a Surinam slave named Kwasi had a tree named after him by Linnaeus, visited the Dutch royal family and ultimately earned his freedom through his botanical expertise. Africans' knowledge continued to stimulate anxiety among masters, however, in no small degree because of the political empowerment afforded by "occult" Afro-Caribbean religious traditions to which they were related. This fear of the political subversiveness of African knowledge is also an important theme of the final essay, by François Regourd. Regourd examines how the practice of animal magnetism, the latest medical trend from Paris, arrived in the 1780s in Saint Domingue, the most profitable plantation colony in the Atlantic world. Mesmerism quickly divided the medical and scientific community, becoming a lightning rod for controversy. Ironically, animal magnetism's popularity on the island prompted the establishment of the *Cercle des Philadelphes*, the first scientific academy in the French colonial world. Intriguingly, French settlers started to describe what they took to be mesmeric practices among Saint Domingue's black population. Whether or not the Vodou practitioners were starting consciously to adopt the trappings of mesmerism remains uncertain; clearly, however, once discredited, mesmerism was used as a label for exorcising white fears of the power of Afro-Caribbean religion and natural knowledge. As in the British colonies described by Parrish, the inscrutability of slaves' religious practices and natural knowledge made figures such as Macandal—a slave who allegedly wielded botanical and magical powers and plotted to poison the planters—into powerful touchstones for fears of catastrophic civil insurrection.

These stories can be read comparatively to see how Europeans and American Creoles competed in projects to make knowledge in relatively discrete zones of territorial and maritime influence. However, ultimately they should also be seen as interlocking parts of a larger history of knowledge, communication, and empire in the Atlantic world—a history that transcends any single national framework and is defined as much by unpredictable crossings as mercantile containment.[33] Two factors came increasingly to characterize most if not all of the Atlantic world in the period 1500–1800: the acceleration and proliferation of commercial oceanic voyages and the development of complex settler societies, to differing degrees hybrid worlds of European, Amerindian, and African. Both powerfully shaped the production of knowledge. Commercial intensification produced a proliferation of itineraries governed by no central mechanism, creating lines of communication around the periphery and back across the ocean. The knowledge created by Atlantic travelers was therefore often the product of commercial contingency rather than of imperial design. At the same time, the formation of Creole American societies made the circulation of knowledge inescapably part of the fabric of imperial relationships between metropolis and colonies, *peninsulares* and Creoles, masters, slaves, and indigenous peoples. Our historical maps of the journeys of American curiosities into European cabinets should begin not with transoceanic circulation but the local American relationships—Creole, indigenous, slave—that made the objects available for shipping in the first place, while also tracing movement in other directions to explore how Creoles, especially in Spanish America, assembled *Wunderkammern* of their own.[34] A vast amount remains to be learned by probing the motives of non-Europeans in exchanging knowledge with colonizers as well. Though "empiricism"—a capacious label applied to a vast array of practices—could act imperially as a filter for appropriating non-European expertise while rejecting indigenous belief systems, gathering practical knowledge on the periphery could also embolden Creole and enslaved actors. The ultimate product of profitable imperial extraction was rebellion and revolution.

Without doubt, the career of empire was highly variable across time and space in this Atlantic world. Formal attempts at coordinating the production of natural knowledge as a branch of imperial policy through the resources of the state occurred at different moments for each of the European powers, as we have noted, and with mixed results.[35]

What the engagement with and ultimate resistance to such projects shows, however, is how the production of knowledge depended not only on networks of communication and the organization of projects from metropolitan centers, but on the political willingness of travelers and settlers to participate in such projects. Empire, like science, was a social project that, despite its aim to command assent, was politically fragile. Empire should form part of any Atlantic narrative because it usefully brings political relationships into focus. However, it should not shape that narrative from the center outward, in national or even comparative terms, because as maritime, territorial, and human space, the Atlantic always outreached the empires that tried to subject it. The far side of the ocean was, after all, a matter of perspective.

Notes

1 Francis Bacon, "New Atlantis," in *The Works of Francis Bacon*, ed. James Spedding, Robert L. Ellis and Douglas D. Heath, 14 vols (London: Longman,

1857–74), 3: 125–66, at 131, 145, 156–64. For recent interpretations, see Bronwen Price, ed., *Francis Bacon's New Atlantis: New Interdisciplinary Essays* (Manchester: Manchester University Press, 2003), esp. Richard Serjeantson, "Natural Knowledge in the *New Atlantis*," 82–105.

2 Ralph Bauer, *The Cultural Geography of Colonial American Literatures: Empire, Travel, Modernity* (Cambridge: Cambridge University Press, 2003), 20–1.

3 On Humboldtian science, see Susan Faye Cannon, *Science in Culture: The Early Victorian Period* (New York: Science History Publications, 1978); Michael Dettelbach, "Global physics and aesthetic empire: Humboldt's physical portrait of the tropics," in David Philip Miller and Peter Hanns Reill, eds, *Visions of Empire: Voyages, Botany, and Representations of Nature* (Cambridge: Cambridge University Press, 1996), 258–92; Marie-Noëlle Bourguet, "Landscape with numbers: natural history, travel and instruments in the late eighteenth and early nineteenth centuries," in Marie-Noëlle Bourguet, Christian Licoppe, H. Otto Sibum, eds, *Instruments, Travel and Science: Itineraries of Precision from the Seventeenth to the Twentieth Century* (London: Routledge, 2002), 96–125; Mary Louise Pratt, *Imperial Eyes: Travel-Writing and Transculturation* (New York and London: Routledge, 1992), 111–97; and *Alexander von Humboldt: Netzwerke des Wissens* (Berlin: Haus der Kulturen der Welt, 1999). On Humboldt's debt to Creole writers, see Jorge Cañizares-Esguerra, "How derivative was Humboldt? Microcosmic nature narratives in early modern Spanish America and the (other) origins of Humboldt's ecological sensibilities," in Londa Schiebinger and Claudia Swan, eds, *Colonial Botany: Science, Commerce and Politics in the Early Modern World* (Philadelphia: University of Pennsylvania Press, 2005), 148–65.

4 For Humboldt's desire to join the expedition to Egypt, see Hanno Beck, "'Ich bereite mich ohne Unterlass auf ein grosses Ziel vor': Die Jugendzeit Alexander von Humboldts," in *Alexander von Humboldt: Netzwerke des Wissens*, 47–9. On itineraries of precision, see Bourguet, et al., *Instruments, Travel and Science*.

5 On heroic rhetoric, see Mary Terrall, "Heroic narratives of quest and discovery," *Configurations* 6 (1998), 223–42; on Pacific voyages and commemorations, see, for example, David Philip Miller and Peter Hanns Reill, eds, *Visions of Empire: Voyages, Botany, and Representations of Nature* (Cambridge: Cambridge University Press, 1996).

6 Bacon, "New Atlantis," 165–6; J. H. Elliott, *The Old World and the New, 1492–1650* (Cambridge: Cambridge University Press, 1970) plays down the impact of "discovery," but compare with Anthony Grafton, *New Worlds, Ancient Texts: The Power of Tradition and the Shock of Discovery* (Cambridge, Mass.: Harvard University Press, 1992), and Anthony Pagden, *European Encounters with the New World: From Renaissance to Romanticism* (New Haven: Yale University Press, 1993).

7 For recent accounts of early modern natural knowledge, see Steven Shapin, *The Scientific Revolution* (Chicago: University of Chicago Press, 1996); Peter Dear, *Revolutionizing the Sciences: European Knowledge and its Ambitions, 1500–1700* (Princeton: Princeton University Press, 2001); William Clark, Jan Golinski and Simon Schaffer, eds, *The Sciences in Enlightened Europe* (Chicago: University of Chicago Press, 1999); and *The Cambridge History of Science*, vol. 3: *Early Modern*

Science, ed. Katharine Park and Lorraine Daston (Cambridge: Cambridge University Press, 2006), and vol. 4: *Eighteenth-Century Science*, ed. Roy Porter (Cambridge: Cambridge University Press, 2003).

8   Steven J. Harris, "Long-distance corporations, big sciences, and the geography of knowledge," *Configurations* 6 (1998), 269–304. On cabinets, see Lorraine Daston and Katharine Park, *Wonders and the Order of Nature, 1150–1750* (New York: Zone, 1998). On the "planetary consciousness" fostered by Newton and Linnaeus, see Pratt, *Imperial Eyes*, 15–37; see also Felicity A. Nussbaum, ed., *The Global Eighteenth Century* (Baltimore: Johns Hopkins University Press, 2003); on the variety of early modern plant taxonomies, see Drayton, *Nature's Government*, 3–25.

9   Martin W. Lewis, "Dividing the ocean sea," *Geographical Review* 89 (1999), 188–214; Philip E. Steinberg, *The Social Construction of the Ocean* (Cambridge: Cambridge University Press, 2001); on explorations of the nineteenth-century ocean, see Helen Rozwadowski, *Fathoming the Ocean: The Discovery and Exploration of the Deep Sea* (Cambridge, Mass.: Harvard University Press, 2005).

10 Fernand Braudel, *The Mediterranean and the Mediterranean World in the Age of Philip II*, trans. Siân Reynolds, 2 vols (New York: Harper and Row, 1972), 1: 224; Alison Games, "Teaching Atlantic history," *Itinerario* 23 (1999), 162–73; David Armitage, "Three concepts of Atlantic history," in David Armitage and Michael J. Braddick, eds, *The British Atlantic World, 1500–1800* (Basingstoke and New York: Palgrave, 2002), 11–27; for an overview of the formation of the Atlantic world, see D. W. Meinig, *The Shaping of America, Volume I: Atlantic America, 1492–1800* (New Haven: Yale University Press, 1986).

11 Robert Hooke quoted in Steven Shapin and Simon Schaffer, *Leviathan and the Air-Pump: Hobbes, Boyle, and the Experimental Life* (Princeton: Princeton University Press, 1985), 36–7. See also Mary Baine Campbell, *Wonder and Science: Imagining Early Modern Worlds* (Ithaca: Cornell University Press, 1999).

12 See David Armitage, *The Ideological Origins of the British Empire* (Cambridge: Cambridge University Press, 2000), 29–36.

13 Anthony Pagden, *Lords of All the World: Ideologies of Empire in Spain, Britain, and France, c.1500–c.1800* (New Haven: Yale University Press, 1995); on Dutch conceptions of empire, see P. J. Drooglever, "The Netherlands colonial empire: historical outline and some legal aspects," in H. F. van Panhuys, ed., *International Law in the Netherlands* 1 (1978), 103–65. As Margaret Jacob rightly points out in her afterword to this volume, much more work needs to be done on the production of natural knowledge in the Dutch Atlantic. See, however, Benjamin Schmidt, *Innocence Abroad: The Dutch Imagination and the New World, 1570–1670* (Cambridge: Cambridge University Press, 2001), and most recently, Harold J. Cook, *Matters of Exchange: Commerce, Medicine, and Science in the Dutch Golden Age* (New Haven: Yale University Press, 2007). On Jesuit scientific networks, see below, note 26.

14 On the problem of "the empire of the seas," see Armitage, *Ideological Origins of the British Empire*, 100–24; Steinberg, *Social Construction of the Ocean*, 68–109;

Mónica Brito Vieira, "*Mare liberum* vs. *mare clausum*: Grotius, Freitas, and Selden's debate on dominion over the seas," *Journal of the History of Ideas* 64 (2003), 361–78; Garrett Mattingly, "No peace beyond what line?" *Transactions of the Royal Historical Society* 13 (1963), 145–62. Bynkershoek's *De Dominio maris* (1702) is discussed in Coleman Phillipson, "The great jurists of the world, IX: Cornelius van Bynkershoek," *Journal of the Society of Comparative Legislation*, new series, 9 (1908), 27–49. See also Peter Linebaugh and Marcus Rediker, *The Many-Headed Hydra: Sailors, Slaves, Commoners, and the Hidden History of the Revolutionary Atlantic* (London: Verso, 2000); Janice E. Thomson, *Mercenaries, Pirates, and Sovereigns: State-Building and Extraterritorial Violence in Early Modern Europe* (Princeton: Princeton University Press, 1994); Richard White, *The Middle Ground: Indians, Empires, and Republics in the Great Lakes Region, 1650–1815* (Cambridge and New York: Cambridge University Press, 1991); Christine Daniels and Michael V. Kennedy, eds, *Negotiated Empires: Centers and Peripheries in the Americas, 1500–1820* (New York and London: Routledge, 2002), 235–65.

15 George Basalla, "The spread of western science," *Science* 156 (May 1967), 611–22; for background on science and empire studies in the 1960s in relation to modernization theory and diffusionism, see Roy MacLeod, ed., "Nature and empire: science and the colonial enterprise," *Osiris* 15 (2000), esp. 2–3; and see also the articles in "Focus: colonial science," *Isis* 96 (2005), 52–87, as well as James E. McClellan, *Colonialism and Science: Saint Domingue in the Old Regime* (Baltimore: Johns Hopkins University Press, 1992); and Sandra Harding, *Is Science Multicultural?: Postcolonialisms, Feminisms, and Epistemologies* (Bloomington: Indiana University Press, 1998).

16 Bruno Latour, *Science in Action: How to Follow Scientists and Engineers Through Society* (Cambridge, Mass.: Harvard University Press, 1987), 215–57; see also Latour, "Circulating reference: sampling the soil in the Amazon forest," in *Pandora's Hope: Essays on the Reality of Science Studies* (Cambridge, Mass.: Harvard University Press, 1999), 24–80; Joseph O'Connell, "Metrology: the creation of universality by the circulation of particulars," *Social Studies of Science* 23 (1993), 129–73; and Kapil Raj, *Relocating Modern Science: Circulation and the Construction of Knowledge in South Asia and Europe, 1650–1900* (Basingstoke and New York: Palgrave MacMillan, 2007).

17 For a critical analysis of the La Pérouse case, see Michael T. Bravo, "Ethnographic navigation and the geographical gift," in David N. Livingstone and Charles W. J. Withers, eds, *Geography and Enlightenment* (Chicago: University of Chicago Press, 1999), 199–235.

18 Richard H. Drayton, "Knowledge and empire," in P. J. Marshall, ed., *The Oxford History of the British Empire*, vol. 2: *The Eighteenth Century* (Oxford and New York: Oxford University Press, 1998), 231–52; Marcus Rediker, *Villains of All Nations: Atlantic Pirates in the Golden Age* (Boston: Beacon Press, 2004), 127–47; on the Casa de la Contratación, see the essays in this volume by Sandman and Barrera-Osorio; on Banks, see John Gascoigne, *Science in the Service of Empire: Joseph Banks, the British State and the Uses of Science in the Age of Revolution* (Cambridge: Cambridge University Press, 1998).

19 Pamela H. Smith and Paula Findlen, eds, *Merchants and Marvels: Commerce, Science and Art in Early Modern Europe* (New York and London: Routledge, 2002). For other recent accounts of early modern commercial networks and science, see: Larry Stewart, "Global pillage: science, commerce, and empire," in Roy Porter, ed., *Eighteenth-century Science*, 825–44; Alison Sandman and Eric H. Ash, "Trading expertise: Sebastian Cabot between Spain and England," *Renaissance Quarterly* 57 (2004), 813–46; Charlotte de Castelnau-L'Estoile and François Regourd, eds, *Connaissances et pouvoirs: les espaces impériaux (XVIe-XVIIIe siècles): France, Espagne, Portugal* (Pessac: Presses universitaires de Bordeaux, 2005); and Daniela Bleichmar, Paula De Vos, Kristin Huffine, and Kevin Sheehan, eds, *Science in the Spanish and Portuguese Empires (1500-1800)* (Stanford: Stanford University Press, forthcoming 2008).

20 On "playing the stranger," see Bruno Latour and Steve Woolgar, *Laboratory Life: The Construction of Scientific Facts* (Princeton: Princeton University Press, 1986 [1979]), 29; Jan Golinski, *Making Natural Knowledge: Constructivism and the History of Science* (Cambridge: Cambridge University Press, 1998), 30–1; and Shapin and Schaffer, *Leviathan and the Air-Pump*, 3–21.

21 Simon Schaffer, "Golden means: assay instruments and the geography of precision in the Guinea trade," in Bourguet, *et al.*, eds, *Instruments, Travel, and Science*, 20–50; see also his "Instruments as cargo in the China trade," *History of Science* 44 (2006), 217–46, and "'On seeing me write': inscription devices in the south seas," *Representations* 97 (2007), 90–122. Dipesh Chakrabarty, *Provincializing Europe: Postcolonial Thought and Historical Difference* (Princeton: Princeton University Press, 2000), is also relevant here.

22 Harding, *Is Science Multicultural?* 155; Judith Carney, *Black Rice: The African Origins of Rice Cultivation in the Americas* (Cambridge, Mass.: Harvard University Press, 2001).

23 On gentlemanly knowers and their dependencies, see Steven Shapin, *A Social History of Truth: Civility and Science in Seventeenth-Century England* (Chicago: University of Chicago Press, 1994); Barbara J. Shapiro, *A Culture of Fact: England, 1550-1720* (Ithaca: Cornell University Press, 2000), esp. 63–85; and Susan Scott Parrish, *American Curiosity: Cultures of Natural History in the British Colonial Atlantic World* (Chapel Hill: University of North Carolina Press, 2006). On American colonization as a story of cultural incommensurability and the construction of alterity, see Tzvetan Todorov, *The Conquest of America: The Question of the Other*, trans. Richard Howard (New York: Harper and Row, 1984), and Anthony Pagden, *The Fall of Natural Man: The American Indian and the Origins of Comparative Ethnology* (Cambridge: Cambridge University Press, 1982); on colonialism as an interactive and hybridizing process, see Serge Gruzinksi, *The Mestizo Mind: The Intellectual Dynamics of Colonization and Globalization*, trans. Deke Dusinberre (New York and London: Routledge, 2002), and Joseph Roach, *Cities of the Dead: Circum-Atlantic Performance* (New York: Columbia University Press, 1996); on the erasure of indigenous presence in/by science, see Pratt, *Imperial Eyes*, and Antonio Lafuente and Nuria Valverde, "Linnaean botany and Spanish imperial politics," in Schiebinger and Swan, *Colonial Botany*, 139. On the collaboration between indigenous agents and

European colonists in British India, see Kapil Raj, "Surgeons, fakirs, merchants, and craftspeople: making L'Empereur's *Jardin* in early modern South Asia," in Schiebinger and Swan, *Colonial Botany*, 252–69.

24 Schiebinger and Swan, *Colonial Botany*; Amy Meyers and Margaret Beck Pritchard, eds, *Empire's Nature: Mark Catesby's New World Vision* (Chapel Hill: University of North Carolina Press, 1998); E. C. Spary, *Utopia's Garden: French Natural History from Old Regime to Revolution* (Chicago: University of Chicago Press, 2000); Antonio Barrera-Osorio, *Experiencing Nature: The Spanish American Empire and the Early Scientific Revolution* (Austin: University of Texas Press, 2006); Bernardino de Sahagún, *Historia General de las Cosas de la Nueva España* (manuscript c.1570); Nicolás Bautista Monardes, *Primera y segunda y tercera partes de la Historia Medicinal de las Cosas que se traen de nuestras Indias Occidentales que siruen en Medicina* (Seville: Alonso Escriuano, 1574); and the essay in this volume by Barrera-Osorio. On "things," see Bill Brown, ed., "Things," special issue of *Critical Inquiry* 28: 1 (Autumn 2001), Lorraine Daston, ed., *Things that Talk: Object Lessons from Art and Science* (New York: Zone, 2004); and Bruno Latour and Peter Weibel, eds, *Making Things Public: Atmospheres of Democracy* (Cambridge, Mass.: MIT Press, 2005).

25 On New World wonder, see Stephen Greenblatt, *Marvelous Possessions: The Wonder of the New World* (Chicago: University of Chicago Press, 1991); and Campbell, *Wonder and Science*; on enlightened utility, see Lissa Roberts, "Going Dutch: situating science in the Dutch Enlightenment," in William Clark, Jan Golinski and Simon Schaffer, eds, *The Sciences in Enlightened Europe* (Chicago: University of Chicago Press, 1999), 350–88, and Lorraine Daston, "Afterword: the ethos of Enlightenment," ibid., 495–504. On the religious context of natural history, see Richard Ligon, *A True & Exact History of Barbados* (London, 1657), 82; Steven J. Harris, "Mapping Jesuit science: the role of travel in the geography of knowledge," in John W. O'Malley, et al., eds, *The Jesuits: Cultures, Sciences, and the Arts* (Toronto: University of Toronto Press, 1999), 212–40; Richard H. Drayton, *Nature's Government: Science, Imperial Britain, and the "Improvement" of the World* (New Haven: Yale University Press, 2000); and Lisbet Koerner, *Linnaeus: Nature and Nation* (Cambridge, Mass.: Harvard University Press, 1999). Unlike Drayton and Koerner, Londa Schiebinger, *Plants and Empire: Colonial Bioprospecting in the Atlantic World* (Cambridge, Mass.: Harvard University Press, 2004) pursues a utilitarian approach to colonial botany that downplays its religious and cosmological significance. Further examples of the scientific role of missionaries in colonial settings other than the Atlantic can be found in: Richard H. Grove, *Green Imperialism: Colonial Expansion, Tropical Island Edens, and the Origins of Environmentalism, 1600–1860* (Cambridge: Cambridge University Press, 1995); and Sujit Sivasundaram, *Nature and the Godly Empire: Science and Evangelical Mission in the Pacific, 1795–1850* (Cambridge: Cambridge University Press, 2005).

26 On the depersonalization of early modern experimental knowledge, see Shapin and Schaffer, *Leviathan and the Air-Pump*; Shapin, *A Social History of Truth*, esp. 243–66; Bourguet, *et al.*, *Instruments, Travel, and Science*; Robert Boyle, *Heads for the Natural History of a Country* (1692), reprinted in Myra Jehlen and

Michael Warner, eds, *The English Literatures of America* (New York and London: Routledge, 1997), 518–21. On travel writing and witnessing, see Juan Pimentel, *Testigos del Mundo: ciencia, literatura y viajes en la Ilustración* (Madrid: Marcial Pons, 2003). On the Jesuits, see Harris, "Mapping Jesuit science," "Long-distance corporations," and Harris, "Confession-building, long-distance networks, and the organization of Jesuit science," *Early Science and Medicine* 1 (1996), 287–318; Harris, "Jesuit scientific activity in the overseas missions, 1540–1773," *Isis* 96 (2005), 71–9; Mordechai Feingold, ed., *Jesuit Science and the Republic of Letters* (Cambridge, Mass.: MIT Press, 2003); Feingold, ed., *New Science and Jesuit Science: Seventeenth-Century Perspectives* (Dordrecht: Kluwer, 2003); Paula Findlen, ed., *Athanasius Kircher: The Last Man Who Knew Everything* (New York: Routledge, 2004); and Luis Millones Figueroa and Domingo Ledezma, eds, *El saber de los jesuitas: historias naturales y el Nuevo Mundo* (Madrid/Frankfurt am Main: Iberoamericana/Vervuert, 2005).

27 Compare Bernard Bailyn, *Atlantic History: Concepts and Contours* (Cambridge, Mass.: Harvard University Press, 2005), 3–56, with Paul A. Kramer, "Empires, exceptions, and Anglo-Saxons: race and rule between the British and United States empires, 1880–1910," *Journal of American History* 88 (2002), 1315–53.

28 Antonello Gerbi, *The Dispute of the New World: The History of a Polemic, 1750–1900*, trans. Jeremy Moyle (Pittsburgh: Pittsburgh University Press, 1973); Jorge Cañizares-Esguerra, *How to Write the History of the New World: Histories, Epistemologies and Identities in the Eighteenth Century Atlantic World* (Stanford: Stanford University Press, 2001); Bauer, *Cultural Geography of Colonial American Literatures*. On eighteenth-century European views of Africans, see Winthrop D. Jordan, *White Over Black: American Attitudes towards the Negro, 1550–1812* (Chapel Hill: University of North Carolina Press, 1968), 216–65; William B. Cohen, *The French Encounter with Africans: White Response to Blacks, 1530–1880* (Bloomington: Indiana University Press, 1980); and Roxann Wheeler, *The Complexion of Race: Categories of Difference in Eighteenth-Century British Culture* (Philadelphia: University of Pennsylvania Press, 2000); on race in the Atlantic world, see Joyce E. Chaplin, "Natural Philosophy and an Early Racial Idiom in North America: Comparing English and Indian Bodies," *William and Mary Quarterly* 54 (1997), 229–52; Colin Kidd, *The Forging of Races: Race and Scripture in the Protestant Atlantic World, 1600–2000* (Cambridge: Cambridge University Press, 2006); Jorge Cañizares-Esguerra, "New world, new stars: patriotic astrology and the invention of Indian and Creole bodies in colonial Spanish America, 1600–1650," *American Historical Review* 104 (1999), 33–68; María-Elena Martínez, "The black blood of new Spain: *Limpieza de Sangre*, racial violence and gendered power in early colonial Mexico," *William and Mary Quarterly* 61 (2004), 479–520; Guillaume Aubert, "'The blood of France': race and purity of blood in the French Atlantic world," *William and Mary Quarterly* 61 (2004), 439–78; on the history of racial thought more generally, see Ivan Hannaford, *Race: The History of an Idea in the West* (Baltimore: Johns Hopkins University Press, 1996).

29 Charles Bonnet, *The Complexion of Nature*, 2 vols (London: T. Longman, 1766), 1: 68–9, quoted in Wheeler, *The Complexion of Race*, 175.

30 John Webber after P. J. de Loutherbourg, *The Apotheosis of Captain Cook* (London: J. Thane, 1794); on this image, see Greg Dening, *Mr Bligh's Bad Language: Passion, Power and Theatre on the Bounty* (Cambridge: Cambridge University Press, 1992), 271–5. On scientific commemoration in eighteenth-century France and Brazil, see Neil Safier, *Measuring the New World: Enlightenment Science and South America* (Chicago: University of Chicago Press, forthcoming); see also the chapter in this volume by Delbourgo.

31 See, for example, James Delbourgo, *A Most Amazing Scene of Wonders: Electricity and Enlightenment in Early America* (Cambridge, Mass.: Harvard University Press, 2006), chaps. 1–3; and the essay in this volume by Bleichmar.

32 See Schaffer, "Golden means"; Beth Fowkes Tobin, *Colonizing Nature: The Tropics in British Arts and Letters, 1760–1820* (Philadelphia: University of Pennsylvania Press, 2005), chap. 2; Schiebinger, *Plants and Empire*; Parrish, *American Curiosity*, chap. 7. See also Vincent Brown, "Spiritual terror and sacred authority in Jamaican slave society," *Slavery and Abolition* 24 (Apr. 2003), 24–53; Karol K. Weaver, *Medical Revolutionaries: The Enslaved Healers of Eighteenth-Century Saint Domingue* (Urbana: University of Illinois Press, 2006); and, in a different context, Stephan Palmié, *Wizards and Scientists: Explorations in Afro-Cuban Modernity and Tradition* (Durham, NC: Duke University Press, 2002).

33 Ian K. Steele, *The English Atlantic, 1675–1740: An Exploration of Culture and Community* (New York and Oxford: Oxford University Press, 1986); C. A. Bayly, *Empire and Information: Intelligence Gathering and Social Communication in India, 1780–1870* (Cambridge: Cambridge University Press, 1996); Kenneth J. Banks, *Chasing Empire across the Sea: Communications and the State in the French Atlantic, 1713–1763* (Montreal and Kingston: McGill-Queen's University Press, 2002). On the question of national Atlantics versus a "pan-Atlantic" perspective, see Jorge Cañizares-Esguerra, *Puritan Conquistadors: Iberianizing the Atlantic, 1550–1700* (Stanford: Stanford University Press, 2006), 215–33.

34 José Luis Maldonado Polo, "El primer gabinete de historia natural de México y el reconocimiento del Noroeste Novohispano," *Estudios de Historia Novohispana* 21 (2000), 49–66; on the Creole collector Pedro Franco Dávila, see Pimentel, *Testigos del Mundo*, 147–78.

35 Juan Pimentel, "The Iberian vision: science and empire in the framework of a universal monarchy, 1500–1800," *Osiris*, 2nd Series, Vol. 15, *Nature and Empire: Science and the Colonial Enterprise*, ed. Roy McLeod (2000), 17–30; James E. McClellan, III, and François Regourd, "The colonial machine: French science and colonization in the ancien regime," *Osiris* 15 (2000), 31–50; Drayton, "Knowledge and empire."

SECTION I
Networks of Circulation

CHAPTER 1

Controlling Knowledge

Navigation, Cartography, and Secrecy in the Early Modern Spanish Atlantic

ALISON SANDMAN

Secrecy is often portrayed as an aberration, a change from the expected course of events. We tend to assume that knowledge was expected to circulate, at least in the absence of compelling reasons otherwise, and indeed that openness and circulation are important concepts in looking at scientific revolution. Though counterexamples are scarcely unknown or undiscussed, much of the recent emphasis has been on intellectual property and the ownership of ideas or on the difficulties involved in transmitting craft knowledge, especially in writing.[1] In discussions of cartography, conversely, the importance of secrets is assumed almost without question—for charts provided information about the locations of places, information that was not only obviously useful for potential competitors but diplomatically sensitive after the division of the world in the treaty of Tordesillas. Even in the face of widespread diffusion of geographic knowledge, worries about the circulation of charts seems to require little explanation. In this essay, I would like to recast these discussions of secrecy and authorship and look instead at attempts to control geographic knowledge—that is, to control what sorts of information were to be revealed, in what forms, and for what purposes.

The problem for Spain arose from overseas exploration—in many ways from the existence of an "Atlantic World." The dilemma faced by Spanish officials, and so by the navigators and cartographers they regulated, was that geographic knowledge needed to be at one time constrained and disseminated. It needed to be constrained to protect the secrets of navigation and so protect Spanish shipping and settlements but, at the same time, ships needed to go back and forth and charts needed to be available for routine use at sea, providing a practical limit on the secrets that could be kept. Furthermore, the need for secrecy was tempered not only by the need for use unsupervised overseas but by the desire to use maps and charts to legitimate territorial claims, a function they could not logically provide without being public.

The paradoxical nature of this attention to secrecy has not gone unremarked. The changes in levels of secrecy surrounding maps have usually been explained in large part as a reaction to changes in the diplomatic situation vis-à-vis Portugal, complicated by the personal interest of Philip II.[2] Both explanations are compelling, at least when applied broadly to the overall interest in geographic information. In this essay, however, I want to turn the question around and look not so much at the *cause* of the interest in secrecy but at some of the *effects*, particularly the effects on navigation and cartography, on attempts to guide ships around the Atlantic world and to gather information about their journeys. The contrasting requirements of keeping geographical information secret while at the same time using it (whether in navigation or diplomacy) encouraged a distinction between two very different types of knowledge about the Atlantic, one championed by cosmographers and the other by pilots, and two very different ideas of expertise.[3]

The cosmographers, especially in their role as chart makers, focused on information about the relative locations of places—latitudes, longitudes, bearings, distances, and even sizes and shapes. This is information that needed to be learned in situ (i.e., by traveling) and required some cosmographic expertise both to gather and to fit together into a coherent system but could be fairly easily disseminated in a chart. The pilots focused instead on how to get from one place to another, which required not only bearings and distances (and potentially latitudes and longitudes) but details of winds and currents and tides and the entrances to ports. This required mastering a wealth of details, best learned through extensive experience at sea, and in practice difficult to teach without that extensive personal experience at sea.

As officials grappled with the difficulties of keeping information secret, I argue that they developed different strategies for the two types of information. As the information valued by the cosmographers was at one time more useful for diplomacy and less useful for navigation compared to the types of knowledge preferred by the pilots, attempts to control it came to depend as much on careful dissemination as on secrecy per se. At the same time, the detailed local knowledge of the pilots, whether held in their persons or in more detailed charts of harbors or headland views, remained secret while being more and more left out of discussions altogether.

However, this strategy itself had side effects. As the knowledge to be drawn on charts and disseminated more widely diverged from the knowledge used by the pilots in their navigation, the pilots became less and less useful as sources of the information to be drawn on the charts. By the end of the century, this led to an increasing emphasis on methods for long-distance control of the pilots, replacing attempts to enlist their expert aid.[4] The conversion of navigation from a craft to an applied science is usually seen either as a victory for theoretical knowledge or as the successful union of theory and practice. I argue here that it grew in part out of an attempt to subdivide geographic knowledge so as to publicize some while keeping the rest secret and that the lack of public attention to the craft aspects of navigation followed the tacit categorization of this knowledge as secret. The universal, theoretical, systematic knowledge of the cosmographers was simultaneously emphasized and publicized, whereas the tacit, experiential, craft knowledge of the pilots was kept secret and so written out of the discussion.

Restricting Knowledge: Secrecy in Theory and Practice

Before looking in more detail at the effects of secrecy on Atlantic knowledges or the subtle compromise forged by the Spanish officials, I would like to first emphasize the ubiquity of secrecy in sixteenth-century discussions of geography, cartography, and navigation. Discussions of all sorts repeatedly emphasized the importance of restricting knowledge and especially of keeping it out of the hands of foreigners. One cosmographer, for example, denounced a Portuguese chart maker living in Seville on the grounds that he could easily copy charts and send them to his family back in Portugal, allowing the Portuguese to know "the navigation to the Indies and the secrets thereof."[5] Another warned that if foreigners had copies of charts, "they would be able to go to the said Indies and make themselves corsairs, and sack towns as they have done and do every day." On the basis of this reasoning, he advised that an Italian cosmographer should not only be forbidden a license to make and sell charts but should have all charts in his possession confiscated.[6] Both cosmographers treated knowledge of geography, cartography, and the details of navigation to the Indies as state secrets, damaging if revealed.

These particular arguments were somewhat self-serving, using ominous warnings to cloak economic interests and try to drive competitors out of the market. Nonetheless, there were indeed genuine concerns about secrecy, justified in large part by the possible uses of charts both as tools for navigators and as authoritative statements about the locations of places. Given proper instruction both in methods of navigation (which were to some extent general) and the specific routes the Spanish used to get to their overseas territories, and given the geographical information underlying those routes (in either textual or cartographic form), foreign pilots could guide ships to the Indies, attacking Spanish shipping and infringing Spanish colonial monopolies. Moreover, in drawing charts and compiling geographic information, the makers necessarily took positions in the ongoing disputes about the exact locations of places. As Spanish justifications of overseas territorial claims were based in large part on their locations with respect to the demarcation line with Portugal, all geographies were potentially politically sensitive. It would not do for the geographic experts to cede to Portugal territories claimed by the diplomats, no more than for them to train and outfit the pilots of competing countries. Such were the concerns of the king and the Council of the Indies.

These fears were not expressed purely as rhetorical flourishes but in an increasingly elaborated body of laws governing pilots, charts, and nautical instruments. In 1510, a new set of rules for the Casa de la Contratación, the newly founded institution in charge of regulating navigation, explicitly forbade anyone to give out sea charts or information on the Indies without permission.[7] Those in positions of trust were subject to further limitations. Part of the formalities for licensing pilots was a formal oath not to "give or sell or lend the sea chart to a foreigner from outside of this kingdom."[8] Chart makers were forbidden to sell their wares to foreigners and, despite complaints of economic hardship, the Council repeatedly refused to grant exceptions.[9] To limit access, the official pattern charts, the exemplars from which all pilots' charts were supposed to be copied, were kept in a locked chest in the Casa; two separate keys kept by different people were required for access.[10]

It is hardly surprising, then, that foreign chart makers came under suspicion. In 1563, the cosmographer Sancho Gutiérrez objected to the presence of a Portuguese chart maker in Seville, in one of the cases cited above. A certain Andrés Freile from Lisbon, son of a Portuguese cosmographer, had formerly made charts in Seville, but when he was forbidden to do so, he returned to his family in Portugal. Now, however, he had returned and married a local woman (allowing him to be naturalized) and was trying to establish himself making charts. This was dangerous, according to Gutiérrez:

> because he had a copy of the pattern charts for the navigation to the Indies, and if he did not have one he could easily make one and send it to his father, and it would be the occasion that the Portuguese know the navigation to the Indies and the secrets thereof, which would be a great inconvenience…[11]

The Council ordered that the Casa officials investigate, and if they found that Andrés Freile was indeed Portuguese, they should seize his charts and forbid him to make others.

A similar objection was raised against Neapolitan cosmographer Domingo de Villarroel (Doménico Vigliarolo) in the 1580s.[12] The pilot major, Alonso de Chaves, refused to examine his charts for approval, explaining that foreigners were barred from the trade.[13] Asked to justify himself further, Chaves warned that "as such a foreigner and acquaintance and associate of foreigners, he [Villarroel] could give and sell the said charts to them."[14] Chaves concluded that not only should Villarroel be forbidden to sell charts but he should not even be allowed to own them, for he could always take the charts and "go with them to foreign realms" where he could make more and sell them. Chaves suggested that the officials seize all of the charts Villarroel had in his possession, both the pattern charts and the copies he had made, and lock them up in the Casa until the matter was resolved. He emphasized that the entire matter was very important "for the security, pacification, and tranquility of the said navigation of the Indies."[15] The Casa officials decided (without comment) to refer the matter to the Council of the Indies. Meanwhile, Villarroel continued to offer charts for sale.

The laws governing pilots were equally clear. As there was ample opportunity to flee the country should any pilot so desire, the laws focused on limiting who was allowed to become a pilot. Only native or naturalized Castilians were eligible to be licensed as pilots, and they had to provide the information about their eligibility before they could even attend cosmography classes, let alone convene a tribunal for the mandatory licensing exams.[16] Cosmographers teaching classes aimed at pilots swore oaths not to share their knowledge with foreigners.[17] Disputes about nationality restrictions were not uncommon; the pilot major questioned the eligibility of men such as a foundling orphan adopted by a prominent ship owner at the approximate age of seven and a man brought from Portugal as a babe in arms and raised in Seville by his father, himself naturalized.[18]

Despite a consistent shortage of licensed pilots, the Council repeatedly ruled that no one without significant family or property ties to the kingdom should be allowed to be a pilot.

This idea that pilots held privileged knowledge extended even to their associates. Instrument makers were also licensed by the Casa, though these licenses were intended for quality control and to limit competition rather than to control information. As French astrolabe maker Pedro Grateo pointed out, there was nothing secret about an astrolabe. Nonetheless, his licensing was opposed on grounds of secrecy. Rodrigo Zamorano, a cosmographer and rival instrument maker and pilot major, argued that technically he could not even inspect Grateo's instruments, because any necessary corrections would violate his own oath not to teach foreigners. He later argued that even if Grateo were wholly competent, licensing him would be a bad idea because it would give him the opportunity to listen to pilots' gossip:

> and so there is no need for Frenchmen or other foreigners to make [instruments], nor to involve themselves in spying the secrets of the navigation to the Indies, which secrets the pilots of the said navigation, without any caution, speak of and confer about in front of the people who make and sell the said astrolabes and other instruments...[19]

Though astrolabes were in no way secret, he argued, foreigners should not be afforded such opportunities to interact with pilots and learn their secrets. No disposition of the case is recorded, but Grateo is not known to have continued selling nautical instruments in Seville.

Despite this continuing rhetoric, however, true secrecy about either the geography of the Indies or the methods for getting there was clearly a lost cause. Throughout the sixteenth century, skilled mariners and cartographers moved freely from country to country, despite occasional complaints, with Ferdinand Magellan and his cartographical entourage being merely an extreme example. Nor was this interchange of knowledge and expertise limited to Iberia, though as Spain and Portugal funded the majority of the explorations, they received most benefit from the international migrations. The first pilot major, Amerigo Vespucci, was Florentine, with earlier experience exploring for both Spain and Portugal. His nephew, Juan (Giovanni) Vespucci, worked for Spain as a chart maker for more than a decade after his uncle's death, though he was eventually dismissed, perhaps because he was suspected of sending information to the Medici.[20] Another pilot major, Sebastian Cabot, worked for England both before and after his time in Spain. Though the English were eager to learn from him Spanish methods of navigation, the Spanish showed little concern about his departure beyond the hassle of having to fill the empty position.[21] The prevalence of foreigners (and their tendency to defect) was most pronounced in the first half of the century, but as late as the 1590s Domingo Villarroel's departure for France aroused little protest; throughout the century, almost all concerns about foreigners were raised by those with a direct economic interest in suppressing their participation in the Enterprise of the Indies.

Though ordinary pilots were subject to more restrictions than were the people regulating them, it seems highly unlikely that this prevented foreigners from learning navigation in Spain. Most pilots learned their trade aboard ships, and the large percentage of foreigners making up most ships' crews was a notorious cause of concern throughout the century. Even after a 1568 regulation limited the number of foreign crewmembers to six per ship, the official records still show 20 percent foreigners in the last quarter of the sixteenth century, and that is almost certainly an underestimate.[22] Though foreigners could not go on to get licensed as a pilot, the nationality checks came at the very end of the process, so that most of the foreigners denied the right to take the licensing exam were already experienced seamen who considered themselves capable pilots; indeed, no one presented himself before the pilot major without being prepared with testimonials to that effect.[23] The group of Bristol merchants who paid to send a pilot with Cabot's 1526 expedition so that he could learn the routes firsthand were simply more explicit about their aims than was the norm later in the century.[24] In the voluminous records surrounding the voyage there is no evidence that anyone objected to their plan.

Once a ship set sail, geographic information was even harder to control, as a brief consideration of piracy makes thoroughly clear. Both charts and pilots were liable to be captured by pirates or enemy warships. The Portuguese pilot, Nuño da Silva, served Drake for more than a year after being captured, but though local authorities barred him from the Indies for life, the Council of the Indies determined that he wasn't at fault and sent him back to Seville with a safe-conduct (though they did ask that he be watched).[25] According to the accounts of his released prisoners, Drake seems to have made a habit of capturing pilots and interrogating them about local conditions; aside from Silva, he also held (for greater or lesser periods) four pilots and a sailor, Juan Pascual.[26] His concerns seem to have been twofold; first he sought people (such as Silva) who could guide him through the strait of Magellan, widely considered exceptionally difficult and dangerous.[27] Once through the strait, he also sought people with local knowledge; though not a pilot, Juan Pascual was described as a person "acquainted with this coast" who knew "the ports in which water and wood were available."

Drake also routinely seized charts from his conquests, though his primary chart, on which he displayed his potential routes for his prisoners, was made for him in Lisbon.[28] He also carried Spanish charts for the coast of Chile, notwithstanding complaints that these were inaccurate, if not deliberately falsified.[29] As one crewmember later complained,

> Wee following the directions of the common Mapps of the Spanyards were utterly deceaued for of a Malitious Purpose they had set forth the mapp false that they might deceave strangers if anny gave the attempt to trauail that way that they might perish by Running ofe to the sea rather than Touch with anny part of the land of America.[30]

Whether these maps were deliberately deceptive (and if they were, this would only emphasize the importance placed on hiding geographical knowledge), they were manifestly ineffective in keeping the English away from territory claimed by the Spanish. Indeed, Drake was certainly not alone in his incursions; pirates from a variety of countries regularly attacked Spanish convoys and coastal settlements. The very existence, much less notoriety, of such raiders made it clear that the ability to navigate the Indies was far from a Spanish monopoly.[31]

Furthermore, pirates and privateers were not alone in their opposition to the Iberian monopolies. The short-lived French attempt to settle Florida (1562–65) indicates the limited value of geographical secrets.[32] Jean Ribault, the leader of the 1562 expedition, was delighted to report that he had pioneered a new route to the Indies, to the north of the traditional Spanish routes but to the south of those taken by the Newfoundland fishermen. This route, he explained, was better than the one taken by the Spanish because it allowed him to bypass not only the Spanish Antilles but the Bahama channel, arriving directly in Florida, where (despite the indications of shipwrecks on his charts) he found good ports and rivers.[33] His route, thus, was not only shorter but it allowed him to avoid contact with the Spanish and their possibly hostile reactions on learning his intentions. He did not make any attempt to claim that he did not know the Spanish routes or could not follow them but celebrated his own bravery and sense of adventure in pioneering an untried route. The next French voyager, however, abandoned Ribault's route across the Atlantic for the traditional one, making it quite clear that by 1562 the outlines of the route were assumed as common knowledge among seafarers. The Spanish were reduced to defending their territorial claims by force of arms rather than by geographical secrecy.

Thus, between pirates, Frenchmen, and foreign pilots of all sorts, no one in Spain could have supposed that geographical information was truly secret. As the author of one book on navigation argued around 1580, the pirates already had the information they needed to get to the Indies, so there was no reason not to allow him to publish his book.[34] However, neither did anyone conclude from this that all information should be freely available, and indeed the book in question was not published, though a fair copy was made for the Council of the Indies. The question was how to balance the conflicting demands.

In this context, the case of Villarroel, the Neapolitan chart maker who sought a license to sell charts in Seville is instructive. Discounting the pilot major's impassioned arguments in favor of keeping foreigners out of the chart trade, the Council of the Indies acceded instead to arguments from the pilot's representatives that charts were becoming scarce and expensive. Despite his nationality, Villarroel was licensed to make and sell charts in Seville. This solution, however, was relatively short-lived. Complaining of infighting and poverty, he left Spain after little more than a decade and took his box of maps with him to France, where he considered employment in La Rochelle and Bordeaux. Despite having seen the pilot major's direst predictions come true, however, Spanish officials showed little concern. The main information on Villarroel's flight comes from letters introduced as evidence by a rival (the same Zamorano who had complained about the French astrolabe maker) accusing him of going to France to give away the secrets of the navigation to the Indies and requesting his abandoned job.[35] No one else mentioned any danger from either his employment in France or the charts he took with him.

Different Types of Knowledge

So to what extent did anyone actually care about secrecy? One tempting explanation of this confused mass of laws and rhetoric surrounding secrecy, given its impracticality, is simply that it was never intended to work—that the rhetoric was simply a screen for attempts to keep rival makers from entering the market, and that their foreignness simply provided additional ammunition against them rather than a distinct motive for opposing their suit. This explains quite well the actions of the cosmographers in Seville, who were concerned with the difficulties of making a living selling charts and instruments and jealous of any competition. However, it fails to explain why their complaints met with such consistent support from the higher authorities, especially the Council of the Indies, and it fails to explain the repeated refusal to publish geography books.

The Council of the Indies did, I think, genuinely desire to keep charts secret, and genuinely support the laws instituted for that purpose but, at the same time, they always had other concerns. First of all, as charts were useful for navigation not only to interloping foreigners but to Spanish pilots, there was an incentive to ensure a steady supply, even if this made charts easier for foreigners to acquire. This was the motive for allowing Villarroel to make and sell charts.

More important, charts were also used as evidence of the locations of places, especially under Charles V, but charts could lend legitimacy to territorial claims only to the extent that they were public. To this end, ornate versions of charts were given as gifts to foreign dignitaries, whereas others (provided by both sides) were cited by commissions meeting to determine international boundaries.[36] However, if charts were to be public, their message needed to be controlled. Thus, every time there was a debate about charts in Seville, the delegated officials checked each rival chart for possible ramifications about territorial claims, and chart makers duly produced witnesses that their charts did not move the line of demarcation.[37] Though charts were still ostensibly secret, the concern that they not misrepresent the line of demarcation implicitly acknowledges that they would become more widely available.

Fortunately, the information that made the charts most useful diplomatically—latitude, longitude, and boundary lines—was not what was considered most useful for navigation. This contributed, I think, to an unstated compromise. Within the context of general laws about secrecy, attempts to limit information focused on detailed local knowledge and so on harbor views, details of fortifications and, most important, the persons of the pilots and their knowledge of currents, weather conditions, harbors, and places to find wood and water. How to cross the Atlantic was common knowledge (within limits), but how to get through the strait of Magellan and where best to go to make repairs was still (perhaps) proprietary. Certainly this was the type of information sought by Drake. These are the sort of details that could be picked up from pilots, even pilots gossiping in front of an astrolabe maker, but would not be included on the planispheres made as gifts for foreign dignitaries or even (ordinarily) written down at all.

Meanwhile, attempts to disseminate geographic knowledge relied on knowledge of a different scale and type—focusing on charts and planispheres that displayed large portions of the earth, usually in terms of latitude and longitude. The information available on the charts used at sea and associated with the Casa de la Contratación did not include the type of details necessary to enter ports, recognize headlands, or avoid even the most common hazards. As experience remained essential to successful navigation, the inadvertent revelation of these charts posed little practical risk.

The attempts at secrecy, then, help to reveal the disjuncture between two very different concepts of knowledge, both of what types of things were useful to know and of how best to gain this knowledge. This is not to say that pilots were not concerned with latitudes, for they were, but they also insisted on the importance of more accidental local factors and so on the inescapable importance of experience. A definition put forth by the Universidad de Mareantes, which acted as a guild for pilots, included knowledge of winds, routes, lands, ports, and tides, finding latitude from the sun and north star, handling the sails, setting the course, estimating distances traveled, and in general having good judgment.[38] The Universidad repeatedly argued that only men brought up at sea could possibly acquire the experience to make a good pilot and that the key role of licensing exams was to screen out imposters without such lifelong experience. They warned of the dangers of letting candidates study cosmography before taking the exam and so being able to hide their lack of experience and true ability behind a thin veneer of learning.[39] Though much of the debate was about who should control navigation, it also revealed a different appreciation of what the job entailed. In modern terms, while cartographers were concerned only with features relevant to deep-sea navigation, pilots were also concerned with ship handling and pilotage. And this sort of experiential detail could be kept secret, even while the planispheres were allowed to circulate.

Gathering and Communicating Knowledge

In the final section of the paper, I wish to argue (a bit speculatively) that these distinctions—between local knowledge (still worth trying to keep secret) and more general knowledge of latitudes and longitudes (that either should not or could not be secret)—helped to shape cartography in the period and especially the attempts to gather information. The charts made at the Casa de la Contratación—the charts that were officially regulated and inspected and sold—emphasized latitudes and longitudes and distances and compass bearings, rather than headland views or careful soundings. That is, they included the details useful to armchair navigators planning trips, to diplomats making territorial claims, and to some extent to the leaders of convoys planning general routes across the Atlantic, but not those important to individual pilots attempting to find a safe anchorage, enter a tricky port, or orient themselves approaching an unfamiliar coast.

This disjunction, however, meant that the pilots and cartographers were increasingly at odds about what counted as useful information—while cartographers emphasized the importance of precise latitudes, pilots were more likely to mention the desirability of naming and identifying all ports on a given island. As a result, over the course of the century, cosmographers began first to discount the expertise of the pilots (because they were not reporting the desired information) and then to work on methods to constrain the actions of the pilots, so that they would provide the necessary information despite themselves. Rather than synthesizing and correcting the information provided by the pilots (as they had done in the middle of the century), the cosmographers attempted to force the pilots to gather the new types of information requested by the cosmographers, primarily latitudes and longitudes.

This new focus is clear in a variety of attempts to gather information in the last decades of the sixteenth century, all of which found the cosmographers asking extremely specific questions. Moving beyond the long-standing requirement for all pilots to report to the cosmographer in residence at the end of the voyage, they instead worked to distribute forms and questionnaires. These forms constrained the actions available to the pilots (and occasionally other agents pressed into service) in two distinct ways. First they worked to standardize the information returned, both in form and in content, so that all was directly comparable and no area was abandoned or forgotten. More important, they required the agents to report the empirical data as directly as possible, leaving all interpretation and calculation for the experts in Spain. As they increasingly distrusted the expertise of the pilots, cosmographers worked to limit their discretion.

Ideally, none of this would have been necessary, and cosmographers could simply have delegated one of their own to make the necessary observations. As there were not, however, a sufficient number of qualified cosmographers to send one with every ship, in practice using cosmographers required a special-purpose expedition. This was tried in the 1580s, when Jaime Juan, a Valencian cosmographer, was sent on a surveying expedition to New Spain and the Philippines.[40] Described as a "man expert in mathematics and astronomical calculations," he was accompanied by assistants (including one specified as a painter), experimental instruments to try out, and detailed instructions.[41] He was supposed to observe latitude and compass declination everywhere possible, make maps of every place he went, copy any other maps he found, observe lunar eclipses when they occurred, and meet up and consult with other cosmographers. He was also asked to teach the pilots on his ship how to use the instruments he carried and to watch carefully to see what sorts of errors they made. The entire project was expected to take at least six or eight years. Unfortunately, however, Juan died before finishing the expedition, and most of his papers were lost. The lack of usable results had a dampening effect on other suggested expeditions, suggesting that alternative approaches might prove more cost effective.

One alternate approach was to enlist people already present in the places to be surveyed, thus avoiding the trouble and expense of a voyage. In the 1570s and 1580s the cosmographer major, Juan Lopez de Velasco, tried to gather information about lunar eclipses this way.[42] His was not a small-scale endeavor—in several different letters, he requested observations for eclipses in 1577, 1578, 1581, 1582, 1584, and 1588—asking for observations from a variety of New and Old World locations, which could (if reliable) be used to fix points on the charts, anchoring the longitude estimates obtained by other means.[43] Copies of the request were sent throughout Spanish territories—the copy of one such letter in the outgoing letter register lists more than twenty-five addressees—and each letter gave the dates of the next lunar eclipses and asked the recipient to arrange for observations. The request was accompanied by a set of detailed instructions on how to make the observations—printed by order of the Council of the Indies, presumably to make sure that everyone received sufficient and identical information about how to proceed.[44]

The topics covered in the instructions indicate their concern with wresting data from the ill-trained.[45] Despite their length, they never gave any overall explanation of what was to be done or what information was to be gathered or why any of this was useful—all of which would be helpful in understanding the instructions. Notably, the instructions gave no procedure for determining from the observations the local time of the lunar eclipse, which was the entire point of the endeavor. Instead, the instructions simply provided step-by-step procedures, strictly limited to those activities that absolutely had to be done in the field.

The instructions indicate the limited expectations Velasco had for his chosen observers, for he made every effort to specify not only what information should be gathered but how it should be recorded. According to the instructions, for example, the observer should carefully measure with dividers the length of the noon shadow, and record that information: "on a sheet, not of parchment but of paper, they are to make two rays or straight lines in ink, one exactly the size of the shadow, no matter how small, and the other equal to the length of the stile … declaring in writing on each of the said lines which is the measure of the shadow, and which of the stile." The care to produce a graphical result (rather than a measurement) is symptomatic of the desire to do all the calculations centrally, rather than relying on distributed expertise. A tracing of the length of the shadow onto paper—the least elastic of the available materials—would have been the closest thing possible to a direct observation, eliminating as much on-site interpretation as possible and leaving both calculation and interpretation for the experts in Seville. The enterprise was not a resounding success. Few observations were sent back to Spain, and most of those came from cosmographers. Andrés García de Céspedes, who made extensive reference to Velasco's results and reprinted the instructions as an appendix to his book on hydrography, nonetheless listed more eclipse observations for Spain (where they added little information) than for the New World.

Despite the disappointing results, Spanish cosmographers did not abandon attempts to gather latitudes and longitudes. In 1593, a committee made up of pilots and cosmographers met under the direction of Pedro Ambrosio Ondériz, who had been sent from court to reform charts, instruments, and navigation practices.[46] They determined that more detailed information was needed about the exact locations of places in the Indies and designed a detailed set of instructions to go out with the next convoys. As had the eclipse instructions, these also tried to minimize the calculations and interpretations required of the pilots.

First of all, the instructions specified that the pilots were to have the newly designed astrolabes, the ones that were large enough to be divided into half-degrees, and so to increase the precision of the pilots' measurements. Using either those astrolabes or their own, the pilots were to find the noon altitude of the sun. The instructions were quite clear that this was the only information the pilots need provide: "he should not take into account declination, nor the rules for taking the latitudes of places, but only what the astrolabe says…"[47] The pilots were not to make any adjustments or calculations, which could introduce errors, but simply to report the raw observations, complete (presumably) with place and date. Note that this measurement should not have been difficult for the pilots—the observation itself, and the calculations necessary for finding latitude from it, had been a mandated part of every pilots' licensing exam for several generations. However, the format represents a distinct demotion in the status of the pilots; the choice between asking whether an island is positioned correctly on the charts and asking for the height of the sun at noon at that island on a specific day marks a profound difference of opinion about the expertise of the observer.

The rest of the instructions consisted largely of a list of the usual routes and the places along them that should be measured—thus, the New Spain fleet was instructed to take observations in the Canaries, La Deseada, Antigua, Guadalupe, Monserrat, Puerto Rico, and many other islands. This is not a trivial number of observations—essentially the pilots were to take observations at every possible stop along their routes. If followed, these instructions would lead to detailed data about the latitudes of most ports in the New World, data that were relatively independent of the understanding of the pilots, as they were responsible only for the raw data, and the processing was to take place in Spain.

Unfortunately, most of the pilots did not do the observations, or at least did not return them. Ondériz complained that he received only three responses, and he did not trust even them.[48] This general lack of interest indicates that despite the stated justification of improving navigation, the desire for latitudes came far more from cosmographers and royal officials than from pilots. When the pilots were asked explicitly about problems in their charts, their responses indicated that they did indeed use latitudes but that their sense of what most needed fixing on the charts was notably different from the concerns of the cosmographers. Though there were a few complaints about mistaken latitudes, the more common suggestions concerned level of detail; several pilots suggested that the charts should be larger, and some described specific danger spots that should be marked.[49] This was the sort of local knowledge that there was no diplomatic advantage to disseminating.

By this time, the cosmographers' preoccupation with latitudes and longitudes had caused their interests to largely diverge from the requirements of navigation. Even the traditional reports of the pilots, cast in terms of distances and bearings, began to be reinterpreted to put the information into the terms preferred by the cosmographers. Andrés García de Céspedes, for example, spent many pages explaining how to reinterpret the pilots' reports of their courses to gain reliable information about longitude, despite his claim that the pilots themselves could not be trusted to correctly perform the calculations or even to understand the changing relationship between longitude and distance at different parts of the earth.[50] Though Céspedes had presided over a general reform of navigation and cartography, his main interest was in longitude and particularly in using new information about longitudes to uphold the validity of Spain's claims in the East Indies. He was far less interested in the ins and outs of getting safely to the Indies.

In conclusion, as the standards of the cosmographers came to predominate, their interest in latitude and longitude and their lack of interest in the sorts of local knowledge learned best through experience came to define navigation. The types of knowledge most important for cartography came to be the only types worth discussing. And the information that was still somewhat secret, the details of ports and currents and sandbanks and reefs, was written out of the discussions of navigation, which instead came to be cast solely in terms of celestial navigation. Professionalized, systematized navigation came to rely on knowledge that could be drawn on charts and safely allowed to circulate. Meanwhile, the secrets of the pilots, learned only at sea, remained secret (insofar as they did) primarily by being unspoken and unwritten.

Notes

1. See especially Pamela O. Long, *Openness, Secrecy, Authorship: Technical Arts and the Culture of Knowledge from Antiquity to the Renaissance* (Baltimore: Johns Hopkins University Press, 2001).
2. See especially Richard L. Kagan, "Arcana Imperii: Mapas, ciencia y poder en la corte de Felipe IV," 49–70 in *El Atlas del Rey Planeta: La "Descripción de España y de las costas y puertos de sus reinos" de Pedro Texeira (1634)*, ed. Felipe Pereda and Fernando Marías (Madrid: Nerea, 2002), 63–8.
3. On early modern ideas of expertise, see Eric H. Ash, *Power, Knowledge, and Expertise in Elizabethan England* (Baltimore: Johns Hopkins University Press, 2004). In addition, my distinction between universal and local knowledge is closely related to the one he draws between theoretical and empirical knowledge.
4. Long-distance control and the difficulties of using agents has of course been discussed in the science studies literature. See especially Steven J. Harris, "Long-distance corporations, big sciences, and the geography of knowledge," *Configurations* 6 (1998), 269–304 and John Law, "On the methods of long-distance control: vessels, navigation and the Portuguese route to India," in John Law, ed., *Power, Action and Belief: A New Sociology of Knowledge?* (Boston: Routledge & Kegan Paul, 1986), 234–63.
5. Seville, Archivo General de Indias (henceforth AGI) Indiferente, 1966, L. 15, fol. 18r, December 16, 1563 letter from the Council of the Indies to the Casa officials.
6. AGI, Patronato, 252, R.77, block 2, fol. 6v, undated 1584 statement from Alonso de Chaves.
7. José Pulido Rubio, *El Piloto Mayor: Pilotos Mayores, Catedraticos de Cosmografía y Cosmógrafos de la Casa de la Contratación de Sevilla*, Expansion of 1923 edition (Seville: Escuela de Estudios Hispano-Americanos, 1950), 382.
8. This order appears in the official record of each licensing exam, many of which for the period following 1568 are preserved in AGI Contratación 54A and B. I have quoted from the earliest licensing exam I have located, the June 22, 1551 *carta de examen* of Benito Sanchez: "… no deis ny vendais ny presteis la carta de marear a estranjero fuera del rreyno…" AGI, Justicia, 836, N. 6, image 795.
9. See, for example, AGI, Indiferente, 741, N. 128, October 30, 1586 consulta of the Council of the Indies on the salary of Domingo de Villarroel, discussing his complaints of poverty.
10. AGI, Contratación, 5554, September 5, 1592 petition from Domingo de Villarroel to inspect the box; upon inspection one lock was found to be broken, and the contents were listed.

11 "[P]or que tiene sacado el treslado de los padrones de la navegaçion de las yndias, y que si no le a sacado le podra sacar con façilidad y enbiarle a su padre y que sera ocasion que los portugeses sepan la navegaçion de las yndias y secretos de ellas y que seria cosa de gran ynconviniente…" AGI, Indiferente, 1966, L. 15, fol. 18r, December 16, 1563 letter from the Council of the Indies to the Casa officials.

12 For a brief biography, see Agustín Hernando, "Los Cosmógrafos de la Casa de Contratación y la Cartografía de Andalucía," in *Miscelanea Geografica en Homenaje al Professor Luis Gil Varon* (Cordoba: Servicio de Publicaciones de la Universidad de Córdoba, 1994), 134–6; for a detailed account of his time in Spain, see Pulido Rubio, *El Piloto Mayor (1950)*, 647–95.

13 AGI, Patronato, 262, R.1, block 2, fols. 9r–10v, August 4, 1584 statement from Alonso de Chaves.

14 "[Y] siendo esto ansi/ como lo es= con justa causa e rrazon se podra y puede mandar e proveer que el dicho domyngo de villarroel siendo como es estrangero ytaliano e tan çerrado de lengua e no conoçido= no haga ny venda las dichas cartas= porque como tal estrangero e conoçido e famyliar de estrangeros les puede dar e vender las dichas cartas para el efeto dicho y aun se podria henobar? la mar de cosarioss y enemigos." AGI, Patronato, 262, R.1, block 2, fol. 9v.

15 "[E] como tal estrangero tenyendo como tiene en su poder los patrones de las cartas desta navegaçion se puede yr con ellos a rreynos estranoss donde las haga e venda e rresulta el mysmo daño y asi conviene que luego se saquen de su poder los dichos padrones e las demas cartas que tuviere fechas e se pongan en la dicha casa hasta tanto que su magestad siendo ynformado desto provea lo que mas convenga a su serbiçio por ser como es negoçio de mucha ynportançia para la seguridad paçificaçion e quietud de la dicha navegaçion de sus yndias" AGI, Patronato, 262, R.1, block 2, fols. 9v–10r.

16 The treatment of foreigners who wanted to be pilots was frequently controversial, since there were rarely enough Castilians to supply the need. See AGI, Indiferente, 1961, L. 3, f. 183rv, December 11, 1534 cedula to Sebastian Cabot clarifying rules for excluding foreigners from pilots' licensing exams; AGI, Patronato, 251, R. 47, August 2, 1547 cedula to Sebastian Cabot discussing foreigners being allowed to take the pilot licensing exam; AGI, Indiferente, 1562, L. 2, f. 57, October 29, 1561 account of treatment of foreigners who applied to be licensed as pilots; AGI, Indiferente, 1966, L. 14, ff. 98r–99v, November 9, 1561 letter to Casa officials, giving new rules for licensing foreigners as pilots.

17 AGI, Contratación, 734, No. 1, fol. 5rv, October 25, 1593 petition from Rodrigo Zamorano.

18 AGI, Indiferente, 2005, 1593 case of Francisco Manuel, foundling of unknown nationality; AGI, Justicia, 1154, N. 6, R. 2, case of Juan Nuñez, portuguese. Manuel was allowed to take the licensing exam, but the Council of the Indies denied Nuñez's appeal.

19 "[Y] assi no ay necesidad de que officiales franceses ni otros estrangeros destos reynos los hagan ni se entremetan a espiar los secretos de la navegacion de las indias, los quales los pilotos de la dicha navegacion sin recato alguno tratan y confieren delante de las personas que les hazen y venden los dichos astrolabios y otros instrumentos, y assi por sola esta causa se confia esto de los cosmografos

de su magestad por ser personas de quien se tiene toda satisfacion." AGI, Contratacion, 734, No. 1, fols. 7v–8r, October 29, [1593] petition from Rodrigo Zamorano.

20 Consuelo Varela, *Colón y los Florentinos* (Madrid: Alianza Editorial, 1988), 78–81.

21 His departure did spark several years of diplomatic interchanges about whether he would be allowed to visit the emperor to reveal certain unnamed secrets. Cabot asked the Spanish to get him permission to leave, while asking the English to protect him. Eventually both sides saw through the charade. Cabot later sent a letter revealing alleged French and English plans to attack Peru through the Amazon; it was forwarded without comment. No one in Spain expressed any worry about what he might reveal to the English. See Alison Sandman and Eric Ash, "Trading expertise: Sebastian Cabot between Spain and England," *Renaissance Quarterly* 57 (2004), 813–46.

22 Pablo Emilio Pérez-Mallaína Bueno, *Spain's Men of the Sea: Daily Life on the Indies Fleets in the Sixteenth Century*, translated by Carla Rahn Phillips (Baltimore: Johns Hopkins University Press, 1998), 54–62.

23 The requirements for a pilots' license included six years of experience at sea, character references, and experienced pilots willing to swear that the pilot was fit. Only at this point would a seaman petition the pilot major for permission to present testimony and attend the cosmography classes. Thus, all foreigners refused by the pilot major can be assumed to have at least six years experience sailing to the Indies.

24 On this group, see E. G. R. Taylor, "Introduction," in *A Brief Summe of Geographie by Roger Barlow*, ed. Taylor, Hakluyt Society, Second Series, vol. 69 (London: Hakluyt Society, 1932).

25 On Silva's reception on his release, including his own depositions and the documents of his inquisition trial, see Zelia Nuttall, trans. and ed., *New Light on Drake: A Collection of Documents Relating to His Voyage of Circumnavigation 1577–1580*, Works Issued by the Hakluyt Society, Second Series, vol. 34 (London: Hakluyt Society, 1914), 242–71 and 295–399. On the decision of the Council of the Indies, see 397–9.

26 The pilots were Juan Griego, Custodio Rodriguez, San Juan de Anton, and Alonso Sanchez Colchero. On Griego, see Pedro Sarmiento de Gamboa, "Relación de lo que el Cosario Francisco Hizo y Robó en la Costa de Chile y Pirú," in *Viajes al Estrecho de Magallanes (1579–1584)*, 2 vols, ed. Ángel Rosenblat (Buenos Aires: Emecé Editores, 1950), 182–3; on Rodriguez, see Nuttall, *New Light on Drake*, 138; on Anton, see pages 163–75; on Sanchez, see pages 183, 187, 195–6; on Pascual, see pages 197 and 335. Drake also captured a pilot named Benito Diaz Bravo who was released after only one day (146).

27 Helen Wallis, "English enterprise in the region of the Strait of Magellan," in *Merchants and Scholars: Essays in the History of Exploration and Trade*, ed. John Parker (Minneapolis: University of Minnesota Press, 1965), especially 208.

28 For a summary of Drake's sources of information, see Helen Wallis, "The cartography of Drake's voyage," in *Sir Francis Drake and the Famous Voyage, 1577–1580: Essays Commemorating the Quadricentennial of Drake's*

Circumnavigation of the Earth, ed. Norman J.W. Thrower (Berkeley: University of California Press, 1984), 131; For primary source references to seizing charts, see Sarmiento de Gamboa, "Relación de lo Que el Cosario Francisco Hizo y Robó en la Costa de Chile y Pirú," 181; Nuttall, *New Light on Drake*, 66, 184, 186, 197; on the chart bought in Lisbon (for 800 cruzados), see Nuttall, *New Light on Drake*, 162.

29 San Juan de Anton reported complaints that the falsified ("falsa") chart was part of the reason the fleet got separated.

30 Quoted in Wallis, "The cartography of Drake's voyage," 131. Wallis thinks the accusation is unjust, and the problem is simply one of longitude and the difficulties of compilation. See also Wallis, "English enterprise in the region of the Strait of Magellan," 204–6.

31 Interestingly enough, the details of Drake's voyage were also kept secret; see Wallis, "The cartography of Drake's voyage," 133–7.

32 For a good account of these settlement attempts, especially the motives of both sides, see John T. McGrath, *The French in Early Florida: In the Eye of the Hurricane* (Gainesville: University Press of Florida, 2000).

33 Jean Ribaut, *The Whole and True Discoverye of Terra Florida* (London: Rouland Hall for Thomas Hacket, 1563).

34 Juan Escalante de Mendoza, *Itinerario de Navegación*, 1575, edited by Roberto Barreiro-Meiro (Madrid: Museo Naval), 13.

35 "… se a ido a frança sin licencia, y llevadose los patrones, papeles y otros secretos de la navegacion de las indias para comunicarlos con los enemigos de nuestra santa fe catolica y destos reynos, y dexó el offiçio desamparado y no ay quien le sirva …" Statement of Rodrigo Zamorano, AGI, Patronato, 262, R. 11, block 3, image 39; see Pulido Rubio, *El Piloto Mayor (1950)*, 692–5. He introduced as evidence two letters sent by Villarroel from France: AGI, Patronato, 262, R. 11, images 41–2, August 6, 1596 letter from Villarroel to Maestro Pedro, clockmaker, sent from Bordeaux; and image 44, undated letter sent from Villarroel to Romulo Folla, surgeon.

36 For a brief list of the charts produced to support the Spanish claims, see Luisa Martín-Merás, "La Cartografía de los Descubrimientos en la Epoca de Carlos V," in *Carlos V: La Náutica y la Navegación*, ed. José Ignacio González-Aller Hierro (Barcelona and Madrid: Lunweg Editores, 2000), 85–6; most are reproduced either in the article or in the accompanying catalog. On cosmographical meetings, see Ursula Lamb, "The Spanish cosmographic juntas of the 16th century," *Terrae Incognitae* 6 (1974), 54–6, reprinted in *Cosmographers and Pilots of the Spanish Maritime Empire* (London, Brookfield: Variorum, 1995), V, 51–64. For later cosmographic meetings, see AGI, Patronato, 49, R.12, account of October 8, 1566 meeting. See also the discussions in David Goodman, *Power and Penury: Government, Technology and Science in Philip II's Spain* (Cambridge: Cambridge University Press, 1988), 56–61.

37 See, for example, the discussion in AGI, Justicia, 1146, N. 3, R. 2, block 3, image 359, undated Spring 1545 statement from Sancho Gutiérrez, and images 149ff (question 13 and answers by pilots).

38 "... el arte de la navegaçion que consiste en conoçer los bientos derrotas tierras puertos cabos baxos lunas mareas y en tomar el altura por el sol o estrella en hechar los puntos en mandar governar y velejar en dar a las naos su andana en considerar las singladures y en tener discreçion y prudençia para acudir a lo que pidieren los casos ...," AGI, Contratación, 5014, July 28, 1586 cedula discussing a complaint from the Universidad de Mareantes that Rodrigo Zamorano is not a sailor and so not fit to be pilot major.

39 The general documents are in AGI, Patronato, 251, R.76. The suggestions were all put forward by Diego de Sotomayor and Andrés de Paz, the deputies of the Universidad de Mareantes, as part of a questionnaire on the office of pilot major. Most of their witnesses agreed. AGI, Patronato, 251, R.76, block 3, folio 2, August 2, 1584. A number of pilots echoed these concerns a decade later when asked about the need for reforms; see AGI, Patronato, 262, R.2.

40 On this expedition, see M. I. Vicente Maroto and Mariano Esteban Piñeiro, *Aspectos de la ciencia aplicada en la España del Siglo de Oro* ([Spain]: Junta de Castillay León, Consejería de Cultura y Bienestar Social, 1991), 403–6, and María Luisa Rodríguez-Sala, "La misión científica de Jaime Juan en la Nueva España y las Islas Filipinas," in *El eclipse de luna: Misión científica de Felipe II en Nueva España*, ed.María Luisa Rodríguez-Sala (Huelva: Universidad de Huelva, 1998), 43–66. Many of the relevant records are in AGI, Indiferente, 740, N. 103, and Filipinas, 339, L. 1, fols. 225r–36v.

41 The draft instructions began, "A Jayme Joan natural valençiano hombre experto en Mathematicas y calculaçiones de Astronomiay que sabra muy bien hazer las observaçiones que se le mandaron y ordenaren tocante a las descripçiones de las tierras provinçias y lugares e islas segun su longitud y latitud ..." AGI, Indiferente, 740, N. 103, Block 3, f. 1r, 1583 instructions for Jaime Juan.

42 See María Luisa Rodríguez-Sala, ed., *El eclipse de luna: Misión científica de Felipe II en Nueva España* (Huelva: Universidad de Huelva, 1998), for a facsimile and transcription of the observations of the November 17, 1584 eclipse, and Clinton R. Edwards, "Mapping by questionnaire: an early Spanish attempt to determine New World geographical positions," *Imago Mundi* 23 (1969), 17–28, 18–22, which includes an English translation of one set of instructions. See also García de Céspedes, *Hydrografía*, fols. 161v-69v, for a near contemporary critique of López de Velasco's methods.

43 AGI, Indiferente, 427, L. 30, fols, 278r–9r, May 25, 1577 cedula to the Virrey of New Spain and others; AGI, Indiferente, 427, L. 30, fols. 374v–5v, June 1, 1587 cedula with a list of twenty-seven recipients. For the results, see AGI, Mapas y Planos, Mexico, 34A-F; reproduced in Rodríguez-Sala, *Eclipse de luna*, 103–62.

44 He was reimbursed for the printing on at least four separate occasions. See AGI, Indiferente, 426, L. 26, f. 37v, June 1577 to pay him 330 reales for printing costs; f. 214v, August 27, 1580 order to reimburse him 88 reales; L. 27, f. 60rv, August 12, 1583 order to pay him 12 ducados; and f. 161v, May 30, 1587 order to pay him 3 ducados.

45 The instructions were reprinted in Andrés García de Céspedes, *Hydrografía* (Madrid: Juan de la Cuesta, 1606), fols. 161v–2v.

46 On these reforms in general, see M. I. Vicente Maroto and M. Esteban Piñeiro, *Aspectos de la Ciencia Aplicada en la España del Siglo de Oro* (Madrid: Junta de Castilla y Leon, 1991), 407–31. On the meetings at the Casa, see AGI, Indiferente, 742, no. 151c, December 22, 1593 report from the Casa officials, and Patronato, 262, R. 2 for the pilots' statements.

47 "En todas las partes a donde llegaren, y particularmente en las que aqui se diran, con los Astrolabios que llevan divididos en medios grados, o con los suyos mesmos tomaran a medio dia, la mayor altura que el Sol tuviere entonces, assi de grados, como de medios y quartos, a buena estimativa, estando presente un escrivano que de fe dello: y donde no vuiere escrivano, lo escriva el mismo Piloto que lo hiziere, el qual no ha de hazer otra cuenta alguna de declinacion, ni de reglas que se tienen para tomar las alturas de las tierras, sino solamente la que se ha dicho del Astrolabio, porque lo que se pretende, es saber la mayor altura que el Sol tenia a medio dia en aquel lugar el dia en que lo haze, lo qual se ha de regular con mucho cuydado." AGI, Indiferente, 742, No. 151c-4, instruction printed December 15, 1593.

48 AGI, Indiferente, 868, July 28, 1595 *consulta* of the Council of the Indies, and Vicente Maroto and Esteban Piñeiro, *Ciencia aplicada*, 420-2 and 448-50 (transcription).

49 AGI, Patronato, 262, R. 2, fols. 19r–61v.

50 I go into this in more detail in "An apologia for the pilots' charts: politics, projections, and pilots' reports in early modern Spain," *Imago Mundi* 56, no. 1 (2004), 7–22.

CHAPTER 2

Vers la ligne

Circulating Measurements Around the French Atlantic

NICHOLAS DEW

> For who can doubt, if Gravity be the cause of the fall of a Stone in *Europe*, that in *America* the cause of the fall is the same?
>
> —Roger Cotes, preface to Newton's *Principia*, 2nd edn (1713)

> Just think, *milord:* without the voyage and experiments of those sent by Louis XIV to Cayenne in 1672 … never would Newton have made his discoveries concerning attraction.
>
> —Voltaire, letter to Lord Hervey (1740), prefacing the *Siècle de Louis XIV*

The Atlantic between the Local and the Universal

For early Enlightenment thinkers, the Atlantic space could be taken to symbolize the relationship between the local and the universal. The phrase "in America as in Europe"—with its echoes of "on Earth as it is in Heaven"—seems to have become a rhetorical shorthand, by this time, for talking about global regularities, whether physical or moral. At the same time, however, the phrase implies a reminder that global laws could be observed or verified only through a series of local instances.

Writing in the mid-1680s, both Newton and Locke invoke the bridging of the gulf between the Old World and the New as an emblem of universal law. At one point in his *Letter concerning Toleration* (written in 1685 and published in 1689), Locke illustrated his argument that the civil magistrate should not intervene in religious matters by saying that even Native Americans, living under European rule, should be free to practice their own religion. "For the reason of the thing is equal, both in *America* and in *Europe*. And neither Pagans there, nor any Dissenting Christians here, can with any right be deprived of their worldly Goods" by a "Court-Church" (Locke was writing this in Holland, as a Whig in exile during James II's reign). "Civil rights," Locke says, should be guaranteed against religious intolerance in all places equally across the globe.[1]

Writing at the same time, but in Cambridge rather than Amsterdam, Newton outlined a series of general methodological "hypotheses," which he later re-dubbed "Rules for Philosophizing" ("Regulae philosophandi"), which appeared in his *Philosophiae naturalis Principia mathematica* (written in 1685–86, published in 1687)—more specifically, at the start of Book Three, which is "On the system of the world." The second of these stated that "*The causes of natural effects of the same kind must be the same.* Examples are the cause of respiration in Man and Beast, or of the descent of a stone in *Europe* and in *America*, or of Light in a Kitchen fire and in the Sun, or of the reflection of light on Earth and in the Planets."[2] This passage was glossed in a preface added to the *Principia*'s second edition (1713) by its editor, Roger Cotes, who took the theme of Rule 2 and amplified it in a series of copious variations:

> For who can doubt, if Gravity be the cause of the fall of a Stone in *Europe*, that in *America* the cause of the fall is the same? If Gravity is mutual between a Stone and the Earth in *Europe*, who will deny that it is mutual in *America*? If in *Europe* the attractive forces of the Stone and the Earth is compounded of the attractive forces of the parts, who will deny that in *America* the force is similarly compounded? If in *Europe* the attraction of the Earth is propagated to all kinds of bodies and to all distances, why should we not say that in *America* it is propagated in the same way? All Philosophy is based on this Rule: inasmuch as, if it is taken away, there is then nothing we can affirm about things Universally. The constitution of singular things can be found by Observations and Experiments; and proceeding from there, it is only by this Rule that we make judgments about the nature of things universally.[3]

Here Cotes sets out what was, and is, a particularly influential idea of how scientific knowledge is made, in which universal statements—the stuff of natural philosophy proper—are derived from singular, local observations. Newton's original reference to stones falling "in America as in Europe" was not merely a commonplace, though. He was more likely alluding to passages that come later on in Book Three. An important section of that book (Proposition 20), dealing with the shape of the earth, discusses the weakening of the earth's gravitational pull at places near the equator, using as evidence the length of a pendulum beating seconds.[4] The key measurements that Newton cited were from the expeditions to Cayenne (in French Guiana), Gorée (just off the Senegambia coast, in West Africa), and the Caribbean islands, that had been carried out by agents of the Paris Académie des Sciences in the 1670s and 1680s. As further results were reported in the decades after the *Principia* first appeared, Newton gathered more evidence to fit his theory that the earth had an equatorial bulge, carefully sifting the data, rejecting those measurements he found unsuitable, and keeping those he needed. In the second and third editions (1713 and 1726), Book 3 Proposition 20 became much longer, as Newton presented the new data along with increasingly elaborate defenses of his interpretation of them. This was because, as we will see, in the years since 1687, the status of those measurements had become increasingly controversial. (Later, in the 1720s and 1730s, the question of the shape of the earth was to become a fierce debate and prompted two international expeditions, to Lapland and to Ecuador, to settle the question.[5] All that, though, was to come later.)

The doublet "in America, as in Europe" functions in different ways for Locke and for Newton, even if both are concerned with the universal and the local. As the (again, later) Dispute of the New World reminds us, the gap between America and Europe could be, for contemporaries, as much physical as cultural.[6] America was, literally, *l'autre monde*. It was precisely the assumed ontological gulf separating the two worlds that allowed the trope "in America as in Europe" to carry such weight. To Locke's readers, the diversity in human institutions was implicit but could be bridged by establishing universal claims on the basis of natural rights; for the readers of Newton and Cotes, the diversity in created nature was implicit but could be bridged by establishing universal regularities through observation and experiment (and because, for Newton, God guaranteed such regularities with His divine will).[7] The aptness of the image of falling stones lay in its encapsulation of Newton's argument: that not only were the reported observations of varying gravity in the Tropics actually true but, in addition, they could be explained by the rule of universal attraction—as long as readers were prepared to accept the idea of a nonspherical earth (which was by no means self-evident).

What both the Locke and Cotes passages have in common, though, is the way in which they invoke Atlantic knowledge as if such a thing were readily available and uncontroversial. Their rhetoric can be taken as voicing a metropolitan myth of science *toute faite*, in which all the hard work that had to go into making such statements possible was nonchalantly glossed over. Nowhere in Cotes's overlong list does he mention real American phenomena or the real experimental work that was actually done in the Caribbean: instead, his statements are all theoretical truths figured out on Newton's desk in Cambridge. At the "front end" of the *Principia* (by 1713 already a monument to British natural philosophy), the tropical labors of a few technicians of the French Académie were—perhaps inevitably, given the mounting Anglo-French rivalry in science as in much else—passed over in silence.

One model for understanding the establishment of "universal" (i.e., portable, durable, translatable) science out of local, particular knowledges is the notion of *metrology*. For measurements to be circulated reliably, shared standards need to be established through the precirculation of specific objects. This was something that Locke and Newton (and all the savants of their day) were acutely aware of—the need for universal standards was a favorite theme of the period, as we shall see—and then as now, measurement standards had to be established and calibrated by the circulation of special artifacts (usually the instruments or objects that provided the standard). Actor-network theorists have taken the notion of metrology further by adding to the literal sense a more metaphorical sense in which the term stands not only for standards of measurement but for all of those techniques that allow scientific information to be relayed around a network: without metrology, there is no science.[8] The technical labors that Cotes glosses over—the work of the French envoys sent to the Tropics— were part of a metrological enterprise (in both the contemporary and the sociological senses) to allow quantitative observations to circulate reliably around the Atlantic space. The aim of this paper is to situate that enterprise within its geographical context and to use the seconds-pendulum as a case study in the construction of scientific authority within the networks of the French Atlantic.

Cassini's Grand Design

Texts such as Newton's "System of the World" (*Principia*, Book 3) reveal how baroque natural philosophy profited from a global network of communication. Newton employs data provided by networks of observers in three important sections of that Book, dealing with pendulums, tides, and comets, observed or measured from points around the world.[9] He could do this because such astronomers as Gian-Domenico Cassini (Cassini-I) in Paris or Edmund Halley in London—but also large numbers of obscure figures, many of them not even named by Newton—were acutely aware of the profit to be gained from collecting information from all over the globe. In the period from around 1670 to around 1720, this tentative global network was slowly being built up. When dealing with the shape and size of the earth, Newton relies on a network of sources that has three distinctive features: the network is mainly French; it is mainly Atlantic (or to use the terms Newton uses, it stretches across the "Atlantic Sea and the Æthiopic Sea");[10] and it is distinctly absent from the historiography of Newton's science. (The first two may well explain the third.) What interests me here is not Newton but the construction of this French scientific network connecting the colonies of Louis XIV.

In the first few years after its founding in 1666, the French Academy of Sciences began a surprisingly ambitious program to make observations in distant parts of the world. This program was largely the brainchild of Cassini, who understood the opportunity that global travel, if combined with the newest instruments and techniques, represented for the advancement of astronomy and geography. Jean Richer's 1670 voyage to Acadia, and then his mission to Cayenne in Guiana in 1672, were the first of the long-distance expeditions organized as part of Cassini's scheme. Many others followed: even before the end of the century, various junior members of the Academy, but also specially trained Jesuits, Minim friars, royal engineers, and hydrography teachers, were sent to Canada, the Caribbean, South America, West Africa and, in the case of the Jesuits, the Cape, India, Siam, and China. The results of their astronomical observations, physical experiments, and botanical and zoological curiosity were collated in Paris and eventually published so they could circulate around the Republic of Letters (which of course included readers in the colonies, too).[11]

Cassini's brilliance, or no small part of it, lay in being able to convince Colbert and Louis XIV that they wanted to fund his astronomical expeditions. He had specific astronomical aims, some of which might not have been guaranteed to interest the king, such as finding the earth's distance from the sun or measuring solar parallax, but he knew that the missions could be made more appealing to his mercantilist masters by tying them firmly to the mast of a national and global mapping project. Cassini's "grand dessein" was to collect longitude and latitude figures from around France and the world—the longitude to be established using his own method of observing the eclipses of Jupiter's moons—and to accumulate coordinates on a floor map in the western tower of the Observatoire. The famous Observatoire map has survived only in the engraved version published in 1696.[12] The Observatoire floor map has been seen as an emblem of a new kind of cartography, one that was able to achieve unprecedented levels of precision, partly because of the adoption of new instruments and new techniques. The Académie's astronomers Adrien Auzout and Jean Picard had been behind the development of quadrants mounted with telescopes and micrometers, allowing for more accurate angular measurement; perhaps more important, the Académie used pendulum clocks designed by Christiaan Huygens.[13] Cassini meanwhile provided his own techniques, including the method of finding longitude by simultaneous observations of the eclipses of Jupiter's moons—not a secret method in itself but one that was only as accurate as the revised tables of the moon's movements that Cassini himself was in the process of producing.[14] The Observatoire became the center of French cartography for the rest of the eighteenth century, and the French national survey, the first of its kind, completed by Cassini's grandson and greatgrandson, set the standard for other countries to follow.[15]

Scientific Travel Before Imperial Networks

Cassini certainly deserves credit for getting this ambitious global mapping project off the ground. However, the older historiography of a bright new age of French cartography (shaped by the rhetorical histories of the Académie)[16] needs to be rewritten, with a firmer grasp of the material practices of Cassini's envoys. As we turn from ideals to reality, we are confronted with an absence of models. Scientific journeys of the period before the high Enlightenment—before the age of La Condamine, before the Transits of Venus—present us with something of a recognition problem. The models that we have to hand come from either an earlier or a later period (either the sixteenth-century Iberian empires or the eighteenth-century Pacific explorations) and from cases in which the alliance between the scientific institutions and the state was much closer than was the case in the late seventeenth century. The fact that Cook, La Pérouse, or Malaspina traveled on specially commissioned vessels should not lead us to overlook that for many scientific travelers in the seventeenth and early eighteenth centuries, securing a passage was often a far more complex business. In the 1670s and 1680s, the Académie des sciences was not able to negotiate special ships for its envoys (something Halley managed in his 1698 voyage), so the early observers could travel only on commercial ships, and securing passages for astronomers and their instruments was by no means easy.[17] More seriously, it was impossible to know with any certainty whether the agents would even get to their destination. Even somewhere close to Europe, such as the Canary Islands, proved impossible to reach in the 1670s and 1680s: the Académie wanted Richer in 1672, and Jean Deshayes and le Sieur Varin in 1682, to visit the Canaries, and both failed to do so because of the unavailability of willing sea captains. In the latter case, this meant that what had been planned as a trip to Tenerife turned into a triangular voyage to Senegal and the Antilles. Likewise, Guillaume de Glos was supposed to go to the Portuguese colony of São Tomé (under the Equator) to make solar observations, but this too failed, and de Glos followed Deshayes and Varin from Senegal to the Caribbean. Even when the envoys did manage to reach their intended destinations, they were not always able to stay there long enough to do their work, nor did the Académie have any power to ensure that the trading company agents treated them well.

If, in this period, the Académie des sciences appears to have been almost powerless over the fate of its envoys once they had left Paris, the same might in fact be said—*mutatis mutandis*—of the French monopoly Companies, and indeed of most branches of the Bourbon state. The nascent French Atlantic trade network was certainly unlike the Spanish or Portuguese empires and was likewise dissimilar to the Dutch or British colonial systems, so much so that to speak of a "French Atlantic" in the period 1660–1713 may even be misleading. In comparison to its competitors, the French triangular network of this period lacked a developed infrastructure, it lacked settler colonies (with the exception of the sparsely populated New France), and it was supplied by a smaller volume of shipping, both across the ocean and between colonies. The ships that Cassini's trainees traveled on in the 1670s and 1680s were owned, officially, by the French Senegal Company, but these ships were quite often "prizes" ("*prises*") taken from the (even smaller) Danish, Courlander, or other trading companies, and they were frequently manned by a motley crew of mixed confessions and nationalities, including French Huguenots and Dutchmen. The boundaries between agents of the state and agents of private enterprise—blurry enough within Europe at this date—were even harder to draw in the colonies, where privateers could sometimes become admirals (and vice versa).[18]

Perhaps the only reliable feature of the Atlantic network was that there would be *some* vessels making triangular voyages. The trading networks existed already, thanks to the perceived profitability of the slave trade and the sugar plantations. What the Académie des sciences was effectively doing, in the late seventeenth century, was to make use of this preexisting network for its own ends—to act as freeloaders, as it were—rather than setting up a network of its own. The fragility of the French trading posts, fortified islands that were repeatedly conquered from the Dutch and English and just as easily lost, and the relative lack of what might be termed civil society in the French islands at this date make it hard to classify the early French voyages as "*colonial* science." By the second half of the eighteenth century, Saint Domingue and Guiana would have more developed Creole societies, with literary associations or botanical gardens; nothing of this kind existed when Jean Richer arrived in Guiana in 1672. For this very reason, even the notion of a "network"—a concept that has been so useful in recent science studies—has to be handled with care. A network implies that the nodes of the net are relatively durable, that agents stay in the remote locations for long stretches, sending back a sustained stream of information to the centre. In the case of the early expeditions, though, the sites visited (the nodes) are visited briefly, before the expedition moves on—which silences the node until the next expedition happens to pass through. Unlike, say, the networks of the Dutch East Indies Company or the Society of Jesus or the Casa de la Contratación—all of which had observers stationed for a relatively long time—the Académie des sciences had only a *virtual* scientific network in the late seventeenth century, one that existed only if a sufficient number of expeditions could be carried out.[19]

Richer's Pendulum

The degree to which the Académie's early expeditions were shaped by the practical limitations of French colonial transport is one example of how fragile the whole enterprise was and how remarkable it was that any results were achieved at all. Even when expeditions were able to reach their destination, though, the trouble with interpreting their results was only beginning. The case of Richer's seconds pendulum measurement will be followed here because it links us back to the issue of metrology, and the question of how local knowledge produced in the colonies relates to scientific interpretations produced in metropolitan centers.[20]

In scientific travel beyond Europe, the role of the observer's location was somewhat paradoxical. On the one hand, the observers needed to be there, in the exotic location, to do the work. This is clearest in the case of natural history, ethnography, or other "field sciences," but it is equally true for astronomical expeditions: Cassini wanted Richer to go to a site near the equator, not primarily to investigate the pendulum question but because observing the sun when it was higher in the sky would reduce the effects of atmospheric refraction on solar observations.[21] However, conversely, the fact that the research was done in a distant location caused trouble for the credibility of the observer. This was partly a question of replication—how could anybody replicate what Richer had done in Cayenne, without taking a passage to Guiana themselves?—but it was also more than that. There was also the fact that Richer, Deshayes, and the other French voyagers had inferior status within the Académie: Richer was an *élève* (assistant) of the Académie, but most of the other envoys did not even have this lowly rank within the institution. In addition, the natural environment of the Tropics was given a special explanatory power by those who interpreted the data from the comfort of their European institutions.

After all, Richer's pendulum finding was at first rejected and remained controversial for many years. Richer's measurement was regarded by Cassini and others at the Académie—Jean Picard, Christiaan Huygens at first, and Philippe de La Hire for somewhat longer—as a mistake, an observational error. This was quite understandable, because the margins of the measurement were so small—Richer had found that a seconds pendulum just over three feet long needed to be shortened by one and a quarter "lines," or twelfths of a Paris inch.[22] However, a more important reason why Cassini and the other senior academicians were suspicious of the quality of Richer's workmanship was that they already had several theoretical reasons for wanting the finding to be wrong. The Academy was engaged in a program to make the length of a seconds pendulum a universal, global standard of measurement that would transcend all frontiers and survive into distant posterity.[23] Gradually, through the 1670s, they had established—with much labor and controversy—that the seconds pendulum had a constant length across all of Europe, from Uraniborg in the north, down to Bayonne in the south of France. In 1679, the Fellows of the Royal Society of London still insisted on finding a different, longer figure, but even they were forced to admit their mistake when the Academy sent a special envoy, Ole Roemer (accompanied by John Locke), who was able to show them that they had not being doing the experiment properly. If the senior members of the Parisian Academy had no fear of telling the Royal Society of London that their experimental technique was faulty, it should not surprise us if they decided to classify as dubious the work of one of their young assistants (*élèves*). To make matters worse, Huygens thought Richer had mishandled his clocks during a previous voyage and therefore refused to trust him.[24] It was because of this cloud of suspicion that Cassini instructed so many subsequent envoys to repeat the pendulum test in extra-European locations.[25] From Cassini's point of view, a hard-won and much-needed metrological network was being built in Europe, and the only obstruction was coming from the colonies. As Richer's finding was, by 1680, the only anomaly left, a one-off finding, its status was extremely questionable: if the irregularity was not due to Richer, it could only be attributed to Nature herself. Or, to use the terms of the period, the pendulum was in danger of becoming a "monster."[26]

Perhaps it *was* a monster after all, at least in the sense of an artificial phenomenon produced by an occasional, and *local*, irregularity? This was the view taken by the academician Philippe de La Hire (1640–1718), who gave a paper to the Academy in 1703 on the question. La Hire argued that all of the numbers produced by Richer and, by this date, the several subsequent expeditions were indeed exact measurements—thus saving the honor of the Academy's own trainees—but explained them by the effects of the local climate on the observers' instruments.[27] La Hire was one of the Academy's keener meteorologists, annually reporting series of readings from the barometers and thermometers at the Observatoire. He claimed that the quality of the air must be the main cause of the Richer phenomenon. The air's resistance to the pendulum must, he argued, be a key factor, depending on the humidity, temperature, and pressure of the atmosphere. Above all, the metal parts of the observers' clock and—crucially—the iron rulers used for measuring the length of the seconds pendulum would surely expand in the hot air of the colonies (thus making the pendulum only *seem* to be shorter). As he put it:

> We could say, then, that near the Line [Equator], and between the Tropics, the heat is very great, and so metals expand and extend considerably more than they do in Europe, perhaps also by means of a particular cause, namely the vapors and exhalations [of the earth] which penetrate them—for, as we know, such vapors and exhalations are very penetrating in those countries—and the metals would expand more at one time than at another, and more in one place than in another. That is why these causes of extension, which are not considerable over here [in Europe] might be very different at Gorée and at Cayenne, and at different times; because we are persuaded that near the Tropics, the heat is much stronger than near the Line [the Equator].[28]

This would also explain why the pendulum had needed even more shortening on the island of Gorée (according to the Deshayes voyage of 1682), even though Senegambia is farther north than Cayenne: because, as La Hire says, "we know" (*on sait*) that near the Tropics the climate is fiercer than at the Equator. So La Hire is happier to give enormous explanatory power to traditional doctrines about the climate of the Torrid Zone than to classify the Senegal measurement as a mistake.[29] In addition, La Hire makes use of the Cartesian notion of the "exhalations of the earth"—a topic that was being debated in the Académie at the time in relation to the theory of earthquakes—to bolster his theory of the climatic expansion of the metal measuring rods.[30]

La Hire was also worried about the material used to make the pendulum cord. Picard's original (and fullest) account of how the experiment was done explains:

> The bob of our pendulum was of brass, one inch in diameter, and made on a lathe. The thread with which the first experiments were made, was of flat silk, but because it extended noticeably at the least dampness in the air, we found that it was better to use a simple piece of *Pite*, which is a sort of flax brought from America…[31]

Silk was tested against *pite*—both exotic commodities made available to Parisians by long-distance trade—and the *pite* was found to perform better. *Pite*, known in English as "silkgrass," was a kind of hemp produced from an agave plant, rather like sisal, and it was used by the Arawak in the Cayenne region to make hammocks.[32] The privileging of such a particular New World substance clearly had implications for the successful replication of the experiment. Despite the apparent superiority of *pite* over silk, in his effort to find reasons for why Richer's finding might be an error, La Hire argued that pendulums made with a cord of *pite* tended to turn slightly in their oscillations, which would disrupt the reading.[33]

As we have seen, in the later editions of the *Principia* (1713 and 1726), Newton reworks the proposition on the pendulum to include a much fuller account of the data from expeditions. Newton also needed to dismiss La Hire's attempt to explain the "Richer effect" away, arguing that the climate, although clearly a factor, could not account for *all* of the many shortenings that had by now been measured.[34] It is worth noting that in theorizing about the seconds pendulum, Newton, Huygens, and La Hire could not help touching on what at first glance seems an eclectic range of topics: the thermal expansion of metals, the central fire of the earth, the causes of earthquakes and springs, the origin of the earth as a swirling ball of fluid, the causes of tropical prevailing winds, the earth's exhalation of subtle vapors, and so on. However, this range of material was needed because an explanation of the Richer effect touched on some of the fundamentals of natural philosophy, not least the contested theory of falling bodies. Moreover, the way in which the pendulum data were handled by metropolitan savants reveals the geography of scientific authority. Even though the pendulum measurements were quantitative findings and apt material for Newton's ambitiously accurate calculations, when such numerical results had to be sifted and classified, like natural-historical reports, local qualitative explanations were often required.

Cassini had carefully designed the working methods of his envoys so as to keep as much control over their data in his own hands. He wanted their astronomical observations sent to him in as "raw" a form as possible, so he could do the working himself, and he also made sure the Académie carefully edited the reports that the envoys produced on their return. In editing the publication of the voyage report, Cassini and the Académie tended to remove interpretations offered by their underlings. Richer's report of the pendulum finding, in its original form, concluded with the line that "The same experiment is very much in conformity with the view of the Copernicans, for, supposing the mobility of the earth, it follows that bodies weigh less near the Equator than at Paris."[35] This interpretative speculation was cut from the published account. Cassini, who held a skeptical (or "fictionalist") position on the Copernican theory, preferred to feign no hypotheses.[36]

How Richer's pendulum finding (and those of the other envoys who came after him) was interpreted clearly made a difference as to whether the length of a seconds pendulum was going to survive as a candidate for a universal standard of measurement. That it was so controversial had also to do with the fact that the pendulum experiment was so difficult to do. It involved careful daily solar observations with a quadrant to regulate a pendulum clock, which provided the reference point for the pendulum beating seconds of mean solar time. The narrow margin of Richer's finding was close to the limits of accuracy of the apparatus used, and perhaps the only reason why Richer was able to produce a figure that was quite close to the modern one is that he spent a period of thirteen months at Cayenne, repeating the experiment every week, and reported an average figure. Moreover, to carry his measurement back to Paris, he brought back a metal rule marked with the length he had established at Cayenne.[37] A similar practice was followed by Deshayes and Varin on their voyage to Gorée and the Antilles a decade later: Deshayes marked his pendulum lengths on a reed attached to the tube of the telescope.[38] This meant that the academicians had access to an object that they themselves could measure, repeatedly if necessary. In fact, in seventeenth-century metrological work, it was normal to send specimens rather than numbers, as an object provided the only guarantee (short of personal travel) that the units had been precisely translated. (Or to put it another way: when trying to verify conversion rates between units, there could be no substitute for moving the physical referents around.) When Picard wanted an example of the English foot, he asked Oldenburg to send a brass rule from London—but he also asked *who* had made the markings, because the credibility of the artifact was closely tied to the craftsmanship of the maker.[39]

Conclusion

Normally, the expeditions of the early French Académie des sciences are seen as inaugurating a new era of precision measurement in astronomy, geodesy, and cartography. The association between the French scientific institutions and long-range networks is strong in the literature on eighteenth-century science, and it has usually been supposed that some of the explanation for this must lie in the relationship between the French Academy and the Bourbon state.[40] However, as we have seen, the state had very little effective power to control Atlantic communications. Looking at these earlier French expeditions from a pragmatic point of view reveals the extent to which the credibility of observations was profoundly shaped by their geography, both at the micro level of rulers and pendulum cords and at the macro level of shipping routes and destinations. The fate of Richer's pendulum finding illustrates the complex geography of precision measurement in the age of Cassini's cartographic designs. Factual information reported from Cayenne and from other sites around the French Atlantic trade network had a peculiarly troubled status in the late seventeenth century, partly because the form of scientific life that was to become familiar in the course of the next century—the "scientific expedition"—was still new, and its "rules" were still unstable. How could factual data from the other side of the Atlantic be trusted? Measurements could be made to travel reliably only if the instruments and people involved could be trusted. However, as we have seen, natural philosophy in the age of Newton and Locke increasingly called for stable values that could traverse the Ocean. The metrological network that the Académie set up allowed Richer's finding to circulate—even as far as Newton's Cambridge—but it was also extremely fragile. The techniques, standards, and instruments that would travel with increasing frequency around the world in the eighteenth century, and which were to become the fabric of "Humboldtian science" a century later, were still in the process of being developed in Richer's period. Rather than circulating references within a standardized scheme, baroque metrologists had to circulate cumbersome (and mutable) specimens, such as reeds or metal rules—and then worry about the effects of climate on them.[41]

Voltaire was fond of quipping that Newton owed his theory of attraction to the envoys of Louis XIV.[42] In one sense, Voltaire was right. But the early French Atlantic voyages can tell us more than their accidental status as sources for Newton might imply (important though that story is). The voyages of the Académie's envoys highlight the importance of the emerging trade network of the French Atlantic for the scientific community of the late seventeenth century, but at the same time they reveal the degree to which scientific networks—like the state's imperial networks on which they relied—were fragile and apt to escape from metropolitan control.[43]

Notes

1 [John Locke], *A Letter concerning Toleration: Humbly Submitted, &c.* (London: Awnsham Churchill, 1689), 35. Locke later argued the somewhat contrasting view that Natives could justly be deprived of their land for "failing" to cultivate it: see David Armitage, *The Ideological Origins of the British Empire* (Cambridge: Cambridge University Press, 2000), 97–8; and James Tully, "Rediscovering America: the *Two Treatises* and aboriginal rights," in *An Approach to Political Philosophy: Locke in Contexts* (Cambridge: Cambridge University Press, 1993), 137–76.
2 Isaac Newton, *Philosophiae naturalis Principia mathematica* (London: Joseph Streater, 1687), 402; compare Newton, *The "Principia": Mathematical Principles of Natural Philosophy*, ed. and trans. I. Bernard Cohen and Anne Whitman, with Julia Budenz (Berkeley: University of California Press, 1999), 795 (Newton's Rules). Newton qualified the phrasing in the second edition (1713) to make it rather more cautious: "the causes *assigned to* natural effects of the same kind must be, *so far as possible*, the same": see Alexandre Koyré, *Newtonian Studies* (London: Chapman and Hall, 1965), 265–6. I have modified the Cohen-Whitman translation slightly in the light of the Latin text, and have restored the original's capitalization and italicization. The Latin texts of the *Principia*'s editions are now readily available via EEBO (Early English Books Online) and ECCO (Eighteenth Century Collections Online).
3 Newton, *The "Principia,"* 391 (Cotes's preface to the 2nd edition).
4 Newton, *The "Principia,"* 826–32 (Book 3, Proposition 20; see also the commentary, 233–8); for the critical edition of the Latin text, see *Isaac Newton's Philosophiae naturalis principia mathematica: The Third Edition (1726) with Variant Readings*, eds Alexandre Koyré and I. Bernard Cohen, with Anne Whitman, 2 vols (Cambridge: Cambridge University Press, 1972), vol. 2, 600–10.
5 Mary Terrall, "Representing the Earth's shape: the polemics surrounding Maupertuis's expedition to Lapland," *Isis* 83 (1992), 218–37; Terrall, *The Man who Flattened the Earth: Maupertuis and the Sciences in the Enlightenment* (Chicago: University of Chicago Press, 2002), 88–172; Rob Iliffe, "'Aplatisseur du monde et de Cassini': Maupertuis, precision measurement and the shape of the Earth in the 1730s," *History of Science* 31 (1993), 335–75; John L. Greenberg, *The Problem of the Earth's Shape from Newton to Clairaut* (Cambridge: Cambridge University Press, 1995); Antonio Lafuente and Antonio J. Delgada, *La geometrización de la tierra: observaciones y resultados de la expedición geodésica hispano-francesa al virreinato del Peru (1735–44)* (Madrid: Consejo Superior de Investigaciones Científicas, Instituto Arnau de Vilanova, 1984); and Neil Safier, *Measuring the New World: Enlightenment Science and South America* (Chicago: University of Chicago Press, forthcoming).

6 Antonello Gerbi, *The Dispute of the New World: the history of a polemic, 1750–1900*, trans. Jeremy Moyle (Pittsburgh: Pittsburgh University Press, 1973); Jorge Cañizares-Esguerra, *How to Write the History of the New World: Histories, Epistemologies, and Identities in the Eighteenth-Century Atlantic World* (Stanford: Stanford University Press, 2001). See also Ralph Bauer, *The Cultural Geography of Colonial American Literatures: Empire, Travel, Modernity* (Cambridge: Cambridge University Press, 2003); and James Delbourgo, "Leviathan and the Atlantic," *History of Science* 43 (2005), 101–7.

7 On Newton's sense of the "diversity of created things, each in its place and time," see the discussion in Koyré, *Newtonian Studies*, 265–6. We might point out the reversal in the positions of nonhuman and human nature that has taken place between the seventeenth century and the twenty-first: then, nonhuman nature was relatively disunited (witness the dispute of the New World), whereas human nature was thought to be unified by Biblical origins and natural law; in modern thought, arguably, the natural world is ontologically much more unified, whereas human cultures are considered radically plural.

8 Bruno Latour, *Science in Action: How to Follow Scientists and Engineers Through Society* (Cambridge, Mass.: Harvard University Press, 1987), 247–57; see also Latour, "Circulating reference: sampling the soil in the Amazon forest," in *Pandora's Hope: Essays on the Reality of Science Studies* (Cambridge, Mass.: Harvard University Press, 1999), 24–80; Joseph O'Connell, "Metrology: the creation of universality by the circulation of particulars," *Social Studies of Science* 23 (1993), 129–73; David N. Livingstone, *Putting Science in its Place: Geographies of Scientific Knowledge* (Chicago: University of Chicago Press, 2003), 140–77. On seventeenth-century metrology, see for example Zur Shalev, "Measurer of all things: John Greaves (1602–1652), the Great Pyramid, and early modern metrology," *Journal of the History of Ideas* 63 (2002), 555–75; and Simon Schaffer, "Golden means: assay instruments and the geography of precision in the Guinea trade," in *Instruments, Travel and Science: Itineraries of Precision from the Seventeenth to the Twentieth Century*, ed. Marie-Noëlle Bourguet, Christian Licoppe and H. Otto Sibum (London: Routledge, 2002), 20–50.

9 Respectively, Book 3, propositions 20 (pendulums); 24 and 37 (tides); and 41 (comets). The largest network is for the comet in proposition 41, where Newton's sources are spread from Boston to Balasore and as far back as the *Anglo-Saxon Chronicle*. This is discussed in Simon Schaffer, "Newton on the beach: the information order of *Principia mathematica*" (forthcoming), a paper to which my argument owes a great deal.

10 The "Ethiopic Sea" was a standard name for what is now termed the South Atlantic in Newton's period. Newton only refers to the Atlantic twice, and always in the form "maris Atlantici et maris Æthiopici"; see *The "Principia,"* 835, 877.

11 On the early French expeditions, and Richer in particular, see: John W. Olmsted, "The scientific expedition of Jean Richer to Cayenne (1672–1673)," *Isis* 34 (1942), 117–28; idem., "The voyage of Jean Richer to Acadia in 1670: a study in the relations of science and navigation under Colbert," *Proceedings of the American Philosophical Society* 104 (1960), 612–34; François Regourd, "Sciences et colonisation sous l'ancien régime: le cas de la Guyane et des Antilles françaises, XVIIe–XVIIIe siècles," doctoral thesis (Université de Bordeaux III,

2000), 233–358, esp. 271–7; Jordan Kellman, "Discovery and enlightenment at sea: maritime exploration and observation in the eighteenth-century French scientific community," Ph.D. dissertation (Princeton University, 1997), esp. ch. 2; Florence C. Hsia, "French Jesuits and the Mission to China: science, religion, history," Ph.D. dissertation (University of Chicago, 1999), ch. 1.

12 Cassini's *planisphère terrestre*, published by Jean-Baptiste Nolin, can be seen on the Bibliothèque nationale's Gallica site (gallica.bnf.fr); a reproduction can be also be found in David Turnbull, "Cartography and science in early modern Europe: mapping the construction of knowledge spaces," *Imago Mundi* 48 (1996), 5–24, at 15.

13 On the work of Picard and Auzout, see Guy Picolet, ed., *Jean Picard et les débuts de l'astronomie de précision au XVIIe siècle* (Paris: CNRS, 1987). Picard presented Colbert with a project to map the whole of France in February 1681: Archives de l'Académie des sciences (Paris), Registres des Procès-Verbaux (henceforth AAS, RPV), vol. 9, ff. 96v–97r.

14 On Cassini's method, see Albert Van Helden, "Longitude and the satellites of Jupiter," in William J. H. Andrewes, ed., *The Quest for Longitude* (Cambridge, Mass.: Collection of Historical Scientific Instruments, 1996), 86–100. Cassini published his first set of tables for Jupiter's moons in 1668, but his revised numbers were not published until 1693. The envoys were using his revised figures (AAS, RPV, vol. 9, ff. 89r–92v, Dec. 14, 1680).

15 See Christian Sandler, *Die Reformation der Kartographie um 1700* (Munich and Berlin: R. Oldenbourg, 1905); Léon Gallois, "L'Académie des sciences et les origines de la Carte de Cassini," *Annales de Géographie* 18 (1909), 193–204, 289–310; Josef W. Konvitz, *Cartography in France, 1660–1848: Science, Engineering, and Statecraft* (Chicago: University of Chicago Press, 1987); Turnbull, "Cartography and science"; Monique Pelletier, *Cartographie de la France et du monde de la Renaissance au Siècle des Lumières* (Paris: Bibliothèque nationale de France, 2001).

16 On which see Mary Terrall, "Heroic narratives of quest and discovery," *Configurations* 6 (1998), 223–42.

17 A good survey of the later voyages is provided by Rob Iliffe, "Science and voyages of discovery," in *The Cambridge History of Science*, vol. 4: *Eighteenth-century Science*, ed. Roy Porter (Cambridge: Cambridge University Press, 2003), 618–45. On Halley, see Alan Cook, *Edmond Halley: Charting the Heavens and the Seas* (Oxford: Clarendon Press, 1998), 61–88, 256–91.

18 The themes of the last two paragraphs are explored more fully in my "Timekeeping and trust in the French Atlantic" (forthcoming). On the French Atlantic in this period, see James S. Pritchard, *In Search of Empire: the French in the Americas, 1670–1730* (Cambridge: Cambridge University Press, 2004); Kenneth J. Banks, *Chasing Empire across the Sea: Communications and the State in the French Atlantic, 1713–1763* (Montreal and Kingston: McGill-Queen's University Press, 2002). Privateers who gained official posts include Jean Du Casse and Jean Bart (see *Dictionnaire de biographie française*, s. v.). My remarks here on the early modern state owe a lot to Janice E. Thomson, *Mercenaries, Pirates, and Sovereigns: State-building and Extraterritorial Violence in Early Modern Europe* (Princeton: Princeton University Press, 1994).

19 On these other famous examples of early modern long-distance networks, see Steven J. Harris, "Confession-building, long-distance networks, and the organization of Jesuit science," *Early Science and Medicine* 1 (1996), 287–318; Harris, "Long-distance corporations, big sciences and the geography of knowledge," *Configurations* 6 (1998), 269–304; and Harris, "Networks of travel, correspondence, and exchange," in *The Cambridge History of Science*, vol. 3: *Early Modern Science*, eds Katharine Park and Lorraine Daston (Cambridge: Cambridge University Press, 2006), 341–62.
20 The material in this section is dealt with more fully in my article, "Gravity in the tropics: global metrology and the seconds pendulum, 1672–1726" (forthcoming).
21 Good accounts of Richer's astronomical work are in Olmsted (art. cited) and John L. Heilbron, *The Sun in the Church: Cathedrals as Solar Observatories* (Cambridge, Mass.: Harvard University Press, 1999), 132–4.
22 A line (*ligne*) was a twelfth of a Paris inch (*pouce*), and measured 2.25575 mm, according to modern reference works; so Richer's pendulum was shortened by 2.83 mm.
23 See Pierre Costabel, "Picard et l'étalon universel de longueur fondé sur le pendule," in Picolet, ed., *Jean Picard*, 315–28; see also George Sarton, "The first explanation of decimal fractions and measures (1585): together with a history of the decimal idea and a facsimile (no. XVII) of Stevin's Disme," *Isis* 23 (1935), 153–244, at 188–94.
24 Olmsted, articles cited. Huygens only really accepted the Richer phenomena after the Dutch voyage of the Alkmaar, which he had had more involvement with. See Eric Schliesser and George E. Smith, "Huygens's 1688 report to the directors of the Dutch East India Company on the measurement of longitude at sea and its implications for the non-uniformity of gravity," *De zeventiende eeuw* 12 (1996): 198–214.
25 Giovanni Domenico Cassini (I), "Instructions generales," in "Les Elemens de l'astronomie verifiez par Mr Cassini par le rapport de ses tables aux observations de M Richer faites en l'isle de Caïenne: Avec les observations de MM. Varin, des Hayes, et De Glos faites en Afriques & en Amerique" ("1684"), in [Académie des sciences,] *Recueil d'observations faites en plusieurs voyages par ordre de Sa Majesté, pour perfectionner l'astronomie et la geographie* (Paris: Imprimerie Royale, 1693), separately paginated, here 55: "seulement à la Caïenne elle s'est trouvée plus courte; mais on doute si cela n'est point arrivé par quelque defaut dans l'Observation."
26 In the seventeenth and eighteenth centuries, singular facts and artificial phenomena could be *suspected* of being "monsters" (i.e., one-off events that were not part of the "ordinary course of nature"): see Peter Dear, *Discipline and Experience: The Mathematical Way in the Scientific Revolution* (Chicago: University of Chicago Press, 1995), 18, 48. We might add that it was precisely because at this time there was no statistical theory of error (not developed until Laplace a century later) that the sorting of "good" measurements from "bad" was necessarily a qualitative, taxonomic operation.
27 Philippe de La Hire, "Remarques sur les inégalités du mouvement des Horloges à Pendule," *Histoire de l'Académie Royale des Sciences. Année* [1703]: *Avec les*

Mémoires de Mathematique & de Physique (Paris: Jean Boudot, 1705), section "Mémoires," 285–99 (paper dated Dec. 15, 1703); cf. commentary by Fontenelle in section "Histoire," 130–5. La Hire writes: "Les observations faites à Cayenne et à Gorée ne laissent aucun lieu de douter qu'elles ne soient trés certaines & trés exactes par toutes les circonstances qui y sont raportées (292) … Mais ne pourroit-on point soupçonner que cette difference de longueur du pendule n'est point réelle, mais seulement apparente, & qu'elle ne vient que de la mesure dont on s'est servi" (293).

28 La Hire, "Remarques," 293: "On pourroit donc dire que vers la ligne & entre les Tropiques où les chaleurs sont fort grandes, les métaux s'étendent & s'allongent tres-considerablement au-delà de ce qu'ils font dans ces païs-ci [i.e., Europe], & peut-être encore par une cause particuliere des vapeurs & des exhalaisons qui les pénetrent, comme on sçait qu'elles sont tres-pénetrantes en ces païs-là; & enfin plus dans un tems que dans un autre, & plus dans un lieu que dans un autre. C'est pourquoy ces causes d'extension qui ne sont pas considerables dans ces païs-ci [Europe], peuvent être tres-differentes à Gorée & à Cayenne, & dans des tems differens; car on est persuadé que vers les Tropiques les chaleurs sont bien plus fortes que vers la ligne."

29 The effects of the climate on the observer had also been mentioned in contemporary accounts of Richer's mission: one writer mentioned that "he found that country too hot to work in, and too full of clouds and vapors to be able to observe": Henri Justel to Henry Oldenburg, Aug. 16, 1673, in Henry Oldenburg, *The Correspondence*, eds A. Rupert Hall and Mary Boas Hall, 13 vols (Madison, Milwaukee, and London: University of Wisconsin Press, 1965–86), vol. 10, 152–3 ("Il trouve ce pais là trop chaud pour y travailler, et trop plein de nuages et de vapeurs pour observer").

30 See Rhoda Rappaport, *When Geologists were Historians, 1665–1750* (Ithaca: Cornell University Press, 1997), 180–9.

31 [Jean Picard], *Mesure de la Terre* (Paris: Imprimerie royale, 1671), 4: "La boule de nostre pendule estoit de cuivre, d'vn pouce de diametre, & faite au tour. Le filet avec lequel les premieres experiences ont esté faites, estoit de soye platte; mais parce qu'elle s'allonge sensiblement à la moindre humidité de l'air, on a trouvé qu'il valoit mieux se servir d'vn simple brin de Pite, qui est une sorte de filasse qu'on apporte de l'Amérique…."

32 A contemporary travel account describes *pite* as follows: "The *Arouages*, the *Araotes*, and most of the other Nations, toward the River *Orenoque*, make their Beds of the Thread of *Pite*, in Net-Work, which they hang up after the Manner the other do their Cotton Hamocks. *Pite* is a kind of Hemp or Flax, but much longer and whiter, of this they make their Cords for the Tackling of the Masts and Sails of their *Canoos*, as for other Occasions; this *Pite* is also much lighter and stronger than Hemp, and nothing near so apt to rot in the Water; they make very fine Thread of it to mend their Arrows, and for other such uses. Perhaps the *Aloe Yuccæ foliis, Catal. Plantar. Jamaic.* p. 118"; from the notes to: "A Journal of the travels of John Grillet, and Francis Bechamel into Guiana, in the Year, 1674. In order to Discover the Great Lake of Parima, and the many Cities said to be situated on its Banks, and reputed the Richest in the World," in [Cristobal

d'Acuña,] *Voyages and Discoveries in South-America* (London: for S. Buckley, 1698), part 3, 57. For "silkgrass," see *OED*; and Robert Hooke, *The Diary of Robert Hooke, M.A., M.D., F.R.S., 1672–1680*, ed. Henry W. Robinson and Walter Adams (London: Taylor & Francis, 1935), 412.

33 La Hire, "Remarques," 297.

34 See *Principia* (1999), 830–2 (end of book 3, proposition 20).

35 "Cette mesme Experience est fort conforme au sentiment des Coperniciens, car posant la mobilité de la terre, Il s'ensuit que les corps pesent moins vers l'Equateur, qu'a Paris"; AAS, RPV, 7, ff. 88r–94v ("Observations physiques faittes en l'Isle de Caïenne par le Sr Richer"), at f. 88v. Compare the published account: *Recueil d'observations faites en plusieurs voyages par ordre de Sa Majesté, pour perfectionner l'astronomie et la geographie: Avec divers Traitez Astronomiques. Par Messieurs de l'Academie royale des Sciences* (Paris: Imprimerie royale, 1693), part 2: "Observations astronomiques et physiques faites en l'isle de Caïenne. Par M. Richer de l'Academie Royale des Sciences," separately paginated, 66; reprinted in *Mémoires de l'Academie royale des sciences: depuis 1666 jusqu'à 1699*, vol. 7, part 1 (Paris: Compagnie des Libraires, 1729), 320.

36 See Heilbron, *The Sun in the Church*, 185–7. Whether this was the principal reason for Cassini's reluctance to admit Richer's finding is not clear.

37 The lack of work on Richer's experimental practice has led to misunderstandings, such as this passage in Newton, *Principia*, ed. Koyré and Cohen, 610: "Richerus autem observationes in Cayenna factas, singulis septimanis per menses decem iteravit, & longitudines penduli *in virga ferrea ibi notatas* cum longitudinibus ejus in Gallia similiter notatis contulit." The 1999 translation has "Moreover, Richer repeated his observation in Cayenne every week during a ten-month period, and compared the lengths he found there for a pendulum *consisting of an iron rod* with its length similarly found in France" (Newton, *The "Principia,"* 832). Clearly, the translation should be "the lengths he found there for a pendulum *marked on an iron rod*" (emphases added). The ruler is a "verge de fer" in the French texts, which Newton's Latin mirrors.

38 See Bibliothèque de l'Observatoire (Paris), ms B.5.2, p. 79 (from the Gorée voyage of 1682–84: Deshayes measures the pendulum cord with a length of straw and attaches this to the tube of the telescope), and p. 337 (from Deshayes's 1700 voyage to Saint Domingue, where he mentions carrying a reed from France marked with the pendulum length).

39 Oldenburg, *Correspondence*, 7: 497 (Francis Vernon to Oldenburg, March 8, 1671, letter no. 1648): Vernon says that Picard wants to know who marked the brass rule that he had received from England, "& whether hee may securely depend upon it as the exact English measure, for hee intends to make use of it in his booke. … he feares there is a mistake either in the observation or the scale… Ease him if you can."

40 See on this theme, James E. McClellan and François Regourd, "The colonial machine: French science and colonization in the Ancien Régime," *Osiris*, new series, 15 (2000), 31–50. It should be noted that McClellan and Regourd are mainly concerned with the later eighteenth century.

41 For similar cases, see Simon Schaffer, "Golden means." For the qualitative, rather than precise, nature of natural philosophy pre-1760, see J. L. Heilbron, *Weighing Imponderables and other Quantitative Science Around 1800*, Historical Studies in the Physical and Biological Sciences: Supplement to vol. 24, part 1 (Berkeley: University of California Press, 1993), 185–242.

42 See the epigraph to this chapter: Voltaire, "Lettre à milord Hervey" (1740), prefacing the *Siècle de Louis XIV*, in Voltaire, *Œuvres historiques*, ed. René Pomeau, "Bibliothèque de la Pléiade" (Paris: Gallimard, 1957), 608–12, at 610; compare this passage from ch. 31 of the *Siècle* (at 999): "On envoie, en 1672, des physiciens à la Cayenne faire des observations utiles. Ce voyage a été la première origine de la connaissance de l'aplatissement de la terre, démontré depuis par le grand Newton; et il a preparé à ces voyages plus fameux qui, depuis, ont illustré le règne de Louis XV."

43 If the French Atlantic space became a more manageable arena for scientific communication in the ninety years separating Richer from the era of the Transits of Venus voyages (the 1760s), this was probably due to the accumulation of data from successive voyages, changes in the relationship between the Académie des sciences and the maritime network, as well as the development of a Creole civil society in the colonies. For the later period, especially as it developed in the French Antilles, see James E. McClellan, III, *Colonialism and Science: Saint Domingue in the Old Regime* (Baltimore: Johns Hopkins University Press, 1992); Regourd, "Sciences et colonisation," part 3; and McClellan and Regourd, "The colonial machine."

CHAPTER 3

Knowing the Ocean

Benjamin Franklin and the Circulation of Atlantic Knowledge

JOYCE E. CHAPLIN

At the heart of early modern empires and early modern science lay an ocean, the Atlantic. That ocean circulated many things—natural specimens, people, commodities, diseases, books—but, perhaps most interestingly, it circulated ideas about its own physical circulation. Yet for almost three centuries, those circulating ideas failed to cohere into a shared vision of what the ocean was or even what people thought it might be. That was mostly the case because the actors with the greatest knowledge of the sea—sailors—had kept themselves (and their knowledge) to themselves. Then, in the eighteenth century, a small number of landsmen managed to publicize what sailors knew. It took a persistent, well-connected, charming, persuasive, curious, and ambitious landsman to discover, embellish, and publish mariners' knowledge of the Atlantic. It took Benjamin Franklin, for instance, to put the Gulf Stream on the map. The story of Franklin's feat is a useful corrective to received ideas about colonial science and about early modern Atlantic worlds. It is customary to see science and empire in terms of distinct geographic zones, usually divided hierarchically into center and periphery; even when the empire strikes back (as when colonized peoples turn out to be the scientific innovators), a geographically expressed power relationship is still the operative framework. However, knowledge about the Atlantic Ocean was different: in this instance, there was no center.

To be sure, the Atlantic Ocean was a center of empire, but if that ocean physically contained and expressed imperial power, European officials had a remarkably haphazard intellectual grasp of it. The maritime communities with the greatest Atlantic expertise were unlikely to have their knowledge come under direct control of any kind. If elements of this hands-off approach could be found in all the European empires around the Atlantic, it was a particularly striking feature of the first British empire into which Franklin was born. England had initially colonized the Americas by granting a remarkable degree of self-government to trading companies and other chartered groups, and these private, localized autonomies affected maritime and colonial communities in the long term. Well into the eighteenth century, oceans remained astonishingly absent from the learned and imperial discussions about nature that were most likely to make their way into print. For these reasons, hydrography, the science of bodies of water, remained remarkably open to a variety of talents and to a range of interpretations.

Colonists had their own distinctive ways of viewing their place within an empire that faced the Atlantic Ocean, as was apparent in Franklin's three succeeding charts of the Gulf Stream, of 1768, 1782, and 1786. His hydrographic work was openly political, far more so than his better-known electrical experiments. Sequentially, he used the Gulf Stream to describe colonial, anti-imperial, international, and American views of the Atlantic Ocean. As he used the Atlantic to define his shifting national characters, his opportunism was as striking as his exuberance: he went through two hydrographic nationalities in as many decades. His blending of personal progress and seaborne investigation was an interesting rehearsal for the romantic, questing science that would characterize later oceanic explorers, including Alexander von Humboldt and Charles Darwin. Throughout his quest, Franklin made clear his dependence on those who still held unparalleled knowledge of the sea; mariners. However, that raises the essential question: why did they discuss their knowledge with him?[1]

* * *

Working people had not always been eager to discuss their trades with outsiders. If scholars generally agree that artisans made substantial contributions to the sciences, they are less certain how that happened.[2] Sailors' substantial, if complicated, contributions to hydrography show one way in which it occurred.

Secrecy had been a hallmark of craftwork during the Middle Ages but, by at least the end of the seventeenth century, crafts were losing their mystery. Erosion of social ranks and the proliferation of printed material both helped to publicize trade secrets. And a new definition of knowledge insisted that more of it should be made public, accessible to anyone; this was supposedly superior to the medieval forms of learning and skill that had been sequestered in monasteries or guilds. Rather than learn about trades through apprenticeship, any person literate in a vernacular language could read about them in newly accessible texts, even if he or she would never actually master what was portrayed.[3]

The occasional leaks in craft secrets became a veritable torrent by the early eighteenth century. The same was true of work in the sciences, some of which fruitfully interacted with artisanal knowledge. More treatises on nature were published in vernacular languages, rather than in Latin; more people (including women and those who lacked formal education) were expected to be interested in the sciences; and more demonstrations of nature's properties were done in public. Compendia, including early encyclopedias (culminating with the French *Encyclopédie*, published from 1751 to 1772), exemplified these trends.[4]

Publication of maritime knowledge, however, proceeded unevenly, which makes it one of the most intriguing areas of natural knowledge in the early modern era. Mariners—like tradesmen on land—belonged to a community whose work was, in part, protected by the corporate ethic of a craft with its particular secrets. Many features of shipboard life emphasized its exclusive nature, as with the ritual humiliation of anyone who crossed the equator for the first time. Even on land, sailors looked different. Their rocking gait, sunburned skin, lash-scarred bodies, outlandish clothing, and (eventually) tattoos gave them away, and their argot was quite deliberately designed to exclude outsiders.[5]

Yet in the eighteenth century, more men than ever were entering this stubbornly separate trade. Supplies of ships and sailors rose as Britain's empire did, and they would rocket up still further during any war. On both sides of the Atlantic, young men from areas that were overstocked with labor (including colonial port cities) had to look seaward for their livelihood. And, tellingly, the Royal Navy became the most powerful branch of the British state, drawing many men into its service.[6]

Indeed, the state was declaring new authority over maritime affairs, including maritime knowledge. Impressment represented the most brutal form of control: in wartime, any man who had any nautical skill could be forcibly "impressed" into naval service. The state offered carrots as well as sticks. In 1677, for example, the Royal Navy instituted an examination for the rank of lieutenant. A successful candidate had to certify that he had spent at least three years at sea as a midshipman and then prove that he had learned the crafts of navigation and gunnery. Published books on mathematics, geometry, and navigation helped highborn men to bone up on nautical skills and make a career in the navy (Figure 3.1). Some ordinary seamen or "tarpaulins" could also study these books (sailors remained fairly literate, as a group) and advance into the officer ranks. These reforms established the navy as a state-controlled profession, which also meant that they eroded the internally based craft ethic of seamanship.[7]

Meanwhile, men of science had their own interest in the sea. Hydrography had existed since ancient times, but it was significantly revived at the end of the seventeenth century, when natural philosophers, such as Robert Boyle and Edmond Halley, took stabs at defining the physical features of the oceans. Such men did their own experiments on the sea, but they also consulted the largely unpublished knowledge that mariners still

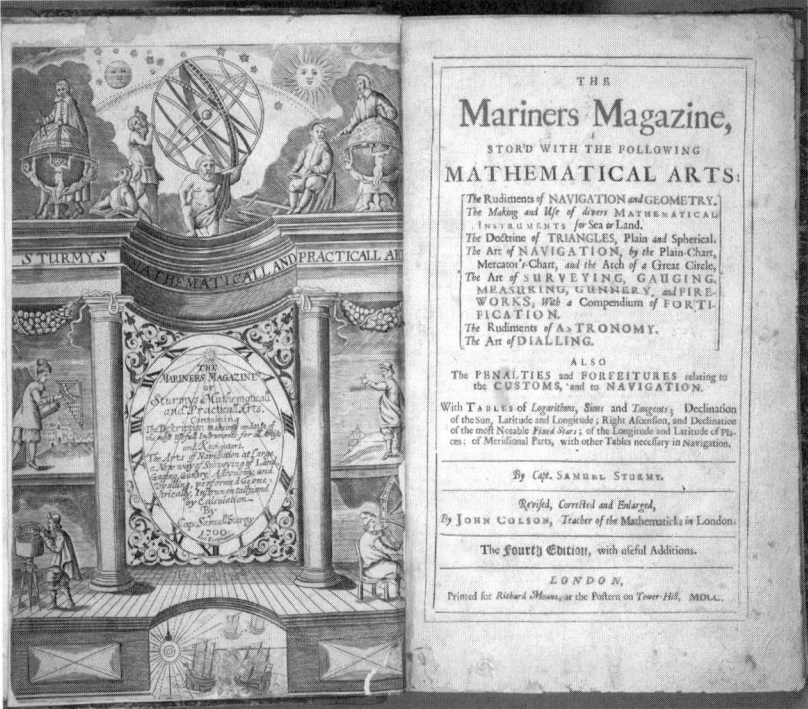

Figure 3.1 Samuel Sturmy, *The Mariners Magazine...*, 4th edn (London, 1700), title pages. This popular seaman's compendium went through multiple editions. Franklin studied an edition of Sturmy when he tried to learn geometry. Reproduction courtesy of the Huntington Library.

kept as trade secrets. The long eighteenth century saw an unprecedented publication of such knowledge, sometimes sponsored by gentleman naturalists (e.g., Halley) but also by the mariners themselves.[8]

The timing of these developments indicates that the expansion of empire was the context for publication of maritime knowledge, yet the state did not always control the publication. Unlike Spain, Portugal, and France, England had never created a central body to administer questions of navigation and hydrography. Not until 1795 (and the Napoleonic Wars) would Britain create a governmental hydrographic office. In the meantime, officials encouraged such work indirectly by permitting private firms to publish books, pamphlets and, especially, charts. Little was censored—by mid-century, quite a lot of maritime information became publicly available, especially in Portsmouth and London. The London market flourished on Tower Hill, a lively nautical nest where press gangs might be snatching men for His Majesty's ships even as customers perused charts in nearby shops. Anyone with enough money—women, landsmen, even foreigners—could buy a British chart of the Atlantic.[9]

Three private agencies dominated British marine cartography. The oldest, Mount and Page, was a grand firm that, under a variety of names, spanned four centuries. By the early eighteenth century, it was sole publisher of the canonical *English Pilot*, a classic sea atlas. The second enterprise, Sayer and Bennett, was newer, dating from 1720 and specializing in charts rather than atlases or books. It scored a coup in 1775 when it obtained the rights to Captain James Cook's surveys of the Canadian coastline, then carried on to publish Cook's cartographic surveys of the Pacific. Last, there was Thomas Jefferys, a newcomer who published two important maritime atlases, *The American Neptune* and *The Atlantic Pilot*. The former represented coastlines and coastal waters in gorgeous detail; the latter was more practical. Yet it was rare for any of these charts to indicate the features of the open ocean.[10]

And even as the maps piled up, cartographers could not agree on basic features for their products. Where should they place the prime meridian, the zero point from which to calculate east–west positions? Such a line established a center for any map, but Europeans (let alone anyone else) could not agree on a common center for the world. A cartographer generally picked a place within his own nation, but even a single nation had several possibilities. English mapmakers could choose London, though many chose the Lizard, the last piece of land a ship would see when sailing away from England.

Nor could cartographers agree on a name for the ocean that divided Britain from British America. The English did not consistently use *Atlantic* to designate the waters that separated Western and Eastern hemispheres until the latter part of the eighteenth century. Before then, they often labeled the North Atlantic, the part they knew best, as either "the Western Ocean" or the "North Sea." These are the names commonly seen on British maps and sea charts through the early part of the eighteenth century. The designation North Sea particularly undermined any sense of a greater Atlantic ocean, or world; what we call the South Atlantic was considered an entirely different ocean, usually called the "Ethiopian Sea." (This tendency was true of Europeans generally, who saw two entirely different oceans where we would see two parts of one, the Atlantic.) These labels—Western Ocean, North Sea, and Ethiopian Sea—simply oriented geography in relation to Britain, which was the whole point.[11]

The term *Atlantic*, of course existed before the late eighteenth century, and European cartographers did sometimes use it. However, the English usually deployed the word to describe the land and sea near the southern Mediterranean or Africa (i.e., near Mount Atlas in Libya). The word *Atlantic* was also useful to talk about anything that was gigantic or mighty, like the mythic figure Atlas himself. Indeed, when in the early nineteenth century William Blake wrote of a mystical moment when "all the Atlantic mountains shook," he probably wanted to convey the old meanings of the term along with the relatively newer reference to the ocean between the Western and Eastern hemispheres. This is another hint that specialized knowledge of the sea—including specific terms for oceans—was not part of public culture until well into the eighteenth century.[12]

How did the knowledge become public? Sailors and nonsailors nudged it into print as the empire expanded though, surprisingly, it was not always people in the first group who were motivated by imperial concerns. This was apparent, for example, in the careers of Edmond Halley and William Dampier.

Halley was a fellow of the Royal Society and would become Astronomer Royal. Best known for his work in astronomy, he predicted the transits of Venus across the sun in 1761 and 1769 and the cyclic return of his eponymous comet. He also investigated tides and the trade winds; the latter he explained according to thermal trends, variations of hot and cold. Halley was also the first English person "to commission and command a naval vessel for the purpose of scientific exploration." Three times, between 1698 and 1701, Halley commanded the *Paramore* to chart the lines of magnetic variation in the north and south Atlantic, a feat that was meant to expedite navigation and was therefore of enormous interest to the state. The Admiralty ordered the navy to support Halley and, extraordinarily, they listed him among the officers to be paid prize monies seized from enemy ships—the first instance when learned status won someone this right. Halley's resulting chart appeared in 1701, published by Mount and Page.[13]

In contrast to Halley, William Dampier came from an ordinary background—he was apprenticed as a sailor and rose through the ranks. He circumnavigated the globe three times and became a privateer who harassed enemy vessels on behalf of the Royal Navy, although he (unlike Halley) was not always in the employ of the state. He is now famous as the captain who abandoned Alexander Selkirk on Juan Fernández Island, an episode that gave Daniel Defoe an idea for a book titled *Robinson Crusoe*; Jonathan Swift and Tobias Smollett also drew on Dampier's much reprinted travel narratives. Dampier was also a maritime expert—and a person of powerful ambition—who capitalized on the new interest in mariners' knowledge. He published an influential *Discourse of Winds, Breezes, Storms, Tides, and Currents* (1699) and became a protégé of the Royal Society, though never its fellow. Dampier is thus called the father of English hydrography—and it is noteworthy that an actual sailor has received this designation.[14]

A comparison of Dampier and Halley indicates a revealing distinction: sailors seemed to use the term *Atlantic* before nonsailors did. Halley's 1701 chart of the variation of the compass referred to "the Western & Southern Oceans," meaning north and south Atlantic. This was true even of the posthumous editions of the map issued in 1760 and 1773. However, Dampier used "Atlantic" fairly consistently to describe both the northern and southern parts of the ocean. He specified that "the Atlantic Ocean" was the one that "parts Africa from America," a designation that indicated both south and north Atlantic. Here is another hint that mariners' terminology and analysis had become, by the turn of the eighteenth century, quite sophisticated but were only beginning to be publicized to wider audiences. The gap between mariners and others was comically apparent in one of Dampier's published charts. He had supplied the title, which uses the term *Atlantick*; his cartographer paid no attention to that and instead resorted to the traditional names of "North Sea" and "Ethiopian Sea" on the chart itself [Figure 3.2].[15]

Soon, however, everyone caught up with Dampier—the term *Atlantic* proliferated over the eighteenth century. The keyword searchable electronic version of Readex's *Early American Newspapers* demonstrates increase in the use of *Atlantic* or *Atlantick*. The words rise from an initial and single use in the 1720s to 760 instances in the 1790s. The key decade for the trend was the 1760s, when American newspapers used the word 147 times, up from a mere 10 incidents in the 1750s.[16]

The timing of the change indicates a stronger public interest in Atlantic and imperial affairs, which itself reflected the growing importance of

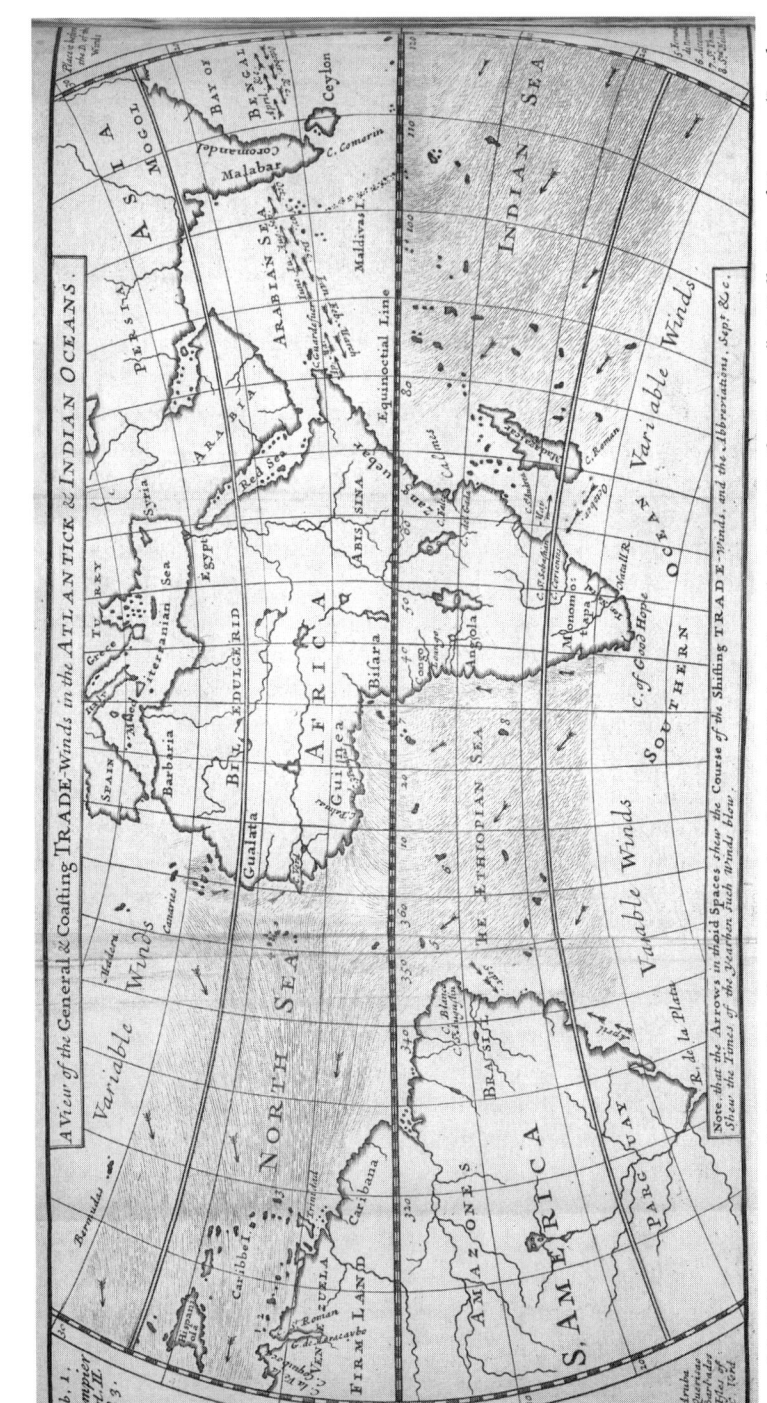

Figure 3.2 William Dampier, "A View of the General Coasting & Trade Winds in the Atlantick & Pacific Oceans," *A Collection of Voyages* (London, 1729). This was an early thematic chart of the world's oceans. Reproduction courtesy of the Huntington Library.

Britain's Atlantic colonies. By 1700, an empire of economically viable English colonies and lucrative trading networks existed; over the eighteenth century, the continental American population would increase with unprecedented speed. Imperial warfare, beginning in 1688 and running through the Napoleonic Wars, made British officials think more seriously about the colonies in the Western hemisphere. They even hoped to get more Atlantic territory, even if it meant seizing it from their Spanish and French rivals. The Seven Years War (1756–63) realized that goal: the war transferred an unprecedented amount of territory from France to Britain. Yet the conflict's financial demands initiated British attempts to tax colonists, leading to the protests and the War for Independence that would tear apart the newly enlarged British empire. That the word *Atlantic* proliferated in the 1760s is prime evidence for these crucial transformations in imperial and in colonial consciousnesses.[17]

Thus, the maritime world captured the interest of all sorts of people. The sea had become a center of action in the modern world, a point of observation for experts in the Royal Society, a means for men of science and ordinary mariners to advance themselves, and a highly mythologized subject of famous English novels. Perhaps the high-water mark of this trend was the establishment of a Parliamentary prize for whoever solved the problem of determining longitude at sea, a contest that would be determined on the Atlantic Ocean, that would engage the energies of people from nearly every rank in British society, and that was set up in 1714, when Benjamin Franklin was eight years old and dying to go to sea.[18]

* * *

Artisan, reader, printer, would-be sailor: Benjamin Franklin rather remarkably embodied all the trends that had recently redefined knowledge, including maritime knowledge, in the eighteenth century. Franklin's first wish, as a boy, was to be a sailor. His half-brother, Josiah, had gone to sea, which perhaps signaled to his sibling a route out of provincial Boston. When Franklin's father quashed the idea (mercifully, one thinks), Benjamin was apprenticed as a printer. Still, producing texts allowed Franklin to publish material relating to the sea. In fact, one of the first things he wrote for public consumption was a ballad on pirates that, as a boy, he hawked in Boston's streets. By the time he made his first oceanic voyage, in 1724 to London, Franklin was primed to see this as an adventure and to regard sailors as interesting workers, if ones whose secrets were no longer quite so secret. In 1726, he sailed back to America and ever after contributed to Atlantic meteorology and hydrography, starting with his famous almanac, *Poor Richard*, and culminating with his work on the Gulf Stream.

Franklin always knew and talked to mariners, not least those within his extended family. The sea trades ran deep on the maternal side. Franklin's mother, née Abiah Folger, came from Nantucket, whose residents, beginning with its Indian peoples, looked to the ocean for their livelihoods and who would become legendary whalers. Via the prolific Folgers, Franklin was related to several Nantucket clans, including the still-extant Coffins and the Starbucks. (Some of the Starbucks ran a store that provisioned whalers, an activity very approximately commemorated in the global chain that sells coffee.)[19] Many Franklins had joined the Folgers at sea, and Franklin had several maritime in-laws. His relatives plied the merchant trades and even served in the Royal Navy. One grim consequence of this maritime engagement was being lost at sea, presumed drowned. This was the fate of Josiah Franklin and of Benjamin's nephew, John, another mariner. No wonder that Franklin became an advocate for the art of swimming.[20]

He was keen on most maritime matters, as is evident from a 1726 diary he kept while crossing the Atlantic from London to Philadelphia. Among other observations of the ocean, Franklin noted the presence of "Gulf" weed. The term *Gulf* (or *Gulph*) had first described maritime things in and around the Gulf of Mexico and was undoubtedly derived from Spanish use. The first printed English source to use the word in this manner dates from 1674, when John Josselyn traveled from England to New England and commented on the "gulf weed" that spread over the warmer parts of the North Atlantic. Likewise, in his 1726 journey, Franklin described how he "took up several branches of gulf weed (with which the sea is spread all over from the Western Isles to the coast of America)."[21]

The 1726 passage, westward, would be the longest Atlantic voyage that Franklin ever endured, an excruciating eighty-three days and a sharp contrast to the forty-nine days it had taken him to cross eastward two years earlier. Franklin began to wonder about the difference. In 1746, he and Cadwallader Colden, New York's noted naturalist, pondered the problem. They considered "Whether the much shorter Voyages made by Ships bound hence to England, than by those from England hither, are not in some Degree owing to the Diurnal Motion of the Earth" and "the more frequently westerly winds." Colden also thought that tides might be a factor and, probably referring to Halley's hydrographic work, pointed out that high tides could be mapped in regular patterns. He estimated that a ship crossing west in thirty days would have the tidal force "lessen'd 1/30 of the time as she approaches" America. An "Equation" to calculate the rates of east and west passages at different times of year should therefore be possible, if all ships were constructed and loaded in exactly the same way.[22]

Tellingly, Franklin and Colden corresponded on the Gulf Stream just after the War of Jenkins's Ear (1739–44) had merged into the War of the Austrian Succession (1744–48). Power over the ocean was political power, and people could not think of the Atlantic Ocean and Gulf Stream without thinking of the politics of empire. In their eighteenth-century attempts to wrest New World territory from the Spanish, the British were preoccupied with access to waterways that would signal their mastery over the Americas.

The Gulf Stream's very name betrayed English obsession with Spanish America. The currents running up from the Florida Straits and across the upper part of the North Atlantic are not in fact caused by the Gulf of Mexico. Yet, given the long Spanish dominance over that Gulf, and therefore over parts of the Atlantic Ocean, it was easy for the British to imagine the current pushing its way out of the Gulf of Mexico, steadily conveying silver to Spain.

Still, "Gulf Stream" remained, compared to "Atlantic," a rare term, as the keyword searchable *Early American Newspapers* makes clear. Its use did increase over the eighteenth century but much more slowly. There was only one instance of "Gulph Stream" in the 1740s, just after the War of Jenkins's Ear—probably no coincidence. Use of the term did not take off until the 1780s and especially 1790s, when the United States, like Britain before it, anticipated new mastery of the North Atlantic. When the phrase made an occasional appearance in the middle of the eighteenth century, it usually did so to describe maritime mishap: storms, wrecks, losing one's way. One Captain Edward reported in 1742 that, when forced "into the Gulph Stream," he saw there a similarly off-course sloop as well as an ominously dismasted brig. The incident could have been a metaphor for the exciting, dangerous course of British empire in the 1740s.[23]

It is notable that, when Franklin began to analyze the Atlantic Ocean, he focused on its dangers, particularly storms and waterspouts. He started his research by asking maritime people for information. It was an old habit—Pehr Kalm, a Swedish naturalist who visited North America in the 1740s, reported that Franklin had, perhaps in boyhood, "heard from sea captains in Boston, who had sailed to the most northern parts of this hemisphere." His early experiences led Franklin to expect good information from mariners. During his work on Atlantic storms from the 1740s through 1760s, for example, Franklin credited "an old Sea Captain" with the information that there was little thunder and lightning on the high seas, yet another "old Sea Captain" with accounts of the interaction of oil and water, and "an intelligent Whaleman of Nantucket" (probably a kinsman and possibly his cousin, Timothy Folger) with testimony that Atlantic whirlwinds occurred to the leeward, as Franklin believed was almost universally the case.[24]

Franklin took the opportunity in 1749 to publish his interpretation of Atlantic storms as a sidebar on Lewis Evans's "Map of Pensilvania, New-Jersey, New-York, And the Three Delaware Counties." Evans, a Welsh-born draftsman and surveyor who had worked in Franklin's print shop, had shrewdly judged the late 1740s a good moment to publish a regional map locally. His map very definitely expressed British-American ambitions to move westward into French and Indian territories. It was a form of British imperialism, yet with independent colonial motivation. Evans underscored Pennsylvanians' divided loyalties by inscribing two prime meridians on his map: at the top, the map measures the distance from London; at the bottom, from Philadelphia's State House. Franklin's business partner, David Hall, printed the map, and Franklin himself gave the map a philosophical flourish by providing his description of how Atlantic storms moved against the prevailing pattern of coastal winds and how they carried atmospheric electricity within them.[25]

Franklin was, at this point, using natural philosophy to make a name for himself. He published his breakthrough *Experiments and Observations on Electricity* two years later, in 1751, and did so in London—clear evidence of his ambition to become known to metropolitan men of science. Yet he also emphasized his insights into American and maritime phenomena. In his first sustained work on hydrography—an analysis of waterspouts read at the Royal Society in 1756 and published in subsequent editions of his *Experiments and Observations*—he was careful to cite both men of science and actual mariners. Franklin believed that contrasts in temperature helped to form waterspouts; the correlated rise of warm air and descent of cold and concomitant disturbance of water created a visible and dangerous vortex. He maintained that the currents ascended rather than descended, violently sucking the water into the air rather than crashing down into it.[26]

In all this, Franklin stressed his reliance on maritime testimony. He later explained that he had decided "to postpone Writing" about his subject owing to "a Desire of producing further Information by Inquiry among my Seafaring Acquaintance." His inquiries led him to that "intelligent Whaleman of Nantucket" as well as to "some Accounts of Seamen," but these were anonymous informants. Franklin named only his published sources. These included the one mariner authority he did cite, William Dampier, and some learned figures, including those published in the *Philosophical Transactions*, as with an essay Cotton Mather had written on whirlwinds.[27]

Franklin turned next to one of the biggest maritime mysteries of his age, the fabled Northwest Passage. In 1745, Parliament had set a prize of £20,000 to anyone who could sail from Hudson's Bay to the South Sea. This was a nakedly imperial race. If a Briton found a passage between the Atlantic and Pacific oceans, his nation would command two oceans and the continent lying between them, a powerful threat to French Canada and to Spanish territories on the Pacific Ocean. The parliamentary prize had smaller but equally pointed goals; for example, it encouraged colonial merchants to encroach on French holdings in sub-arctic Canada and beyond.[28]

Many colonists responded to the challenge, including Franklin. In 1752, he and other investors from the mid-Atlantic region outfitted the optimistically named *Argo*, which was to find the Northwest Passage. The investors petitioned for any trading rights that would not infringe on the Hudson Bay Company's patent—another sign that colonists had their own and local ambitions within the greater imperial expansion. Franklin prepared for the venture by reading narratives of voyages into Hudson's Bay, many of which claimed that the flow of water into and out of the bay proved something or other about a fresh- or saltwater passage through the continent. However, in two succeeding years—1753 and 1754—the *Argo* was blocked by ice.[29]

Franklin had nevertheless made a promising contact. The *Argo*'s captain, Charles Swaine, left his manuscript notes and charts with Franklin before he departed for his second voyage. Franklin eventually kept Swaine's journals from both of the journeys, the relevant "Charts," and "a Number of Letters and Papers" about the northern venture. Swaine is, accordingly, the first of Franklin's maritime informants whom we know by name. Franklin seems to have regarded him, moreover, as an equal and a collaborator.[30]

Not so Franklin's other maritime contacts. Around the time he did his work on hydrography, he undertook some revealing acts of charity toward sailors. He was personally responsible for admitting Francis Buckley, "a poor Sailor … in a very bad Condition with sore Legs," to the Pennsylvania Hospital in 1753; Franklin stood "Security" for all charges for Buckley's treatment and maintenance. He would send a gift in 1761 to Charles Hargrave (or Hargrove) who was at the famous seaman's hospital in Greenwich, England. Hargrave had rendered frequent service to Franklin, as when he carried to Philadelphia David Hall, Franklin's business partner and relation by marriage. His charity underlines how Franklin, now retired from the printing trade and an acknowledged natural philosopher, treated sailors as social and intellectual inferiors. A gentleman who was considerate toward workers, he publicized what sailors knew but not the mariners themselves.[31]

His consultations with anonymous maritime workers nevertheless allowed Franklin to stress his own intellectual humility. A former workman, he appeared to sympathize with those who continued to work with their hands, and he extolled the practicality and immediacy of their forms of knowledge. Franklin concluded his letter on waterspouts by announcing that he had "not [as] with some of our learned Moderns disguis'd my Nonsense in Greek, cloth'd it in Algebra, or adorn'd it with Fluxions." He must have realized that the sentence might offend the philosophers whose ranks he had joined, so he omitted it when he published the letter. Without the phrase, Franklin's praise of ordinary and "intelligent" seafarers gave credit to the unlearned without insulting the learned.[32]

This is not to say that it was easy for Franklin—or anyone else—to get information out of sailors, who probably resented landsmen's attempts to break into their craft. At several points in his life, Franklin noted that shipboard officers dismissed his opinions when he consulted them. For example, in 1726, when Franklin had first studied the Atlantic and thought he saw signs of shallower water, he insisted that "the water is now very visibly changed to the eyes of all except the Captain and Mate, and they will by no means allow it; I suppose because they did not see it first." In 1757, when Franklin joined the fleet bound for battle with the French in Louisbourg and asked his captain a question, his "Answer he gave me with an Air of some little Contempt." Though elsewhere Franklin noted useful contributions to knowledge that he received from several sea captains, these positive encounters were only part of the story.[33]

Yet perhaps because he had been a working man and maybe because he was kinsman to sailors, Franklin managed to get plenty of mariners to tell him things, and he stressed their knowledge. When he first wrote out the phrase *"Gulph Stream"* in 1762, as part of an argument for the Northwest Passage, he emphasized that the current was so called "by Seamen." The eighteenth century's other keen observer of the Gulf Stream, the British engineer William Gerard De Brahm likewise admitted, in 1765, that the current was "Vulgarly" or "commonly" known as "Gulph Stream," meaning that was what ordinary sailors called it. So *Gulf Stream*, like *Atlantic*, began as a trade-based term and only slowly made its way outward to landsmen.[34]

When Franklin gave his first long description of the Gulf Stream in 1768, he made even clearer that his insights depended on those of American seamen whom imperial officials would do well to acknowledge. By this point, when colonial feelings had been ruffled by the Stamp Act (1765), soothed by the Act's repeal (1766), and then newly affronted with the Townshend Duties (1767), Franklin felt himself personally vulnerable. He was a well-known colonial critic of imperial officials. Even his loyal service to the British Post Office (he had been a colonial Postmaster General since the 1750s) was under attack from those who believed it would be better performed in Philadelphia, not in London, where Franklin preferred to live.

So Franklin used maritime knowledge to prove himself a dutiful servant of empire. He wrote a letter to Anthony Todd, Secretary of the Post Office, to answer a question the Board of Customs had raised in Boston about "the long passages made by some Ships bound from England to New York" as compared to their swifter passage eastward. The westbound ships were fighting against the Gulf Stream, Franklin explained, whereas at least some of the eastbound traffic benefited from the current, thus the different rates of travel across the North Atlantic. He named the American who had explained it all to him: "Captain [Timothy] Folger a very intelligent Mariner of the Island of Nantuckett." Folger knew the Nantucket whalers who in turn knew "that the Whales are found generally near the Edges of the *Gulph Stream*, a strong Current so called which comes out of the Gulph of Florida, passing Northeasterly along the Coast of America, and the[n] turning off most Easterly running at the rate of 4, 3 ½, 3 and 2 ½ Miles an Hour."[35]

Familiarity with or ignorance of the Gulf Stream depended, in the Folger-Franklin explanation, on Americans' maritime knowledge. Nantucket whalers had to "Cruise along the Edges of the Stream in quest of Whales," thereby becoming "better acquainted with the Course, Breadth, Strength and extent of the same," than did "Navigators" who "only cross it in their Voyages to and from America." Those navigators sailed west in the eastward path of the Gulf Stream, which slowed them down considerably. They did so for good reason: to avoid running into "Cape Sable Shoals, Georges Banks or Nantuckett Shoals," fear of which kept ship captains southward and smack in the middle of the current. Lacking charts or other guides to indicate a clear passage between the shoals and the Gulf Stream, they had no other choice.[36]

It seems incredible that British captains were entirely ignorant of the Gulf Stream. Yet it is true that, for the period from 1765 to 1770, the Post Office neither ordered nor requested packet captains to gather information about navigation or give advice about charts—again, the hands-off style of British imperial governance is striking. And the journals of one very able seaman, James Cook (later the Captain Cook of Pacific exploration) bear out Franklin's claim. Cook surveyed Newfoundland from 1764 to 1765 under the guidance of Surveyor General Samuel Holland and as part of an effort to map the area to regulate the French who continued to fish there. (Cook's charts were published individually by both Mount and Page and Thomas Jefferys, and then gathered into *The Newfoundland Pilot* [1769].) Cook indeed kept an eye out for currents, but after he "Try'd the Current but found none" on May 26, 1765, he stopped looking for them, despite pestering local fisherman about "the hidden dangers" of the tricky Newfoundland coast.[37]

Franklin thought that ship captains should be able to navigate the North Atlantic more precisely, steering an optimal path between the shoals and the Gulf Stream. He asked Folger to mark "on a Chart, the Dimentions Course and Swiftness of the Stream from it's first coming out of the Gulph, where it is narrowest and strongest; till it turns away to go to the Southward of the Western Islands, where it is Broader and weaker." In addition, he solicited from Folger "Written directions" about avoiding both the stream and the dreaded banks and shoals. Folger evidently marked the current on the chart in red, and Franklin sent it to the Post Office to engrave and distribute to

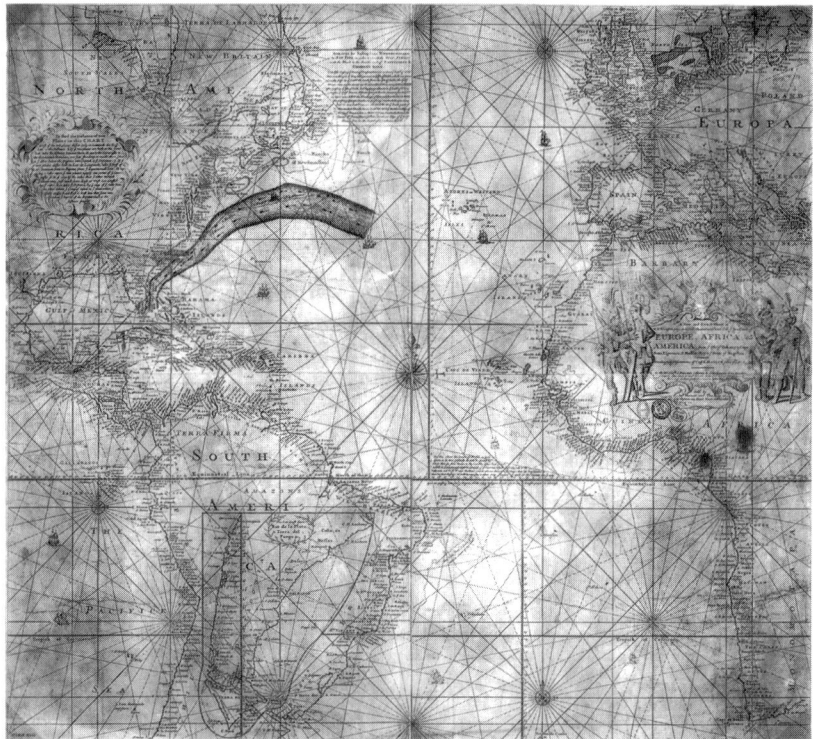

Figure 3.3 Benjamin Franklin and Timothy Folger, chart of the Gulf Stream ([London, 1768]). This is the first known chart of the Gulf Stream. Reproduction courtesy of the Library of Congress.

packet boat captains. The Post Office dutifully printed the chart, using their contract stationer, Mount and Page, which thriftily reused a page from its *English Pilot* to generate the new image of the Atlantic (Figure 3.3).[38]

Thus, the first chart of the Gulf Stream was a fortuitous product of several complicated trends: the publication of maritime knowledge, the collaboration between landsmen and mariners, the growing pains of the British empire, and an emphatically colonial perspective on everything. It was a remarkable example of direct solicitation of information from a mariner, albeit one whose cousin was the most famous North American of the period. It was a landmark contribution to maritime cartography, which was only just beginning to chart the features of the open ocean.

Yet the chart also showed the slowness with which Atlantic knowledge took modern form. The ocean in question is entirely unlabeled, not yet having the canonical name of Atlantic. And, whatever the colonial sensibilities that generated the chart, its prime meridian still lies in England, through the Lizard.[39]

* * *

Franklin's next two Gulf Stream charts, of 1782 and 1786, showed his evolving political allegiances. After he renounced Britain, he abandoned a colonial perspective on anything. His 1782 chart revealed a more openly American view of the Atlantic, yet also one that acknowledged the international rivalries his new nation had to negotiate. The final chart, of 1786, was the most confidently American of all. In creating these images of the Atlantic Ocean, Franklin gave credit and thanks to his nation's French allies, to his own genius, to nearly everyone but the maritime informants who had originally helped him. He had his reasons.

Franklin produced the 1782 chart during the peace negotiations that ended the War for Independence. He may have intended it to assist a Franco-American packet boat service but even more to solidify relations between the allied nations—France and the United States—that had just defeated the British empire. French cartographer Georges-Louis Le Rouge simply copied the northwest corner of the 1768 Mount and Page chart of the Gulf Stream, retaining the prime meridian through the Lizard and still not labeling the ocean at the center. Le Rouge did relabel parts of the map's landscape. He defiantly called the lands to the north of the United States "NOUVE^{LE} FRANCE," perhaps hoping that Britain would yield the Canadian territories they had won from France in the Seven Years War. And he removed any trace of British authority over the coastal American states that had won their independence. As far as Franklin and Le Rouge were concerned, the Gulf Stream was a conduit between the two nations that had significantly challenged British authority in the North Atlantic, nature's link between the French monarchy and the American republic.[40]

For this chart, Franklin still depended on Timothy Folger's collaboration, perhaps even as part of an uneasy quid pro quo. Folger seems to have delivered his manuscript chart of the Gulf Stream and instructions for navigating it when he received an American passport from his cousin in Paris; Franklin made the swap even though Folger was a loyalist, as many Nantucketers were. In preparing the final chart, Franklin noted the contributions of a "Nantucket sea-captain of my acquaintance" without naming him and included the sailing instructions, translated from Folger's 1768 guide into a French so literal that one hopes no trusting navigator ever tried to follow them. However, after Folger was accused of trading with the British, using the passport Franklin had issued as his cover, Franklin renounced his cousin. Their correspondence ends, and Franklin never again had a close intellectual association with a mariner.[41]

Alone, Franklin proceeded to elaborate his major contribution to descriptions of the Gulf Stream: he charted its position according to thermal variation, using three sets of data he had gathered in 1775, 1776, and 1785 on his final Atlantic voyages. In so doing, he brilliantly reversed a tradition in which instruments of precise metrology, including thermometers and barometers, had traced an outward and imperial investigation of the colonies. In Franklin's hands, thermometers now traced a current running out from America, a hydrographic feature that underscored American independence from Europe—it was much harder for Europeans (whether friendly French or jealous Britons) to cross the North Atlantic to America than for Americans to cross over to the Old World.[42]

Franklin published his final Gulf Stream chart (Figure 3.4), his data, and a letter to a French friend as parts of an essay later nicknamed "Maritime Observations," which appeared in the second volume of the American Philosophical Society's *Transactions* (1786). Despite a nod to his nation's French allies (including his correspondent), Franklin celebrated American mastery of nature, particularly oceans. When he had produced his first chart of the Gulf Stream, he had assured Anthony Todd that colonists had maritime skills that the British should respect. Now, Franklin published a longer and more overtly nationalist version of this claim. He did not name Timothy Folger, but his cast-off loyalist cousin might have marveled (or fumed) to see himself ventriloquized to rebuke British captains: "We have informed them that they were stemming a current, that was against them to the value of three miles an hour; and advised them to cross it and get out of it; but they were too wise to be counselled by simple American fishermen."

Figure 3.4 Benjamin Franklin "A Chart of the Gulf Stream," from "Maritime Observations," *American Philosophical Society Transactions* 2 (1786). This is Franklin's third and final Gulf Stream chart. He is at bottom right, conversing with Neptune. Reproduction courtesy of a private collector.

Nature itself pushed Britons away from America—yet they were either too dim to sense the tremendous natural force set against them or too stubborn to take American advice about avoiding it.[43]

Franklin also defined the thermometer as a new hydrographic instrument. His "Maritime Observations" presented three pages of thermal data, advertising that Franklin had been among the very few naturalists who first used thermometers to study the sea's surface and that he had more and more elaborate measurements of temperature than anyone else. Rather than follow the common pattern of using instruments to measure temperature in the air, he innovated by placing them in the sea. Henceforth, he claimed, "the thermometer may be an useful instrument to a navigator."[44]

This upped the ante. No longer would mariners' working knowledge of the stream's location (the variations in its color, the sudden appearance of squalls or whales, the need to take off or put on clothing, the smell of the air) suffice. Technology, including chronometers and newly detailed charts, would enable observers to read the state of nature more quickly. Thermometers would also, Franklin implied, lessen the need for experience or careful training—dip one in the water, and even a landsman knew where he was. With instruments and data, philosophers gave hydrography the precision that began to turn it into oceanography.[45]

It was superbly appropriate that Franklin published his 1786 "Maritime Observations" with the "American Philosophical Society, Held at Philadelphia, for Promoting Useful Knowledge." The Society embodied the eighteenth century's shift from divided, jealously guarded trade groups to unified, publicly accessible knowledge. Franklin's role in publicizing the Gulf Stream outlines this shift, and the publication of his "Maritime Observations" with the Society draws a line under the historic transformation in the production of knowledge. That he added his experimental protocol for measuring the temperature of Atlantic water and his resulting data made clear that he had promoted himself from a would-be sailor to sailors' intellectual and social superior.

The chart of the Gulf Stream in the "Maritime Observations" includes a representation of Franklin himself, in the bottom right corner, conversing with Neptune, even pointing an emphatic finger at the god of the sea (Figure 3.4). The illustration falls within a rich tradition of popular images of Franklin, sometimes in the realm of the gods, often with his index finger pointing at something or someone. The divine ruler of oceans stands in for all the sailors who had talked to Franklin, the mere mortals who nevertheless proved essential to definitions of the Atlantic. The ordinary sailors have been erased from the picture, even as their name for the featured ocean, "Atlantic," finally labels it. Those floating, hard-to-regulate, tricky-to-interrogate mariners might have been pushed to the margins, but their achievement in naming an ocean was now front and center, a lasting reminder of their importance to Atlantic knowledge.[46]

Acknowledgments

The editors of this volume, an anonymous reviewer, and David Armitage gave the author very useful comments on her chapter; Andrew Miller helped greatly with the illustrations.

Notes

1 Some themes of this essay are more generally treated in Joyce E. Chaplin, "The Atlantic ocean and its contemporary meanings," in *Atlantic History: A Critical Reappraisal*, eds Jack P. Greene and Philip D. Morgan (Oxford University Press,

forthcoming); idem, *The First Scientific American: Benjamin Franklin and the Pursuit of Genius* (New York: Basic Books, 2006).

2   Paolo Rossi, *Philosophy, Technology, and the Arts in the Early Modern Era*, trans. Salvator Attanasio, ed. Benjamin Nelson (New York: Harper and Row, 1970), 1–62; Dena Goodman, *The Republic of Letters: A Cultural History of the French Enlightenment* (Ithaca: Cornell University Press, 1994), 23–52, ch. 6; Steven Shapin, *A Social History of Truth: Civility and Science in Seventeenth-Century England* (Chicago: University of Chicago Press, 1994), ch. 8; Simon Schaffer, "Experimenters' techniques, dyers' hands, and the electric planetarium," *Isis* 88 (1997), 456–83.

3   Elizabeth L. Eisenstein, *The Printing Press as an Agent of Change: Communication and Cultural Transformations in Early Modern Europe* (New York: Cambridge University Press, 1979), esp. ch. 6; William Eamon, *Science and the Secrets of Nature: Books of Secrets in Medieval and Early Modern Culture* (Princeton: Princeton University Press, 1994), 94–126; Adrian Johns, *The Nature of the Book: Print and Knowledge in the Making* (Chicago: University of Chicago Press, 1998).

4   Jan Golinski, *Science as Public Culture: Chemistry and Enlightenment in Britain, 1760–1820* (Cambridge: Cambridge University Press, 1992); Ann B. Shteir, *Cultivating Women, Cultivating Science: Flora's Daughters and Botany in England, 1760–1860* (Baltimore: Johns Hopkins University Press, 1996); Rudolph M. Bell, *How to Do It: Guides to Good Living for Renaissance Italians* (Chicago: University of Chicago Press, 1999); Frank A. Kafker, *Notable Encyclopedias of the Seventeenth and Eighteenth Centuries: Nine Predecessors of the "Encyclopédie"* (Oxford: Voltaire Foundation, 1981); John Lough, *The "Encyclopédie"* (London: Longman, 1971).

5   N. A. M. Rodger, *The Wooden World: An Anatomy of the Georgian Navy* (New York: Norton, 1996), 45; Marcus Rediker, *Between the Devil and the Deep Blue Sea: Merchant Seamen, Pirates, and the Anglo-American Maritime World, 1700–1750* (Cambridge: Cambridge University Press, 1987), 155–6, 163–9, 186–93.

6   N. A. M. Rodger, *The Command of the Ocean: A Naval History of Britain, 1649–1815* (London: Allen Lane in association with the National Maritime Museum, 2004), 205–6, 636–9; Rediker, *Between the Devil and the Deep Blue Sea*, 30–3, 80–1, 290; Daniel Vickers, "Beyond Jack Tar," *William and Mary Quarterly*, 3rd ser. (hereinafter *WMQ*) 50 (1993): 418–24.

7   J. D. Davies, *Gentlemen and Tarpaulins: The Officers and Men of the Restoration Navy* (Oxford: Clarendon Press, 1991); Richard Ollard, "The navy," in *The Diary of Samuel Pepys*, vol. 10, *Companion*, ed. Robert Latham (Berkeley: University of California Press, 1983), 285–7; Daniel A. Baugh, *British Naval Administration in the Age of Walpole* (Princeton: Princeton University Press, 1965), 100–2; Rediker, *Between the Devil and the Deep Blue Sea*, passim; Ira Dye, "Early American Merchant Seafarers," APS *Proceedings* 120 (1976), 331–60; Rodger, *Command of the Ocean*, 127–30, 312–16.

8   Margaret Deacon, *Scientists and the Sea, 1650–1900: A Study of Marine Science* (London: Academic Press, 1971), chs 4–8.

9   Nicholas Rogers, "Liberty road: opposition to impressment in Britain during the American War of Independence," *Jack Tar in History: Essays in the History of*

Maritime Life and Labour, eds Colin Howell and Richard J. Twomey (Fredricton, New Brunswick: Acadiensis Press, 1991), 57.

10 Mary Blewitt, *Surveys of the Seas: A Brief History of British Hydrography* ([London]: Macgibbon and Kee, 1957), 29; W. R. Chaplin, "A seventeenth-century chart publisher…," *American Neptune* 8 (1948), 310-11; Thomas R. Adams, "Mount and Page: publishers of eighteenth-century maritime books," *A Potencie of Life: Books in Society, The Clark Lectures, 1986-1987*, ed. Nicolas Barker (London: British Library, 1993), 147-53; Victor Suthren, *To Go Upon Discovery: James Cook and Canada, from 1758 to 1779* (Toronto: Dundurn Press, 2000), 153.

11 Ian K. Steele, *The English Atlantic, 1675-1740: An Exploration of Communication and Community* (New York: Oxford University Press, 1986), 14-15.

12 *Oxford English Dictionary*, s. v., "atlantic."

13 Norman J. W. Thrower, ed., *The Three Voyages of Edmond Halley in the "Paramore," 1698-1701* (London: Hakluyt Society, 1981), 15-27, 58, 321.

14 Anton Gill, *The Devil's Mariner: A Life of William Dampier, Pirate and Explorer, 1651-1715* (London: Michael Joseph, 1997); Anna Neill, "Buccaneer ethnography: nature, culture, and nation in the journals of William Dampier," *Eighteenth-Century Studies* 33 (2000), 165-80.

15 Thrower, ed., *Voyages of Halley*, 368-9; Dampier, *Discourse of Winds, Breezes, Storms, Tides, and Currents* (London, 1699), 3.

16 *Early American Newspapers*, www.readex.com, s. v. "Atlantic" and "Atlantick" for each decade from 1690 to 1799.

17 David Armitage, *The Ideological Origins of the British Empire* (Cambridge: Cambridge University Press, 2000), ch. 4; Fred Anderson, *Crucible of War: The Seven Years War and the Fate of Empire in British North America* (New York: Knopf, 2000).

18 Rupert T. Gould, *The Marine Chronometer, Its History and Development* (London: J. D. Potter, 1923), 1-70; Dava Sobel, *Longitude: The True Story of a Lone Genius Who Solved the Greatest Scientific Problem of His Time* (New York: Walker, 1995).

19 Nathaniel Philbrick, *Away Off Shore: Nantucket Island and Its People, 1602-1890* (Nantucket: Mill Hill Press, 1994), 79 (on Starbuck store), ch. 11 (Folgers); Edward Byers, *The Nation of Nantucket: Society and Politics in an Early American Commercial Center, 1660-1820* (Boston: Northeastern University Press, 1987); Edouard A. Stackpole, *The Sea-Hunters: The New England Whalemen During Two Centuries, 1635-1835* (Philadelphia: Lippincott, 1953), 37 (Folgers); on Benjamin Franklin's genealogy, *The Papers of Benjamin Franklin* (hereinafter *PBF*), ed. Leonard W. Labaree et al. (New Haven: Yale University Press, 1959-), I, liv, lv.

20 Peter Franklin to Benjamin Franklin (hereinafter BF), Feb. 21, 1765, *PBF*, XII, 77; *The Autobiography of Benjamin Franklin*, ed. Leonard W. Labaree (New Haven: Yale University Press, 1964), 79; *PBF*, I, lvii (Josiah), lix (brother and brother-in-law), lx (Isaac All, captain from Newport); *PBF*, VII, 120 and illustrations on three following pages.

21 *Oxford English Dictionary*, s. v., "gulf-weed"; BF, "Journal of a voyage," 1726, *PBF*, I, 93-5.

22 BF to Cadwallader Colden, [Feb. 1746], *PBF*, III, 67–8; Colden to BF [Feb. 1746], ibid., 69, 70. See also Steele, *English Atlantic*, Table 9.3, pp. 314–15.
23 www.readex.com; *American Weekly Mercury*, Apr. 1, 1742. See also *Connecticut Journal*, Apr. 12, 1771.
24 Adolph B. Benson, ed., *Peter Kalm's Travels in North America: The English Version of 1770*, 2 vols (New York: Dover, 1937), cited in *PBF*, IV, 56; BF to John Mitchell, Apr. 29, 1749, ibid., III, 376; BF to John Perkins, Feb. 4, 1753, ibid., IV, 431, 441; BF to William Brownrigg, Nov. 7, 1773, *PBF*, XX, 465.
25 Walter Klinefelter, "Lewis Evans and His Maps," APS *Transactions*, n. s., 61 (1971), 5–31; William E. Lingelbach, "Franklin and the Lewis Evans map of 1749," *APS Society Yearbook 1945* (Philadelphia, 1945), 63–73. A digital version of the map is available on the Library of Congress website: http://hdl.loc.gov/loc.gmd/g3790.ar103500 (Franklin's contribution is at upper left).
26 BF to Perkins, Feb. 4, 1753, esp. pp. 433–7. Also read at the Royal Society on June 24, 1756 and printed as Letter XX in *Experiments and Observations*, 1769 and 1774.
27 BF to Perkins, 429, 430n, 431, 431n, 439, 441; on Dampier, see *PBF*, IV, 442–5 (including notes 7, 8).
28 Glyndwr Williams, *Voyages of Delusion: The Quest for the Northwest Passage* (New Haven: Yale University Press, 2003).
29 BF to William Strahan, July 4, 1744, *PBF*, II, 410; Peter Collinson to BF, June 14, 1748, *PBF*, III, 300; *PBF*, X, 85–6; *PBF*, IV, 380–3.
30 *PBF*, IV, 382–3; BF to John Pringle, May 27, 1762, ibid., X, 99.
31 Pennsylvania Hospital: Report of the Weekly Committee, Nov. 24, 1753, *PBF*, V, 116–17; Charles Hargrave to BF, Mar. 6, 1761, *PBF*, IX, 282 and 282n.
32 BF to Perkins, Feb. 4, 1753, 442 and 442n.
33 BF, "Journal of a voyage," 95–6; BF to Brownrigg, Nov. 7, 1773, 465.
34 BF to Pringle, May 27, 1762, *PBF*, X, 94. On De Brahm's work, see Louis De Vorsey, "Pioneer charting of the Gulf Stream: the contributions of Benjamin Franklin and William Gerard De Brahm," *Imago Mundi* 28 (1976), 105–20, quotation on p. 112. De Vorsey was wrong in thinking that De Brahm's chart was the first of the Gulf Stream. See Philip Richardson, "Benjamin Franklin and Timothy Folger's first printed chart of the Gulf Stream," *Science* 207 (1980), 643–5.
35 Lloyd A. Brown, "The river in the ocean," *Essays Honoring Lawrence C. Wroth*, ed. Frederick R. Goff (Portland, ME: Anthoensen, 1951), 69–84; Commissioners of Customs (Boston) to Lords Commissioners of His Majesty's Treasury, May 12, 1768, Treasury 1/465/60–1, National Archives, London; Anthony Todd to Thomas Bradshaw, June 4, 1768, Treasury Letters, Post 1/9, Post Office Museum and Archives, London; BF to Anthony Todd, Oct. 29, 1769 [1768], *PBF*, XV, 246–7.
36 BF to Todd, Oct. 29, 1769 [1768], 247.
37 Instructions to Deputies, Packet Captains and Surveyors, 1763–1811, Post 44/1, Post Office Museum and Archives; Suthren, *To Go Upon Discovery*, 57–60, 97–102, 126–8, 147 (quotation), 154–5; James Cook, master, "A journal of the Grenville," June 14, 1764–Dec. 31, 1765, 111r, 115r, 127r (quotation), ADM

52/1263, National Archives, London. These entries match those in Cook's log book; later log books and journals are silent on currents.
38 BF to Todd, Oct. 29, 1769 [1768], 247; Richardson, "First printed chart of the Gulf Stream"; Chaplin, "A seventeenth-century chart publisher …," 309, 310.
39 To see the map in detail, consult the digital version on the Library of Congress website: http://hdl.loc.gov/loc.gmd/g9112g.ct000753.
40 Ellen R. Cohn, "Benjamin Franklin, Georges-Louis Le Rouge and the Franklin/Folger chart of the Gulf Stream," *Imago Mundi* 52 (2000), 24–42.
41 Ibid., 135–6.
42 Jan Golinski, "Barometers of change: meteorological instruments as machines of enlightenment," *The Sciences in Enlightened Europe*, ed. William Clark, Jan Golinski, and Simon Schaffer (Chicago: University of Chicago Press, 1999), 69–93; Marie-Noëlle Bourguet, Christian Licoppe and H. Otto Sibum, eds, *Instruments, Travel, and Science: Itineraries of Precision from the Seventeenth to the Twentieth Century* (London: Routledge, 2002).
43 "A letter from Dr. Benjamin Franklin … containing sundry maritime observations," APS *Transactions* 2 (1786), 314–15.
44 Ibid., 316; Deacon, *Scientists and the Sea*, 186–8, 202.
45 Cohn, "The Franklin/Folger chart of the Gulf Stream," 135; "Maritime observations," 314, 325–8.
46 Charles Coleman Sellers, *Benjamin Franklin in Portraiture* (New Haven: Yale University Press, 1962), 2, 6, 7, 8, 30, 32, 33. On the continuing interaction between sailors and scientists, see Helen M. Rozwadowski, "Small world: forging a scientific maritime culture," *Isis* 87 (1996), 409–29.

SECTION II
Writing the American Book of Nature

CHAPTER 4

A New World of Secrets

Occult Philosophy and Local Knowledge in the Sixteenth-Century Atlantic

RALPH BAUER

Introduction

In 1492, the year of Christopher Columbus's first transatlantic voyage, Pope Innocent VIII was succeeded as the spiritual head of Christendom by Alexander VI, a Spaniard born near Valencia in 1431. One year after his elevation to the Holy See, Alexander famously issued a bull that divided up Europe's New Worlds, assigning all territories west of a vertical line running about 100 miles west of the Cape Verde Islands to Spain and all territories east of the line to Portugal. In the same year, on June 18, Alexander issued another, less-well-known bull in which he absolved Giovanni Pico della Mirandola, the famous Italian philosopher and scholar who had been in trouble with the Roman Inquisition for dabbling in astrology and magic. Alexander's elevation marked an important (if temporary) reversal in ecclesiastic policy with regard to magic at the turn of the fifteenth century. Whereas the Church Fathers and the medieval church had condemned magic as dangerous and blasphemous, this period saw a reassessment of the relationship between Christianity and magic, especially natural magic, reflected not only in a new philological interest in the Hermetic tradition but also in the (re-)publication of many texts by medieval magi, such as Albertus Magnus and Roger Bacon, and of new syntheses of occult knowledge, such as Heinrich Cornelius Agrippa von Nettesheim's (1486–1535) magisterial *De occulta philosophia* (1533), which fused Aristotelian science, Ptolemaic astrology, neo-Platonist philosophy, and Hermetic and Cabbalistic traditions with Christian doctrine into a coherent system of Renaissance magic.[1]

If Renaissance Europe was already seeing a revival of magic during the late fifteenth century, the European discovery and conquest of the New World, revealing the existence of vast lands and numerous peoples previously unknown, as well as an abundance of new species of plants and mineral wealth, both contributed to and was inspired by this renewed scientific hope of controlling nature through magic. Yet, while recent New Historicist scholarship has been increasingly attentive to the "darker," geopolitical sides of Renaissance cultural productions, investigating the relationship between European colonialism in the New World and the history of modern Western branches of knowledge, such as ethnography, natural history, or cartography,[2] less is known about the relationship between the early modern European expansionist endeavours in the New World and the rise of "Occult Philosophy" in the sixteenth century.[3] This is perhaps owing in part to some lingering anachronisms that have governed the historiography of both early modern science and the European encounter with the New World. On the one hand, in the "Whiggish" tradition that has long dominated the history of science, the significance of the European discovery of the New World has often been seen in terms of a "hard fact" over which medieval European minds stumbled only to be awakened to modernity, inductivism, and scientific progress, whereas the occult sciences have often been seen as inherently regressive and resistant to change.[4] On the other hand, in postcolonial and New Historicist scholarship on the early Americas, particularly, the European encounter with the New World has all too often been seen in terms of a clash between Western rationalism on the one hand and Native American shamanism and magical thinking on the other, as the confrontation between two worlds that were utterly incommensurable.[5] As a result, literary scholarship on the enchanted texts of the discovery and conquest has largely been informed by a post-Baconian disavowal of their epistemological and ideological underpinnings. From this "enlightened" perspective, the sixteenth-century magical quests in the New World for the Philosopher's Stone, elixirs of life, the Fountain of Youth, the Early Paradise, and most of all for the legendary El Dorado, have appeared as mere Quixotic aberrations into "madness" from a general historical path in which the "New World" would become homologous with the "New Science."[6]

More recently, however, historians of science have called into question the alleged reactionary character of Renaissance occult philosophy and instead paid attention to its close connections with the history of modern science. Especially the modern idea of "science as *venatio*" (hunt) for the secrets of nature through experiment and "cunning" (*mêtis*) owes much, as William Eamon has recently shown, to the empirical bent of sixteenth-century occult philosophers.[7] At the same time, historians and anthropologists of the early Americas have shown that the view that European knowledge "sat, in its purity, like a layer of oil" over Native American magic is a "highly misleading one," as the Christian religion and European knowledge were themselves "intermingled with a great deal of magic." Far from dismissing Native American magic as mere superstition, the early explorers and conquerors frequently attributed efficacy and power to Native American empirical knowledge of the environment and its local spiritual forces.[8] Serge Gruzinski, for example, has asserted that it is "no longer possible to describe the intellectual clash between Renaissance chroniclers and Amerindian societies as a merciless duel between the truths of reason and the mistaken ways of primitive societies." Rejecting an older historiographic model in which cultures encountered but never met, Gruzinski emphasized the "mélanges" that occurred in the sixteenth-century Atlantic world, arguing that "the borders [between the indigenous world and the world of the conquistadors] were so intertwined that they became indissociable" even though these "mestizo mechanisms" were frequently the product of intercultural misunderstandings and "gaps in communication."[9]

Gruzinski's work serves as a reminder not only to historians of science but to literary scholars that early modern forms of knowledge, such as the seventeenth-century Baconian "new philosophy" or "the novel" did not emerge exclusively as European phenomena in the process of cultural expansion but rather as the product of complex geographic interactions with various forms of local knowledge in a sixteenth-century globalizing world. I have elsewhere argued, for example, that the "mercantilist" division of intellectual labor in Baconian "new" natural philosophy responded not only to the sociopolitical imperative of incorporating (and subordinating) social groups previously excluded from the production of scientific knowledge (such as merchants, sailors, and artisans) but to the geopolitical imperative of incorporating (and subordinating to the metropolis) local forms of "creole" knowledge in the seventeenth- and eighteenth-century settler empires.[10] Here, I would like to turn back to the sixteenth century to inquire into the role that the language of "Occult Philosophy" played in the sixteenth-century discourse of discovery and in the encounter with local forms of indigenous knowledge in the New World.[11] On the one hand, I want to suggest that Occult Philosophy functioned ideologically and rhetorically in the early historiography about the New World by synthesizing the mercantile values underlying the projects of exploration and conquest, typically conducted by the nonaristocratic sectors of European society, with the aristocratic values of the courts, on whose sponsorship or favors early transoceanic expansionism critically depended. On the other hand, however, I also want to suggest that it would be a mistake—all too common in recent New Historicist scholarship on the literature of the discovery and conquest—to dismiss the language of Occult Philosophy in this literature as mere European fantasies rationalizing material plunder and economic gain. Rather, in their emphasis on the "productive" role of human cognition in the act of unlocking the secrets of nature,[12] occult philosophical traditions provided important epistemological venues in an early modern discourse of "discovery" not only for the apprehension of empirical phenomena that seemed to be at odds with the more authoritative sources of the scholastic canon but for an interaction with local (often oral) forms of knowledge that would otherwise seem utterly incommensurable with the text-based norms that predominated in the European academies.[13] To "discover" meant, in Occult Philosophy as in scholastic epistemology, not so much to "obtain knowledge of (something previously unknown) for the first time"—as it would in the modern sense of the word "to discover" (*OED*, d. 8)—but rather to "expose to view" or "make manifest" something that is already known—as in the pre-modern sense of the word (*OED*, 3a). Unlike in scholastic epistemology, however, in Occult Philosophy to "discover" meant *not* to reveal something that's *generally* known to be true by authority or reason but rather something that is "secret." In the interstice between the known and the unknown, between the old and the new, the category of the secret in Occult Philosophy opened a unique epistemological space of legitimacy not only for apprehending exotic or preternatural phenomena but for the inscription of local forms of knowledge, a unique space of legitimacy foreclosed on by both the modern notion of discovery—predicated on the assumption that whatever it is that is being discovered has not already legitimately been known by someone who is already "there"—and by scholastic philosophy—predicated on the assumption that whatever it is that is being discovered must either make manifest what is already known or otherwise be an error.

It is within this "occult" structure of knowledge, I want to argue, that much of the sixteenth-century discourse of the discovery and conquest of

America must be understood. My primary textual example here will be Walter Ralegh's account of his quest for El Dorado, *The Discoverie of the Large, Rich, and Bewtiful Empyre of Guiana* (1596), but I want to place this text within the larger contexts of both English New World historiography and the larger El Dorado corpus. As I will argue, the originally Native American traditions surrounding El Dorado could capture the sixteenth-century European imagination across generational and national boundaries precisely because they powerfully resonated with the sixteenth-century revival of European occult traditions, such as Neoplatonism and Hermeticism. For sixteenth-century explorers and conquerors, such as Ralegh, New World experience was a series of encounters with occult signs that held the key to an esoteric knowledge and fused both empirical and mystical elements. By resorting to a language rich in alchemical motifs and imagery when writing about his quest, Ralegh constructed his authority as vested not only in the actual (albeit often fragmented) appropriation of local forms of knowledge but (and perhaps more importantly) by his appeal to their status as a holder *and withholder* of local secrets. In other words, his authority as a travel writer about the New World was constructed not through a rhetoric of transparency and self-effacement (as would be called for by the Baconian project of knowledge production) but rather through a rhetoric and secrecy, opacity, and self-investment.[14]

Secret New Worlds

The importance of Occult Philosophy in the European expansionist project is manifest when we consider some of the earliest treatises about the New World in English. Thus, most English readers would have first learned about the New World primarily in the translations from Spanish and continental cosmographers undertaken by Richard Eden, a staunch promoter of English colonialism working under the reign of Mary. Still emphatically pro-Spanish, Eden called on Englishmen to emulate the Iberian conquests in the New World by translating tirelessly the histories and chronicles of the Spanish and Portuguese exploits in east and west "India." Andrew Fitzmaurice has recently made an argument for the important role that humanism played in the translation enterprise of Eden particularly and the intellectual history of English expansionism in the New World more generally. He takes issue with critics who have emphasized the economic aspects of English colonialism—an emphasis he finds "misplaced"—and calls for an investigation of the "moral philosophy" and "*studia humanitatis*" in England's colonial enterprise.[15] However, while surely humanism was one of the most important currents in the reform of *scientia* during the Renaissance[16] and while some English propagandists of empire, such as Eden and Richard Hakluyt, were doubtlessly influenced by the Humanist movement, it would be misguided to see *studia humanitatis* as the founding epistemology of English colonialism. Indeed, it would lead us to apprehend England's colonial world as did the "honest and old counsellor" Gonzalo in Shakespeare's *The Tempest*, who is shipwrecked at Prospero's New World island and whose philosophizing about the "golden age" in which he believes he has arrived is perceived as noble but also as naïve and out of touch with the empirical realities surrounding him on the island. His knowledge confined to the world of books and classical learning, he is oblivious to and manipulated by the island's invisible and secret forces that Prospero's art has conquered. Prospero, not Gonzalo, epitomizes the Renaissance conqueror who manipulates and commands the secret forces of nature on his island through his magic art learned from secret books on the one hand and his cunning intelligence on the other. His new life on the New World isle combines the *vita contemplativa* with the *vita activa*. His epistemology has synthesized the scholar's occult book learning—in "being transported/ And rapt in secret studies"—with the traveler's empirical knowledge learnt in part by simply *going there* and in part by the intelligence gained from his enslaved New World savage, Caliban—who "show'd thee all the qualities o' th' isle" (including the spirit Ariel). Unlike in Baconian science, which would strive to force a wedge between book learning and empirical knowledge, in Prospero's "art" book learning and empiricism complement one another, giving him power both over the mere empiricist Caliban and the Humanist Gonzalo and the shipwrecked court.[17]

To see the English (as well as the European) encounter with the New World in its proper intellectual context, it is important to take a closer look at the textual record of the New World and the intellectual traditions that are employed to contextualize it. For example, one of the first works translated by Eden into English was the monumental and influential *Cosmographiae* (Basel, 1550), written by the German cosmographer Sebastian Münster. Dividing his book into two sections—one on "east India" and one on "west India"—Münster had written one of the earliest and most comprehensive cosmographies to incorporate all the empirical knowledge gained with the new voyages of discovery. A scholar who had learned Hebrew and studied Jewish science, becoming "one of the great Judaic scholars of his day,"[18] Münster focused especially on the secret properties of the natural and mineral wealth of the newly discovered exotic places. Thus, the first section, which begins with a summary of the Portuguese circumnavigation of Africa and the arrival at Calicut, details the virtues of "the diamond stone" and the secret uses to which it is put by Indian healers. The most precious among the stones, Münster argued, the diamond "not onely refuses the forte of Iron but also resisteth the power of fyre, whose heate is so farre unable to melte it… and is also rather made purer thereby then anye wayes defiled or corrupted." Its secret virtues counteract "poisons, and to frustrate the opperacion thereof, and [is] therefore greatly esteemed of kinges and Princes." The second section, on the "West Indies," provides an account of Columbus's first two voyages and then of Magellan's circumnavigation, returning to Columbus's third voyage and the four voyages of Vespucci, and concludes with a theoretical discourse on "whether under the aequinoctial circle or burninge lyne (called Torrida zona) be habitable regions," in which he finds "no sufficiente causes why under that [equinoctial] line should be no habitable regions," citing diverse Greek, Roman, Alexandrian, and medieval textual authorities that appear to back up the empirical evidence derived from the recent explorations but which contradicted Aristotle.

Significantly, in the prologue to his English translation of Münster's text (originally published in Latin), Eden placed Münster in the tradition neither of Aristotle and the scholastic canon nor of Cicero and the humanist canon but rather in that of the German Dominican Albert Graf von Bollstädt (1193–1280), better known as Albertus Magnus, whose occult knowledge was rumoured not only to have made his garden bloom in the middle of winter but even to have constructed an artificial man, who was then allegedly destroyed by one of Albertus's most brilliant students, Thomas Aquinas, who warned his teacher against indulging in such "promethean ambitions."[19] Not only had Albertus already refuted Aristotle regarding the latter's claim of the uninhabitability of the equatorial regions of the earth but his knowledge even exceeded that of St. Augustin, Eden argued, who fell into "erroure in the science of Astronomie in which he travayled but as a stranger."[20]

What are we to make of Eden's invocation of the occult tradition in his attempt to write the first English histories of the New World? Michael Nerlich's account of the transformation of an ideology of "adventure" during the early modern periods provides a useful historical and sociological context. He argues that the reason for the proliferation of magical, fairy tale–like romances of chivalry, "characterized by dragons and sorcerers, giants and fairies," in fifteenth- and sixteenth-century Europe was the social decline of the knightly class and the loss of the "real meaning of knighthood,"[21] as commercial adventure stood in defiant opposition to the courtly ideology of adventure. Whenever possible, therefore, commercial adventure would have to be cloaked in the forms of courtly ideology itself to be persuasive to the aristocracy. Though the aristocratic world of the courts became increasingly aware of its dependence on the world of commerce and trade and on the empirical knowledge on which this world functioned, the bourgeoisie began to "imitate the nobility, bought land, and withdrew onto its holdings, out of the *vita activa* (active life), out of trade." In early modern novels, such as the *Fortunatus*, tropes such as that of the "magic purse" thus obviated the difficult explanation of economic interconnections and allowed for a fairy-tale justification of emerging bourgeois economic relations.[22] Similarly, the historian of science Pamela Smith, in her study of the German alchemist, Johann Joachim Becher, has offered a sociopolitical model for understanding the resurgence of alchemy in the early modern period (though more particularly for seventeenth-century continental Europe). She argues that the increasing importance accorded to empirical knowledge both "opened up possibilities for individuals who did not have an intensive text-based education that would have allowed them to call themselves learned in the old style" while also resulting in the formation of an elite of natural philosophers who tried to distance themselves from rude mechanics. Alchemical philosophers such as Becher thus aimed to mediate between the immovable values of the landed nobility in the courts and the practices and movable values of the money economy of the commercial world. They framed the "commercial projects in the traditional idiom and gesture of noble court culture and translated the commercial values into court culture."[23]

The language of the occult in the texts written by the early explorers and conquerors of exotic New Worlds, such as Columbus, Cortés, or Ralegh, functioned in ways similar to those observed by Nerlich and Smith about the courtly magicians patronized or entertained by the landed aristocracy. Tales neither of traveling merchants nor of crusading knights in the medieval tradition, their accounts about the sixteenth-century quests for New Worlds forged mystical connections between material gain and spiritual renewal. "Gold is most excellent," Columbus wrote from his third transatlantic voyage; "of gold there is formed treasure and with it whoever has it may do what he wishes in this world and come to bring souls into Paradise."[24] Not unlike the alchemical opus, the New World journey appears in these accounts as a plot not only of material transformation (or "ex-change") in trade and plunder but of a spiritual transformation and renewal. The New World explorers and conquerors hereby frequently fashioned themselves in the mercurial image of the alchemist whose quest is not merely to transform base metals into gold but also to penetrate the mystical secrets of nature. Thus, Cortés wrote in his letters from Mexico, which were translated into virtually all Western European vernaculars and into Latin in the course of the sixteenth century, that he was resolved on finding out "the secrets of these parts."[25]

Apart from these ideological connections, however, there are also epistemological connections between what William Eamon has called an early modern "'conquistatorial' attitude" toward nature and Occult Philosophy.[26] Thus, Eden would most likely have been familiar with one of the numerous English translations of Albertus's writings on the magic properties of plants, such as *The Boke of Secrets of Albertus Magnus of the vertues of herbes, stones, and certaine beasts*, the fruit of his experiments with "herbes, stones, and certaine beasts" that have "great vertues" derived from "the influence of the planets ...[as] every one of them taketh their vertue from the higher naturall powers."[27] As Richard Drayton has pointed out, medieval magi, such as Albertus Magnus and Roger Bacon in England, had in part been inspired by "the influence of new and purified classical texts and of Arab example" that reached Europe via Muslim Spain. Aristotle, Plato, Muslim medicine and alchemy "offered motives and methods for the research of all living and dead things" and a naturalistic mysticism, which "sought religious insights in the study of creation, was joined to a mystical Naturalism, which sought to uncover magical means of working with nature." Thus, Bacon believed that unlocking the secrets of stars, stones, and herbs would ultimately allow men to conquer disease and death. The transmutation of base metals into gold was hereby only one of the aims of medieval alchemy. Its main goal was the prolongation of life by way of discovering an elixir that would make the magus immortal.[28] However, though Albertus and Bacon were important precursors for the Renaissance magi, for the most part there remained during the Middle Ages, as Gonzalo Aguirre Beltrán has observed, essentially two medical traditions in Europe: on the one hand, academic medicine that followed the classical theories of Hippocrates, Galen, Dioscorides, and Avicenna, and, on the other, what he calls "folk medicine" (*medicina-folk*), which flourished outside the official academies, in the streets. Though the professors of the former despised the practitioners of the latter, the latter had a tendency toward empiricism and general irreverence with regard to the ecclesiastical and scholastic authorities.[29] During the sixteenth century, however, an increasing number of figures emerged that bridged the gap between the academy and local "folk medicine," men such as Martín del Rio in Spain and Paracelsus in Switzerland.[30]

Paracelsus (1493–1541) represents a good example of this "vibrant counterculture" to academic scholasticism in sixteenth-century natural philosophy. Particularly, his "chemical philosophy" was, as Walter Pagel has written, a "*scientia*" quite different from that which could be "learnt from books or by logical deduction." It was more akin to "empirical and experimental research, to testing, probing and 'knocking at the door' of nature," inspired by a deep distrust of the power of scholastic rationalism. Despite Paracelsus's essentially premodern emphasis on correspondences between microcosm and macrocosm and his tendency to see similarities rather than differences in all natural phenomena, his Occult Philosophy contained "important proto-scientific ideas" by breaking out of text-bound knowledge and by participating in "those trends of skepticism and empiricism which were soon to contribute to the foundation of modern science." His *scientia* combined a highly eclectic and esoteric book learning with a popular and local folk knowledge passed down orally from generation to generation and the empirical knowledge gained from his everyday experience as a surgeon in the mines. However, precisely for this "artisan attitude to life and medicine," Paracelsus was frequently shunned by the academics, despite his fame for producing effective remedies through his alchemical experiments that continued to secure him employment in the courts of local nobilities. Though in the academic medical sciences of his day the active interference at the sick bed was seen as the task of the surgeon, who was considered to be merely a craftsman, this task was seen as beneath the dignity of the scholarly physician. Paracelsus's natural philosophy, by contrast, defied this institutionalized division between the physician and the surgeon, exemplifying what Drayton has called "that peculiar Renaissance conjunction of the worlds of religion, philosophy, magic, and experimental science."[31]

The early Spanish expeditions to the New World not only were motivated by much of the same spirit as Paracelsus's alchemical quests to find out and dominate the secret forces of nature by drawing on both Neoplatonic and occult textual traditions but they combined their book learning with a similar "artisan attitude" and cunning in the acquisition of local knowledge.[32] Gold was hereby only one among a number of natural mysteries that enflamed the early explorers' imagination about the New World. Here was a whole new world of natural secrets, of hitherto unknown plants, stones, heavens, peoples, and local knowledge. Exotic stones, plants, and woods similarly held out the promise of magical transformation through the knowledge and control of their secret properties. Though European explorers did occasionally state their belief that Native American magic was powerless against European weapons of war, the early explorers, conquerors, and settlers—unlike the modern anthropologist—professed their beliefs in the efficacy of magic generally and in native magic particularly. Thus, detailed descriptions of native healing practices can be found in many of the chronicles of the Conquest, including Cortés's letters to Charles V, Bernal Díaz del Castillo's *Historia verdadera*, and the natural histories written by Gonzalo Fernández de Oviedo and José de Acosta. The earliest treatise specifically devoted to the subject was composed by the Sevillano Nicolás Monardes. Although Monardes was never himself in the New World, he diligently collected specimens and firsthand information brought back by the returning explorers and conquerors. In his *Historia medicinal de las cosas que se traen de nuestras Indias Occidentales*, he described many magic native healing practices involving the use of herbs and stones. He reported, for example, that the returning Spaniards were bringing from New Spain "two stones of great virtue" and describes in detail their use in native healing rituals:

> The stone must be wet in cold water, and the sick man must take it in his right hand, and from time to time wet it in cold water. This is the way in which the Indians use them, for they hold it for certain that touching this stone unto a bleeding body part restrains the blood. They have great trust in this because they have witnessed the effects. We have also seen the great effects of this stone in staunching of blood. Those who suffer from the hemorrhoidal flow have cured themselves by moving this stone across the body in a circular motion and by wearing them continually upon their fingers.[33]

In the 1540s, the Franciscan missionary Bernardino de Sahagún had encouraged the students at his Colegio de Santa Cruz at Tlatelolco to collect and record native knowledge of medicinal plants and minerals. One of the most remarkable products of this collaboration was the *Libellus de medicinalibus Indorum herbis*, now commonly referred to as the *Badianus Manuscript*, an herbal written in Nahuatl by Martinus de la Cruz, a teacher of native medicine, and translated into Latin in 1552 by Juannes Badianus, also a student at the *colegio*.[34] Despite the distinctiveness of Mexica cosmology in which the herbal obviously partakes, the treatise displays remarkable similarities to some of its medieval and early modern European counterparts, such as those written by Bacon or Agrippa, in its interest in the magical properties of animals, stones, and plants, suggesting, as Henry Sigerist has noted, that "in the sixteenth century patients were treated along the same lines on both continents."[35] Soon, the news of the rich medical knowledge of the Mexicans came to the attention also of Philip II, who subsequently appointed his court physician, Francisco Hernández, as "Protomédico de Indias" and, in 1570, sent him to the New World charged with the task of collecting native plants and recording their medicinal uses in the natural magic of the natives of New Spain not to gather ethnological information about other cultures but rather to learn the secrets and powers of native healing practices for his own practical use.[36]

Thus, the early Spanish historians of the New World drew on their ethnographic information about native natural magical practices because they themselves held them to be efficacious and because they could presume that the display of this local knowledge would enhance their own authority as men of science. To be sure, although sixteenth-century Europeans were interested in the medicinal properties of native plants and minerals, many of them—and the missionaries especially—professed distrust toward the native *curanderos* (healers) and the spiritual dimension of native healing practices. However, if they often remained hostile to the particular spiritual significance attributed to them in native cosmologies, they did so not because they were dismissive of their magical powers but rather because in the monotheistic cosmology of the Christians, non-Christian cosmologies had to be attributed to the Devil.[37] Especially during the seventeenth century, occult traditions, and especially non-Western ones, would increasingly become the target of censorship and branded as "heathen superstitions" in post-Tridentine Habsburg Spain and Reformation Stuart England.[38] However, during much of the sixteenth century, Europeans were curious about local indigenous knowledge,[39] and treatises about the New World, such as Monardes's *Historia medicinal*, were translated into most Western European languages, including English as *The Joyful Newes out of the new founde worlde, wherein is declared the rare and singular virtues of diverse and sundrie hearbes, trees, oyles, plantes, and stones* (London, 1577). These treatises promised to reveal, as Richard Drayton has observed about their European counterparts, a botanical philosopher's stone with which to transmute disease to health, along with other alchemical techniques which might unlock hidden virtues.[40]

Walter Ralegh and the "Golden King" of Manoa

One of Monardes's most prominent English counterparts and students in the study of the secrets of "materia medica Indiana" was Sir Walter Ralegh, who frequently cited the Spanish doctor in *The Discoverie of the Large, Rich, and Bewtiful Empyre of Guiana*, his account of his first search for El Dorado in 1596. Ralegh reports, for example, taking mineral samples from Guiana, such as a "kinde of white stone (wherein golde is engendered) [of which] we sawe divers hils and rocks in everie part of Guiana."[41] He describes it as "a kinde of greene stone, which the Spaniards call Piedras Hijadas, and we use for spleene stones," and he reports that "every Casique hath one, which their wives for the most part weare, and they esteeme them as great jewels" (146). Similarly, he was interested in the Indians' art of preparing a most deadly poison with which they equip their arrows in warfare and in their knowledge of antidotes:

> There was nothing whereof I was more curious [he writes] than to finde out the true remedies of these poisoned arrows, for besides the mortalitie of the wound they make, the partie shot indureth the most insufferable torment in the world, and abideth a most uglie and lamentable death, somtimes dying starke mad, somtimes their bowels breaking out of their bellies, and are presently discolored, as blacke as pitch, and so unsavory, as no man can endure to cure, or to attend them. (170–1)

This remedy, commonly called *curare* by the Spaniards, was a mixture prepared from the Strychnos root and bark and the subject of many fables among the European colonists. Though it is impossible to know today whether this native secret was actually disclosed to Ralegh, he makes great rhetorical use of it by emphasizing that, unlike the Spaniards, who failed in their attempt of attaining this knowledge through violent extortions, he was able to penetrate this secret, which was so secret that even most of the natives did not know it. "But every one of these Indians know it not," he writes, "no not one among thousands, but their southsaiers and priests, who do conceal it, and onely teach it but from the father to the sonne." Yet, he continues, "I was more beholding to the Guianians, than any other," and "they told me the best way of healing as well therof, as of al other poisons" (171). Among these remedies were many plants, minerals, and stones with magical properties, including one that the natives call *takua* and that played an important role in native healing ceremonies and cosmology.

In its interest in local indigenous knowledge of the botanical pharmacopoeia, Ralegh's account is, as Neil Whitehead has noted, "very much in tune with current hopes that the bio-diversity of the rainforest may eventually reveal many useful medicines, through the search for the 'secret' or 'esoteric' native knowledge."[42] Whitehead therefore goes on to argue that Ralegh's account of his quest for El Dorado is both an "ethnological" *and* an "enchanted" text; to redeem Ralegh from a long tradition of historians and anthropologists who have disparaged him as a source of ethnological information about sixteenth-century native South American culture, Whitehead writes not one but two separate introductions to his recent edition of the *Discoverie* in which he tries to disentangle the magical and literary from the scientific and ethnographic—one entitled "*The Discoverie* as enchanted text" and the other "*The Discoverie* as ethnological text." However, from a historical point of view, it would be anachronistic to attempt to separate the ethnological from the magical when reading

Ralegh's *Discoverie*, as there was for him no such distinction. Unlike the modern "detached" anthropologist, Ralegh was interested in native secrets not for their status as ethnological "information" about indigenous cultures but for their power and efficacy. In this text, as in many other early European accounts of "discovery" about the New World, the ethnological and the magical inform each other in its occult epistemological structure. Thus, in his account of the New World voyage, Ralegh presents himself in the image of a magus who, like Prospero, has been initiated into the secrets of the New World and learned the art of the native shamans.

Couched as they are in the account of his quest for El Dorado in South America, Ralegh's ethnographic descriptions of native natural magic are suggestive with regard to the connections between the occult sciences and Elizabethan colonialism.[43] It is therefore important to keep in mind that the *Discoverie* was composed only *after* his return from South America—in part to redeem himself from the disgrace in which he had fallen with Elizabeth I after marrying one of her ladies-in-waiting without the Queen's approval and in part to promote English initiatives at colonization that, Ralegh felt, were dismally lagging behind those of other nations, mainly Spain. Addressed to two different audiences, court and public, Ralegh's text evidences, as Shannon Miller has observed, two strategies for colonization: fealty and plunder. Indeed, as Miller also notes, the two prefatory letters with which the *Discoverie* was published in 1596—one addressed to "Charles Howard, knight of the Garter ... [and] Robert Cecil, knight, Counceller in her Highnes privie Councels" and the other addressed "To the Reader"—narrate "completely different version of Guiana, highlighting the conflicting conditions through which this project was pursued." It is in this ideological gulf between two opposed systems, royal patronage and private enterprise, Miller argues, in which Ralegh introduces the idea of trade as a "third term" meant to negotiate the ideological contradictions.[44] Yet, though Ralegh's does compare the English interactions with the Native American in trade favorably to those of the Spanish conquest, his account of his search for "Manoa," the city of the legendary ruler known as El Dorado, is only partially invested in the mercantile language of trade; indeed, his account takes pains to emphasize that his endeavor was ultimately *not* directed at attaining private profit: "I could have returned a good quantity of gold readie cast, if I had not shot at another marke, than present profit" (165). This "other marke" was local knowledge, particularly the "secrets" that the Indians had so far concealed from the Spaniards. It is Ralegh's language of Occult Philosophy and that of alchemy in particular, I would argue, that amalgamates the material with the spiritual, the mercantile with the mystical, the bourgeois with the courtly, and European with local indigenous knowledge.

Beginning in the first pages of the *Discoverie*, Ralegh's descriptions of the New World landscape are intensely mystical, as all observable appearances are interpreted as "signes" of something "secret" beneath the surface that he already knows to be there, of something "hidden" and yet to be "discovered." In Guiana, he writes,

> all the rocks, mountains, all stones in the plaines, in woodes, and by the rivers sides are in effect thorow shining, and appeare marveylous rich, which …. Are trew signes of rich mineralles, but are no other then *El madre del oro* (as the Spanyards terme them) which is the mother of golde, or as it is saide by others the scum of gold. (125)

The "mother" or "scum" of gold is a sign that "the mine was farther in the ground" (176), as gold (like medicinal plants) was known to "grow" in the tropics more rapidly than anywhere else in the world.[45] Though scholars have sometimes been befuddled by some of the curious imagery that Ralegh employs in his descriptions of the New World landscape and strange native cultural practices (towers, golden kings, castles, gardens, and so on), the images are in fact very carefully chosen and draw from a long European tradition of alchemical symbolism. For example, one recurrent image that Ralegh resorts to in his descriptions of the waterfalls of Guiana is that of the church tower. Thus, he writes,

> In this branch called *Cararoopana* were also many goodly islands, some of six miles long, some of ten, and some of twenty. When it grew towards sunset, we entered a branch of a river that fell into *Orenoque*, called *Winicapora* ; where I was informed of the mountain of crystal, to which in truth for the length of the way, and the evil season of the year, I was not able to march, nor abide any longer upon the journey. We saw it afar off; and it appeared like a white church-tower of an exceeding height. There falleth over it a mighty river which toucheth no part of the side of the mountain, but rusheth over the top of it, and falleth to the ground with so terrible a noise and clamour, as if a thousand great bells were knocked one against another. I think there is not in the world so strange an overfall, nor so wonderful to behold. (188)

Earlier, he had reported seeing "some ten or twelve overfalls in sight, every one as high over the other as a church tower" (176). The image of the tower more generally, and that of the church tower particularly, has a long tradition in alchemical literature as a symbol for the *athanor* or philosophical "furnace." This metaphoric connection was not only familiar to well-known English mystics, such as Edward Kelly, close associate of John Dee in his alchemical work, but also to such skeptics and satirists as Ben Jonson.[46] In Ralegh's account, alchemical images such as this one function to construct a metaphoric analogy between the New World journey in quest for El Dorado and the alchemist's opus, which aims not only to transform base metal into gold but to attain the Philosopher's Stone (the "Golden King"), bringing about transformation, spiritual renewal, and insight into the occult relations between microcosm and macrocosm.

In part, the mystical aspects of the *Discoverie* must be seen in the context of Ralegh's connection to a circle of English alchemists and dabblers in the occult science. Thus, the distiller John Hester had even dedicated the publication of a book to Ralegh entitled *114 Experiments and Cures*, which claimed to be an English translation of the works of Paracelsus, who was becoming increasingly popular among such English occultists as Robert Fludd.[47] That Ralegh was indeed well versed in the European occult and hermetic corpus is evident from his massive *History of the World*. There, he embarked on a large-scale defense of the European occult corpus and the theory of astrological correspondence between microcosm and macrocosm. Citing "Mirandula," "Albertus," "Dionysius," Apuleius the Platonist," and "Mercurius Trismegistus," he argued that

> it cannot be doubted, but the stars are instruments of far greater use, than to give an obscure light, and for men to gaze on after sunset; it being manifest, that the diversity of seasons, the winters, and summers, more hot and cold, are not so uncertained by the sun and moon alone, who always keep on and the same course, but that the stars have also their working therein. And if we cannot deny, but that God hath given virtues to springs and fountains, to cold earth, to plants and stones, minerals, and to the excremental parts of the basest living creatures, why should we rob the beautiful stars of their working powers?[48]

The confluence between alchemical mysticism and English imperialism in the *Discoverie* was possibly inspired also by Ralegh's courtly connections, via his half-brother, Humphrey Gilbert, with John Dee, who shared Ralegh's and Gilbert's interest not only in the British expansionist project but in the occult sciences. Dee had one of the largest libraries of the day, which held no fewer than ninety-two editions of works by Paracelsus and by such other occultists as Adam von Bodenstein, Alexander von Suchten, Gerhard Dorn, Leonhardt Thurneyesser zum Thurn, and Konrad Gesner.[49] Dee's own works, such as the *Monas Hieroglyphica*, the (now lost)

Mercurius Coelestis, the "Mathematical Preface" to the English *Euclide*, and especially his *General and rare memorials pertayning to the perfect arte of nauigation*, promoted English overseas exploration and expansion in terms of a political and spiritual renewal at home, a renewal—perhaps inspired by his Rosicrucian leanings—that was for him epitomized in the figure of Queen Elizabeth who, according to Tudor mythology, was the heir to an Arthurian, mythical, and mystical "British Empire" whose divine destiny it was to serve as the bulwark against the Catholic Hispano-Papal attempts at universal domination.[50] Thus, he presented Queen Elizabeth with a map of America on the back of which he outlined Elizabeth's "Title Royall to … foreyn Regions," especially in "Atlantis" (or America), by the right of discoveries allegedly made by ancient Britons such as "Lord Madoc, Sonne to Owen Gzynedd Prynce of Northwales," who, Dee argued, founded a "Colonie and inhabited Terra Florida, or thereabouts," and Arthur, who allegedly "not only Conquered Iseland, Groenland, and all the Northern Iles compassing unto Russia, But even … the North Pole." It was an imperial program that rested on the dual purpose of geographic discovery and historical recovery.[51]

However, whereas Dee saw England's ultimate imperial destiny in finding a Northwest passage to China—"to set furth Ships, for a Northwest Discouery"[52]—Ralegh's primary interests had, by the 1590s, firmly focused on the Americas, having in 1587 received a patent to colonize Virginia and now was setting out to wrest South America from the control of Spain. However, as Guiana lay, unlike Virginia, squarely within territories already claimed by Spain, Ralegh invented his own version of "recovery": unlike the Spanish conquest, which had been a violent extortion of the New World's mineral and natural wealth for purely material gain, the English would restore El Dorado, the Golden King, as legitimate ruler of South America. For El Dorado was none other, Ralegh reasoned, than the descendant of the Inca, who fled Peru after the Spanish invasion and founded his city, Manoa, deep in the South American jungles. Thus, he relates learning of certain "prophecies in *Peru*, at such time as the empire was reduced to the Spanish obedience, in their chiefest temples, amongst divers others which foreshadowed the loss of the said empire, that from *Inglatierra* those *Ingas* should be again in time to come restored, and delivered from the servitude of the said conquerors" (199). This plot again finds its origins in an alchemical motif in which the redeeming or rescuing of the "King" was a metaphor for the "conquest" (from the Latin root *conquaerere*, to "bring together") of the corrupt and divided nature of matter that precedes the release of the "quintessence," which must be "forced," as the Elizabethan alchemist William Bloomfield wrote, "out of chaos dark."[53] Thus, the imperial project of restoring the South American "Golden King" would bring renewal and transformation to England's monarchy.

However, though John Dee's imperial mysticism constitutes an important intellectual background for reading Ralegh's *Discoverie*, another equally important one is its relationship with both local Spanish and indigenous sources of knowledge in South America. Despite the professions of his fierce anti-Catholicism and anti-Spanish sentiments, Ralegh was enchanted with the Spanish chronicles of the discovery and conquest of the New World, quoting and translating extensively not only from Monardes but also from Francisco López de Gómara, Pedro Cieza de León, and others (136–7). More important, much of his knowledge was based on unpublished local sources. As he puts it, "Many yeares since, I had knowledge by relation, of that mighty, rich, and beawtifull Empire of Guiana, and of that great and Golden City, which the Spanyards call El Dorado, and the naturals Manoa, which Citie was conquered, reedified, and inlarged by a younger sonne of Guainacapa [Huayna Capac] Emperor of Peru" (121–2). Thus, during an earlier reconnaissance mission to South America in 1594, Ralegh had seized Pedro Sarmiento de Gamboa, an official of the Spanish crown who had been charged with the fortification of Spain's South American main and who first told Ralegh of the legend of El Dorado. Also, during a raid on Trinidad en route for South America, Ralegh captured another Spaniard, Don Antonio de Berrio, the governor of New Granada who had led a number of earlier expeditions into the interior and who related to Ralegh what he had learned about the location of El Dorado from local Indian groups. Berrio also related how one Juan Martínez, a soldier who had been abandoned in a boat during one of these expeditions as punishment for his negligence in the explosion of a powder store, had been captured by Native Americans and brought to a golden city called "Manoa," which allegedly stood on a lake up the Orinoco River and had been founded by El Dorado, the Golden King. According to Martínez, every year El Dorado "carouseth with his captains, tributaries, and governors" as they are "stripped naked and their bodies anointed all over with a kind of white *Balsamum* (by them called *Curcai*) …When they are anointed all over, certain servants of the emperor, having prepared gold made into fine powder, blow it thorough hollow canes upon their naked bodies, until they be all shining from the foot to the head." (140) Also during this time, Ralegh questioned the local Spanish colonists at Margarita and Cumaná, where local traditions had it that the founders of Manoa were really descendants of the Incas who had invaded this land after the destruction of their empire in Peru.[54]

These local traditions encountered by Ralegh were themselves the product of several decades of intercultural hybridization of European and

Native American historio- and geographic traditions. On the one hand, the Spanish conquerors who inquired after native geographical and historical information frequently interpreted the intelligence they obtained in terms of their own classical myths and historical imagination.[55] European traditions to which the Spanish conquerors hereby made frequent recourse include Ovid's account of Golden Age in his *Metamorphoses*, the Biblical Ophir or golden mines of King Salomon, and the land of giants familiar from medieval travel literature and romances of chivalry. However, there was perhaps no story that better narrativized the ambitions and hopes of the Spanish conquest than the tradition of Jason and his Argonauts, who obtained the Golden Fleece by drugging the dragon guarding it as it hung in a sacred grove in Colchis and through the aid of Aetes' famous enchantress daughter, Medea.[56] Essentially an alchemical fantasy, the *Argonautica* lent itself like no other classical epic as a narrative model for narrativizing a modern scientific project of *venatio* and the New World accounts of discovery.[57]

Charles Nicholl, in his imaginative analysis of Ralegh's expedition, has interpreted the Spanish and Ralegh's "alchemical quest" for the "Golden King" primarily in Jungian terms as a struggle toward inner "wholeness" and thus as a "projection" of an "image thrown out from the mind" and over the "emptiness of *terra incognita*."[58] It would be a mistake, however, to see Ralegh's account only in terms of European knowledge. As modern anthropologists have found, the root of the legend has a basis in Native American oral traditions about a probably actual cultural practice among the Muisca, an Indian tribe living in the highlands of Bogotá, in which on an appointed day each year the tribal chief was stripped naked by his attendants and smeared from head to foot with a sticky layer of balsam gum and then powdered with gold that was puffed through tubes of cane until the cacique appeared like a living statue of gold. A Muisca golden statue recovered from a lake in the highlands even depicts El Dorado himself as he paddled his raft on his annual ritual journey to the center of the lake.[59] The earliest recorded version of the story of a rich kingdom or empire lying beyond South America's jungle and the coastal mountains seems to originate with Diego de Ordás's expedition to the Orinoco river in 1529 and was based on bits of gold and information that Ordás obtained from local Indians.[60] Having participated in the conquest of Mexico under Cortés, Ordás had hoped that this kingdom, to which he referred as "the Land of Meta," would equal that of Tenochtitlan in size and splendor. After the Spanish discovery and invasion of the Inca Empire in 1532–33 and the occupation, in 1534, of the northern Inca city of Quito by Sebastián de Benalcázar, one of Pizarro's lieutenants, Spanish soldiers began to gather more details from indigenous informants about this fabulous golden place. It was in these native accounts that the legendary figure known as "El Dorado," a native lord who supposedly covered himself with gold dust and dropped pieces of gold and other precious stones into the waters of a lake surrounding the city, first emerged.[61] These circulating accounts had subsequently inspired numerous expeditions into the South American jungle, including the 1541 expedition led by Gonzalo Pizarro and Francisco de Orellana to "La Canela" (the "land of cinnamon"), which resulted in the discovery of the Amazon; the 1559 expedition by Pedro de Ursúa, which ended in disaster after the rebellion of Lope de Aguirre; and various expeditions under the leadership of the Welser German banking dynasty.[62]

Besides the main motif of the Golden King himself, the various ancillary elements in the El Dorado corpus—including the existence and location of his city and the tradition of the Amazons—are similarly the products of a hybridization of various indigenous and European traditions and often the product also of mutual misunderstandings, cultural de-contextualizations, and lapses in communication. Thus, on the one hand, such place names as "Manoa" or "Manua" are, as Enrique Gandía has suggested, derivations of the Quechua word "manu," which means "debt" or tribute payment due from the Amazonian tribes to the Inca emissaries. Traditions about exotic peoples, such as the Amazons, were most likely based on shreds of local knowledge about distant Peru, particularly the Incan Virgins of the Sun, whose tale had journeyed long and far by word of mouth.[63]

Thus, the "magical" sixteenth-century accounts of discovery, such as the El Dorado corpus, were complex literary hybrids that fused both European and local indigenous forms of knowledge. Similarly, Ralegh's account is replete with native geographic terminology and historiographic traditions, his text (as printed in sixteenth-century editions) littered with italic font in which all native and Spanish words are set. Arguably, it is precisely these exotic importations of local knowledge that construct his authority as a New World travel writer. Though he necessarily translates native into European concepts (such as "empire," "king," or "queen"), the land that he describes is neither a "New England"—a tabula rasa for English inscription—nor the domain of the Devil and heathen superstitions, but rather a place of native "secrets" and local knowledge that hold the key to power and transformation.

In her account of the connections between English colonialism and the history of early modern science, Joyce Chaplin has shown how the rise of mechanical philosophy during the seventeenth century would increasingly cast native knowledge in greater relief to European knowledge—as heathen "superstition."[64] Her account concurs with my contention here that during the greater part of the sixteenth century, Europeans did not perceive native magic in this way but rather deemed it to be powerful and efficacious. However, whereas Chaplin explains this sixteenth-century European attitude by suggesting that early English colonists were weak, dependent on their Native hosts—changing only after they established their own settlements as self-sufficient—I have here attempted to place the early encounters between native and European knowledges in the context of the sixteenth-century emergence of "Occult Philosophy" and to suggest that the reason why sixteenth-century Europeans did not dismiss native magic was not so much that they were weak but rather that sixteenth-century European knowledge was itself pervaded by mystical and magical practices that frequently hybridized with local forms of indigenous knowledge in the early colonial encounters. Thus, though New World travelers who fashioned themselves in the image of the self-imposing magus and claimed to possess native local "secrets"—as did Ralegh in his *The Discoverie of the Large, Rich and Bewtiful Empyre of Guiana*—would later be discredited by the post-Baconian reforms of natural philosophy, which aimed to police and control empirical local knowledge centrally from the imperial metropolis, during the sixteenth century their occult discourse could wield considerable rhetorical power. This is owing, as I have suggested, to the important epistemological, ideological, and geopolitical role that the occult sciences played in an age of rapid cultural expansionism and globalization during the sixteenth century. On the one hand, Occult Philosophy, and especially the language of alchemy, played an important ideological role by fusing mercantile with courtly languages and interests on which the early modern colonial project critically depended. On the other hand, Occult Philosophy provided important conceptual venues for apprehending empirical phenomena and local forms of knowledge that seemed to contradict the more authoritative sources and norms of the scholastic and humanist canon. Particularly in its emphasis on the discovery of "secrets" and its tendency to absorb nonacademic local or artisan knowledge, Occult Philosophy was well suited in an age of rapid cultural expansionism in a way that Aristotelian or humanist epistemology was not. Even though textual traditions continued to play an important role in Occult Philosophy, it accorded relatively greater importance to empirical knowledge than had either humanist philology or scholastic epistemology. Thus, in their empirical bent, epistemological eclecticism, mystical utopianism, and "conquistatorial" cunning, the enchanted travel histories about the New World, such as Ralegh's *Discoverie*, played an important transitional role in the history of modernity that has not been sufficiently appreciated by either historians of science or literary critics. In the interstice between the known and the unknown, between the old and the new, the category of the secret in Occult Philosophy opened a unique epistemological space of legitimate inquiry not only for apprehending exotic things but also for the inscription of local forms of knowledge.

Notes

1. On Pope Alexander and Pico, see Frances Yates, "The Hermetic Tradition in Renaissance Science," in *Art, Science, and History in the Renaissance* (Baltimore: Johns Hopkins Press, 1967), 255–74, 257. On Agrippa, see William Newman, *Promethean Ambitions: Alchemy and the Quest to Perfect Nature* (Chicago: University of Chicago Press, 2004), 34–114. Generally, the literature of the Renaissance revival of Occult Philosophy is vast. For just a few classic landmarks, see D. P. Walker, *Spiritual and Demonic Magic from Ficino to Campanella* (London: Warburg Institute, 1958); Frances Yates, *Giordano Bruno and the Hermetic Tradition* (London: Routledge and Kegan Paul, 1964), *The Occult Philosophy in the Elizabethan Age* (London: Routledge and Kegan Paul, 1979), and *The Rosicrucian Enlightenment* (London: Routledge and Kegan Paul, 1972); Wayne Shumaker, *The Occult Sciences in the Renaissance: a Study in Intellectual Patterns* (Berkeley: University of California Press, 1972); for some recent scholarship, see Valerie Flint, *The Rise of Magic in Early Medieval Europe* (Princeton: Princeton University Press, 1991); William Newman and Anthony Grafton, eds, *Secrets of Nature: Astrology and Alchemy in Early Modern Europe* (Cambridge, Mass.: MIT Press, 2001).

2. For a study of early modern cartography and linguistics in this light, see Walter Mignolo, *The Darker Side of the Renaissance: Literacy, Territoriality, and Colonization* (Ann Arbor: University of Michigan Press, 1994); on comparative ethnology, see Anthony Pagden, *The Fall of Natural Man: the American Indian and the Origins of Comparative Ethnology* (Cambridge: Cambridge University Press, 1982); also *European Encounters with the New World: from Renaissance to Romanticism* (New Haven: Yale University Press, 1993), and Stephen Greenblatt, *Marvelous Possessions: the Wonder of the New World* (Chicago: University of Chicago Press, 1991); on Renaissance classicism in this light, see Jeffrey Knapp, *An Empire Nowhere: England, America, and Literature from Utopia to "The Tempest"* (Berkeley: University of California Press, 1992). For the most comprehensive study of the relationship between English imperialism in the New World and the development of modern science, see Joyce E. Chaplin, *Subject Matter: Technology, the Body, and Science on the Anglo-American Frontier, 1500–1676* (Cambridge, Mass.: Harvard University Press, 2001); for studies of ethnography, see Talal Asad, ed., *Anthropology and the Colonial Encounter* (London: Ithaca Press, 1973), Johannes Fabian, *Time and the Other: How Anthropology Makes its Object* (New York: Columbia University Press, 1983), Peter Hulme, *Colonial Encounters: Europe and the Native Caribbean* (New York: Methuen, 1986); on natural history, see Ralph Bauer, *The Cultural Geography of Colonial American Literatures: Empire, Travel, Modernity* (Cambridge: Cambridge University Press, 2003).

3 For a notable exception here, see Jorge Cañizares-Esguerra, "New Worlds, New Stars: Patriotic Astrology and the Invention of Indian and Creole Bodies in Colonial Spanish America, 1600–1650," *American Historical Review* 104 (1999), 33–68. For the most part, "America" or the "New World" is absent from recent discussion of the occult in the history of science.

4 See Brian Vickers, "Introduction," in Vickers, ed., *Occult and Scientific Mentalities in the Renaissance* (Cambridge: Cambridge University Press, 1984), 8. For earlier dismissals of the occult in the evolution of modern science, see Herbert Butterfield, *The Origins of Modern Science: 1300–1800* (London: G. Bell, 1957); Charles Schmitt, "Reappraisals in Renaissance Science," *History of Science* 16 (1978), 200–14.

5 For examples of "encounter" scholarship informed by the earlier paradigm of incommensurability, see George McLelland Foster, *Culture and Conquest: America's Spanish Heritage* (New York: Wenner-Gren Foundation for Anthropological Research, 1960); see also Stephen Greenblatt, ed., *New World Encounters* (Berkeley: University of California Press, 1993); and Pagden, *European Encounters*.

6 For one example of this approach in modern historical scholarship, see Germán Arciniegas, *The Knight of El Dorado: The Tale of Don Gonzalo Jiménez de Quesada and his Conquest of New Granada, now called Columbia*, trans. Mildred Adams (New York: Viking Press, 1942), 9–30; also Timothy Severin, *The Golden Antilles* (New York: Knopf, 1970); for an early example, see Pedro Simón's account in *The Expedition of Pedro de Ursua and Lope de Aguirre in Search of El Dorado and Omagua in 1560–1*, trans. William Bollaert (New York: B. Franklin, 1971); for an example from literature, see V. S. Naipaul, *The Loss of El Dorado: A History* (New York: Knopf, 1969). On the "mythification" of America in the discourses of the conquest, see Beatriz Pastor, *The Armature of Conquest: Spanish Accounts of the Discovery of America, 1492–1589*, trans. Lydia Longstreth Hunt (Stanford: Stanford University Press, 1992); on the modern homology "New World/New Science," see Denise Albanese, *New Science, New World* (Durham, NC: Duke University Press, 1996).

7 William Eamon, *Science and the Secrets of Nature* (Princeton: Princeton University Press, 1994), 266–300; on "cunning intelligence" in the Western tradition, see Marcel Detienne and Jean-Pierre Vernant, *Cunning Intelligence in Greek Culture and Society* (Brighton: Harvester Press, 1978). Other historians who have explored the importance of magic during the Renaissance and for the emergence of modern science include Yates; also Lynn Thorndike, *A History of Magic and Experimental Science,* vol. 1 (New York: Macmillan Co., 1923); Shumaker, *Occult Sciences in the Renaissance*; Walter Pagel, *Paracelsus: an Introduction to Philosophical Medicine in the Era of the Renaissance*, 2nd edn (Basel and New York: Karger, 1982); A. G. Debus, *The English Paracelsians* (London: Oldbourne, 1965); Paolo Rossi, *Francis Bacon: From Magic to Science* (London: Routledge and Kegan Paul, 1968); Brian Levack, ed., *Renaissance Magic* (New York: Garland, 1992). For a review of this debate, see Patrick Curry, "Revisions of Science and Magic," *History of Science* 23 (1985), 299–325.

8   Fernando Cervantes, *The Devil in the New World: The Impact of Diabolism in New Spain* (New Haven: Yale University Press, 1994); see also Nicholas Griffith and Fernando Cervantes, eds, *Spiritual Encounters: Interactions between Christianity and Native Religions in Colonial America* (Lincoln: University of Nebraska Press, 1999), especially the editors' introduction and the essay by Osvaldo F. Pardo, "Contesting the power to heal: angels, demons, and plants in colonial Mexico" (163–84); William Christian, *Local Religion in Sixteenth-Century Spain* (Princeton: Princeton University Press, 1982); on history, see Catherine Julien, *Reading Inca History* (Iowa City: University of Iowa Press, 2000); also Susan Niles, *The Shape of Inca History: Narrative and Architecture in an Andean Empire* (Iowa City: University of Iowa Press, 1999); and Jorge Cañizares-Esguerra, *How to Write the History of the New World: Histories, Epistemologies, and Identities in the Eighteenth-Century Atlantic World* (Stanford: Stanford University Press, 2001).

9   Inga Clendinnen, for example, quotes a passage from Antonio de Ciudad Real (1588), in which the Spaniards, having just arrived on the coast of Mexico, named the land "Yucatan" because the Natives from whom they inquired the name of the land had answered: "uic athan" (which means "what do you say or speak, that we do not understand you"); see *Ambivalent Conquests: Maya and Spaniard in Yucatan, 1517–1570* (Cambridge: Cambridge University Press, 1987). Gruzinski therefore compares the situation of Amerindians and Europeans encountering one another as that of prisoners in a maze, in which "the groups most directly involved in the Conquest had to learn to rely henceforth only on local, partial knowledge"; see *The Mestizo Mind: the Intellectual Dynamics of Colonization and Globalization*, trans. Deke Dusinberre (New York: Routledge, 2002), 27, 31, 43, 50–1. For an excellent discussion of *mestizaje* with regard to medical knowledge, see Gonzalo Aguirre Beltrán, *Medicina y Magia: el proceso de acultaración en la estructura colonial* (Mexico: Instituto Nacional Indigenista, 1963), especially 115–39.

10 Bauer, *The Cultural Geography of Colonial American Literatures*; Francis Bacon, *The Works of Francis Bacon*, ed. James Spedding, Rober Leslie Ellis, and Douglas Denon Heath, 14 vols (London: Longman, 1857–74), 4: 254.

11 The eminent Mexican historian Edmundo O'Gorman has pioneered this line of inquiry with his brilliant (though now dated) *The Invention of America* (Westport: Greenwood Press, 1972), although he did not address the question of the occult specifically nor the role that indigenous traditions played in the European "invention" of America. See also Fernando Ainsa, *De la edad de oro a El Dorado: Génesis del discurso utópico americano* (Mexico: Fondo de Cultura Económica, 1992).

12 On the role of "productive" knowing from Antiquity to the early modern period, see Antonio Pérez-Ramos, *Francis Bacon's Idea of Science and the Maker's Knowledge Tradition* (Oxford: Clarendon Press, 1988).

13 For an excellent account of the role (and shortcomings) of Aristotelianism in the New World encounter, see Pagden, *Fall of Natural Man*. For an account of Humanism, see Andrew Fitzmaurice, *Humanism and America: An Intellectual History of English Colonialism, 1500–1625* (Cambridge: Cambridge University Press, 2003).

14 On the ideal of the self-effacing traveler in Baconian natural philosophy, see Julie Solomon, "'To know, to fly, to conjure': situating Baconian science at the juncture of early modern modes of reading," *Renaissance Quarterly* 44 (1991), 513–58, 539–41; see also her *Objectivity in the Making: Francis Bacon and the Politics of Inquiry* (Baltimore: Johns Hopkins University Press, 1998).
15 Fitzmaurice, *Humanism and America*, 13, 8.
16 See Pamela Long, "Humanism and Science," in *Renaissance Humanism: Foundations, Forms, and Legacy*, ed. Albert Rabil (Philadelphia: University of Pennsylvania Press, 1988), 486–514.
17 William Shakespeare, *The Tempest*, ed. Stephen Orgel (Oxford: Oxford University Press, 1987), 2: 2. For a more extensive discussion of *The Tempest* specifically in this light, see Frances Yates, *Shakespeare's Last Play* (London: Routledge, 1975); for a discussion of the alchemical imagery in *The Tempest* from a Jungian perspective, see Noel Cobb, *Prospero's Island: The Secret Alchemy at the Heart of "The Tempest"* (London: Coventure, 1984); for discussions of *The Tempest* in connection to New World colonialism, see Peter Hulme, *Colonial Encounters: Europe and the Native Caribbean* (New York: Methuen, 1986); Eric Cheyfitz, *The Poetics of Imperialism: Translation and Colonization from "The Tempest" to "Tarzan"* (New York: Oxford University Press, 1991). No one has, to the best of my knowledge, investigated the connection between Occult Philosophy and imperial ideology in *The Tempest*.
18 Anthony Grafton, *New Worlds, Ancient Texts: The Power of Tradition and the Shock of Discovery* (Cambridge, Mass.: Belknap, 1992), 103.
19 On Albertus Magnus and Aquinas, see Newman, *Promethean Ambitions*, 44–52.
20 Richard Eden, "To the Reader," in Sebastian Muenster, *A Treatyse of the Newe India, as it is Known and Found in these our days*, trans. R. Eden (London: S. Mierdman for Edward Sutton, 1553), unpaginated.
21 Michael Nerlich, *The Ideology of Adventure*, 2 vols (Minneapolis: University of Minnesota Press, 1995), 1: 29.
22 Nerlich, *Ideology of Adventure*, 1: 29, 43, 60, 71.
23 Pamela H. Smith, *The Business of Alchemy: Science and Culture in the Holy Roman Empire* (Princeton: Princeton University Press, 1994), 5–6.
24 Christopher Columbus, *The Four Voyages*, ed. J. M. Cohen (Harmondsworth: Penguin, 1969), 300.
25 Hernán Cortés, *Letters from Mexico*, trans. and ed. Anthony Pagden (New Haven: Grossman Publishers, 1986), 26–8.
26 Eamon, *Science and the Secrets of Nature*, 416 n. 24.
27 *The boke of secretes of Albertus Magnus of the vertues of herbes, stones, and certayne beastes: Also, a boke of the same aucthor of the maruaylous things of the worlde: and of certaine effectes, caused of certayne beastes* (London: Wyllyam Seres, 1570), sig. Biii.
28 Richard Drayton, *Nature's Government: Science, Imperial Britain, and the "Improvement" of the World* (New Haven: Yale University Press, 2000), 5.
29 Aguirre Beltrán, *Medicina y magia*, 23–4.
30 Martín del Rio, *Investigations into Magic*, ed. and trans. P. G. Maxwell-Stuart (Manchester: Manchester University Press, 2000).

31. Charles Gunnoe, "Erastus and Paracelsianism: Theological Motifs in Thomas Erastus' rejection of Paracelsian Natural Philosophy," in Allen G. Debus and Michael Walton, eds, *Reading the Book of Nature: the Other Scientific Revolution* (Kirksville: Sixteenth Century Journal Publishers, 1998), 45–66, 46; Pagel, *Paracelsus*, 15, 24, 51; Drayton, *Nature's Government*, 9.

32. As Anthony Grafton has pointed out, the world of the early explorers and conquerors was the "world of craft and trade, not that of books, [that] produced the forces that really revolutionized the European and then the entire world" (*New Worlds, Ancient Texts*, 61). The discoverers and conquerors were not elite schoolmen but rather merchants, artisans, sailors, soldiers from the middling or even lower classes, only occasionally coming from the impoverished petty nobility. While it is true that, as Grafton points out, these explorers also had a culture of the book, that book culture was "not identical" with that of the scholars (65).

33. Nicolás Monardes, *Historia medicinal de las cosas que se traen de nuestras Indias Occidentales* (Seville: Alonso Escrivano, 1574), 19.

34. *The Badianus Manuscript (Codex Barberini, Latin 241) Vatican Library: An Aztec Herbal of 1552*, introduction, translation, and annotations by Emily Walcott Emmart (Baltimore: Johns Hopkins University Press, 1940).

35. Henry E. Sigerist, "Preface," in *The Badianus Manuscript*, xi.

36. His texts have recently been compiled, studied, and translated in Simon Varey, ed., *The Mexican Treasury: the Writings of Dr. Francisco Hernández*, trans. Rafael Chabrán, Chynthia Chamberlin, and Simon Very (Stanford: Stanford University Press, 2000).

37. See Daniela Bleichmar, "Books, Bodies, and Fields: Sixteenth-Century Transatlantic Encounters with New World Materia Medica," in *Colonial Botany: Science, Commerce, and Politics in the Early Modern World*, ed. Londa Schiebinger and Claudia Swan (Philadelphia: University of Pennsylvania Press, 2005), 83–99, 94; also Aguirre Beltrán, *Medicina y Magia*.

38. See, for example, Hernando Ruiz de Alarcón, *Treatise of the Heathen Superstitions that Today Live Among the Indians Native to New Spain* [1629], trans. and ed. J. Richard Andrews and Ross Hassig (Norman: University of Oklahoma Press, 1984).

39. For a detailed discussion of this, see Aguirre Beltrán, *Medicina y magia*.

40. Drayton, *Nature's Government*, 4–5, 13–14.

41. Walter Ralegh, *The Discoverie of the Large, Rich and Bewtiful Empyre of Guiana*, ed. Neil Whitehead (Manchester: Manchester University Press, 1997), 126. All further quotations from this text refer to this edition and are henceforth given parenthetically in the text.

42. See Whitehead, "The Discoverie as enchanted text," 57 n. 20.

43. Published independently in four separate editions in 1596 before being included in Richard Hakluyt's *Principal Navigations*, the *Discoverie* has long been regarded by historians as one of the most important texts of Elizabethan imperialism; see, for example, Robert H. Schomburgk, "Introduction," in Ralegh, *The Discovery of the Large, Rich and Beautiful Empire of Guiana*, ed. Schomburgk (London: Hakluyt Society, 1848); David Beers Quinn, *Ralegh and the British Empire* (London: Macmillan, 1947); and Vincent Harlow, "Introduction," in *The Discoverie of Guiana, by Sir Walter Ralegh* (London: Argonaut Press, 1928).

44　Shannon Miller, *Invested with Meaning: the Ralegh Circle in the New World* (Philadelphia: University of Pennsylvania Press, 1998), 154.
45　As Ralegh would have read in the *Historia natural y moral de las Indias* (1590) written by the Jesuit natural historian José de Acosta, "minerals grow in the same manner as plants, not because they have real vegetable and inner life, for this is true only of real plants, but because they are produced in the bowels of the earth in such a way, by the virtue and efficacy of the sun and the other planets, that over a long period of time they gradually grow and almost, one might say, propagate." See José de Acosta, *Historia natural y moral de las Indias*, 2nd edn (Mexico City: Fondo de cultura económica, 1962), 142–3.
46　Thus, Edward Kelly, in his *The Theatre of Terrestrial Astronomy*, refers to the gates of the "furnace" as open "towers"; see *Two Excellent Treatises on the Philosophers Stone, together with the Theater of Terrestrial Astronomy* (London, 1676), 141. In Ben Jonson's *The Alchemist*, Subtle says to Ananias: "O are you come? 'Twas time. Your threescore minutes / Were at the last thred, you see, and downe had gone / *Furnus acediae, Turris circulatorius* …. Had all been cinders" (3. 2. 1.-5). For a discussion of the symbol of the tower in alchemical literature, see Lyndy Abraham, *A Dictionary of Alchemical Imagery* (Cambridge: Cambridge University Press, 1998), 203–4.
47　For a discussion of this, see Charles Nicholl, *The Creature in the Map: a Journey to El Dorado* (London: Jonathan Cape, 1995), 278–87.
48　Walter Ralegh, "The History of the World," in *The Works of Sir Walter Ralegh* (New York: B. Franklin, 1965), 8 vols, 2: 3, 4, 27, 28.
49　Nicholas H. Clulee, "John Dee and the Paracelsians," in Debus and Walton, eds, *Reading the Book of Nature*, 111–32, 113. See also Peter French, *John Dee: the World of an Elizabethan Magus* (London: Routledge and Kegan Paul, 1972), 40–61; and William Sherman, *John Dee: the Politics of Reading and Writing in the English Renaissance* (Amherst: University of Massachusetts Press, 1995), 29–52, as well as 40 and 165 on the connections between Ralegh, Gilbert, and Dee; also Deborah Harkness, *John Dee's Conversations with Angels: Cabala, Alchemy, and the End of Nature* (Cambridge: Cambridge University Press, 1999), 124.
50　E. G. R. Taylor, *Tudor Geography* (London: Methuen, 1930), 119–38; also French, *John Dee*, 180; on Dee's "Christian Cabbalism," see Yates, *The Occult Philosophy in the Elizabethan Age*, 76.
51　John Dee, *General and Rare Memorials* (London: 1577), sig. A2; see also French *John Dee*, 197; and Sherman, *John Dee*, 150–1.
52　John Dee, *General and Rare Memorials Pertayning to the Perfect Arte of Nauigation annexed to The Paradoxal Cumpas, in Playne: Now First Published, 24 Yeres, After The First Inuention Thereof* (London: John Daye, 1577), 2.
53　See Nicholl, *The Creature in the Map*, 322.
54　For a discussion of the Spanish sources of much of Ralegh's *Discoverie*, see Pablo Ojer, *La formación del oriente venezolano* (Caracas: Universidad Católica Andrés Bello, 1966), 496–7.
55　Thus, when Cortés, after the conquest of Mexico, had intelligence about an "island" (actually a peninsula) to the west of the American mainland, he famously named this island "California"—after the Amazon queen Calafia, according to

Rodríguez de Montalvo's romance of chivalry *Sergas de Esplandián*. See Irving Leonard, *Books of the Brave: Being an Account of Books and Men in the Spanish Conquest and Settlement of the Sixteenth-Century New World* (Cambridge, Mass.: Harvard University Press, 1949), 38.

56 Ramos states that the image of a country of gold lying on the shores of a lake (the Black Sea in the myth) would have been a familiar reference among the conquering soldiers of the sixteenth century, who fully expected to find a corresponding reality in the newly-discovered lands of South America (*El mito del Dorado*, 404). For further discussion on this topic, see Gil Munilla, *Descubrimiento del Marañón*, 175; Manuel Ferrandis Torres, *El mito del oro en la conquista de América* (Valladolid: Talleres, 1933), 158–70; and Enrique de Gandía, *Historia crítica de los mitos y leyendos de la conquista americana* (Buenos Aires: Centro Difusor del Libro, 1946 [1929]). On the significance of the Argonautic myth for Habsburg imperial ideology and on Philip II's reconstruction of the Argo, see Mary Tanner, *The Last Descendant of Aenas: The Hapsburgs and the Mythic Image of the Emperor* (New Haven: Yale University Press, 1993), 9.

57 On the *Argonautica* and alchemy, see Antoine Faivre, *The Golden Fleece and Alchemy* (Albany: State University of New York Press 1993). On the *Argonautica* and modern science, see Bronislaw Malinowski's introduction to his 1922 classic, *Argonauts of the Western Pacific: an Account of Native Enterprise and Adventure in the Archipelagoes of Melanesian New Guinea* (New York: Dutton, 1960).

58 Nicholl, *The Creature in the Map*, 319–20.

59 See Neil Whitehead, "The Discoverie as ethnological text," 72–91.

60 See Demetrio Ramos Pérez, *El mito del Dorado: su génesis y su proceso* (Caracas: Academia Nacional de la Historia, 1973), 28 and 38; Gandía, *Historia crítica de los mitos*, 125–6; and John Hemming, *The Search for El Dorado* (New York: Joseph, 1978).

61 For a summary of these indigenous accounts, see Demetrio Ramos Perez, *El mito del Dorado*, 293–304.

62 Both of these expeditions produced their corresponding written accounts, Gaspar de Carvajal's *Descubrimiento del Río de las Amazonas* and Francisco Vázquez's *Relación de Omagua y Dorado* respectively. The best sixteenth-century account of the Welser search for El Dorado is Nicholas Federmann, *Indianische Historia: Ein Schöne Kurtzweilige Historia Niclaus Federmanns des Jüngeren von Ulm Erster Raise so er von Hispania und Andolosia ausz in Indias des Occeanischen Mörs Gethan hat, und was ihm allda ist begegnet bisz auff sein widerkunfft inn Hispaniam, auffs kurtezest beschriben, gantz lustig zu lessen* (1558; Stuttgart, 1859).

63 Gandía, *Historia crítica del los mitos*, 90; also ch. 6.

64 See Chaplin, *Subject Matter*, 321–65.

CHAPTER 5

Tropical Empiricism

Making Medical Knowledge in Colonial Brazil

JÚNIA FERREIRA FURTADO

Medical Books: Paradise Lost

By analyzing several medical treatises written during the seventeenth and eighteenth centuries, this paper will discuss the experience that the authors of these texts accumulated by practicing medicine in colonial Brazil. What do these empirical medical practices tell us about Brazil as a colonial society and about the production and communication of knowledge within the larger Atlantic world? These medical authors inaugurated the field of what would come to be called "tropical medicine," concerned as they were not only with knowing the specificities of diseases and their local treatments but with how to extend this knowledge about the various kingdoms of the natural world into other realms as well.[1] They went to great lengths not only to collect information but to classify and order the world and its natural elements. Their books discuss the specific aspects of diseases that in this period were often little known. Many of these diseases were tropical in origin, and these authors attempted to use local products, including plants, animals, and minerals, in their treatments. Collectively, the medical specimens employed in this practice were known as *materia medica*.[2]

The discovery of the New World put European colonizers into contact with a new and unknown nature.³ In general, the native populations were the ones that showed Europeans how to use this natural bounty, especially in relation to medicinal cures. Empirical gathering and testing of plants, drugs, and other objects in a tropical environment required colonizers to enter into material exchange not only with indigenous populations but with African slaves as well.⁴ It was often the case that European savants discarded the intellectual framework provided by their indigenous counterparts. This article sets out to track the vestiges of some of these forms of knowledge and how knowledge was shared between and published by Europeans during the seventeenth and eighteenth centuries. Put simply, the practice of colonial empiricism led to abundant encounters with native knowledge. It is in this sense that empiricism in the Atlantic (as opposed to within Europe) involved and indeed required extended interactions with native cultures. These men and their books became intermediaries between indigenous and European systems of knowledge, and the cultural encounter between these radically different populations raised unparalleled problems of translation, travel, access, and trust. Their books also reflected the relationship between the entrepreneurial structure of the careers of physicians who traveled between Portugal, Holland, and Brazil and indigenous knowledge and practices (including, significantly, the institution of slavery): a kind of local empiricism in which European and indigenous knowledge and practice merged before ultimately finding their way into print within the larger community of intellectuals and savants on both sides of the Atlantic.⁵

Taken from this perspective, America becomes a center rather than a periphery: a place where new ideas are formed and tested. The caravels that incessantly traversed the seas became caravels of culture, not only because they carried men and objects that reproduced European values and culture but because they took back with them to Europe a new perception of the world that, in turn, made its own mark on the culture of the colonizer. European scholars placed this new knowledge into a new cognitive framework whose structure was increasingly based on empiricism and rationality.⁶ This analysis thus requires a reassessment of the notions of center and periphery within the Portuguese empire itself and in relation to the knowledge systems of Europe and the Americas.

These transoceanic movements and the expansion of Portugal's overseas empire from the fifteenth to the eighteenth centuries brought with them an empirical renovation of knowledge that was the result of observations by sailors, merchants, clerics, administrators, and doctors, among others. In particular, this new maritime knowledge revolutionized the maritime sciences, including the construction of boats and maps. Medical knowledge increased as well, especially in fields that concerned the pharmacopoeia and medical botany.⁷

Maritime voyages to the East and to Brazil unveiled a strange and diversified nature to the Portuguese. Knowledge gleaned from new elements found in these exotic environments—the so-called medicinal simples that, when combined in different ways, formed compound elements—would later transform medicinal practice in Portugal, the East, and in South America, and would force authors to adopt new classification schemes, which in turn stimulated the extensive production of literature mixing medicine and natural history.⁸

From an early stage, then, medicine and natural history in Portugal were characterized both by experimentation and the incorporation of knowledge from the direct observation of nature. This practice led to the incorporation of new medicines in treatments, many of which had been prepared by including elements found overseas. It was often the case that the forms of preparation and their uses were learned through contact with native populations. Medical knowledge within Portugal and throughout its overseas dominions was heir to a medieval tradition in Europe that divided medicine into two separate branches: a learned tradition practiced by doctors trained in universities and another, more practical tradition carried out by surgeons, midwives, and barbers (who, among other things, performed bloodletting and extracted teeth). Because there were no universities in the Portuguese colonies, all doctors and surgeons were required to pursue their studies in Europe, even if they had been born in Brazil. More often than not, Luso-Brazilian doctors studied in Coimbra and Évora. Nevertheless, many were trained abroad, most frequently at the Universities of Alcalá and Salamanca in Spain and at the University of Paris.⁹

In the more practical branches of medicine, the tradition of the trade guilds was already in force. In this system, masters passed on their accumulated knowledge to their apprentices. Aspiring apprentices in the area of surgery, for example, had to present a certificate of at least four years of apprenticeship with a recognized surgeon before training for an additional two years with a licensed surgeon. Because of its more practical character and the empirical apprenticeship it required, the practice of surgery was considered a lesser art, peripheral to the practice of medicine and solely performed by the surgeons or barber-surgeons. The *Hospital Real de Todos os Santos*, founded in Lisbon in 1492, became an important center for the study of surgery, and only in the eighteenth century were courses in this field created within Portuguese universities.¹⁰

As was the case in the rest of Europe, Galenic thinking predominated within Portuguese universities up until the first half of the eighteenth century. Galen's theories were not necessarily opposed to experimentation, however. In fact, empiricism was the basis for the formulation of medical knowledge from ancient times, and many Portuguese books were written based on descriptions of observations from actual patients and from the treatments provided them. During the early modern period, increased medical knowledge required many changes to be made to Galen's theoretical framework. Some of these changes were superficial, but others were more profound. Many Portuguese doctors and surgeons extended their medical knowledge through the publication of medical treatises. For many of these doctors, experience overseas was a fundamental component of the renewal of their medical practice and for the incorporation of new elements to the pharmacopoeia they used for treating patients.[11]

From its earliest stages, Portuguese maritime expansion forced early modern authors to question the ancient belief that the Torrid Zone was uninhabited. Very much to the contrary, the New World revealed itself to navigators as a place of paradisiacal qualities, with a mild climate, exuberant vegetation, and an incredible diversity of animal life. Between the sixteenth and seventeenth centuries, many chroniclers of Portuguese colonization in the Americas made this connection between Brazil and the earthly paradise explicit. Observations about Brazil's temperate climate, the land's fertility, the diversity of fauna and flora, and the abundance of natural wealth frequently dot the narratives and contemporaneous descriptions of the discovery and subsequent colonization of Portuguese America. This literature, produced by clerics, travelers, administrative officers, seamen, doctors, surgeons, and others remains a crucial source for the study of nature and medicine in the New World.[12]

Many well-known Portuguese doctors touted the primacy of empiricism and practical learning over established knowledge, and their works were known throughout Europe. For many of these authors, contact with the floral and faunal diversity of New World nature was essential not only in transforming the practice of medicine but in overturning the centuries-old forms of natural classification that had been inherited from antiquity, strongly marked by the influence of Aristotle, Dioscorides, and Pliny. An extensive literature came to extol the virtues of Brazilian flora and fauna and how to use them in medical treatments, and this literature, in turn, contributed significantly to changes in taxonomy. Many names can be cited to illustrate those Portuguese who first described Brazil's medicinal plants and tropical diseases. Abraão Zacuto Lusitano (1575–1642) inaugurated the medical bibliography of Brazilian tropical disease; Aleixo de Abreu (1568–1630) is considered one of the first authors on tropical medicine; Simão Pinheiro Mourão (1618–85) described many of the same tropical diseases and scurvy, which he observed in Recife; in 1694, João Ferreira da Rosa wrote the first book to describe yellow fever in detail; and Miguel Dias Pimenta studied a very common disease, the *achaque* or *mal do bicho*, also known as *corrupção* or *maculo*, which attacked the large intestine and sometimes even caused death. Their writings revealed a great deal of information about animal and plant diversity as well, such as the existence of *jaborandi*, tobacco, *ipeca*, *canafistula*, and many other medicinal plants from Brazil.[13]

The Dutch invasion of Brazil (1624–49) has traditionally been seen as the period in which the greatest amount of new knowledge about Brazilian diseases was produced. Likewise, most scholars still insist that only Dutch authors were capable of producing knowledge that was, at its base, empirically and experimentally derived from Brazilian sources. This affirmation has been reinforced by the assertion that, during this period, the Iberian Peninsula remained entirely cut off from the epistemological changes taking place throughout the rest of Europe, a characteristic most frequently blamed on both the oppressive power of the state and the Inquisition.[14]

I hope to show, by contrast, that there was a regular and vigorous exchange of ideas between Portugal and the rest of the European continent and that this exchange did not only travel in one direction. Though scholars from many different countries throughout Europe were responsible for the transformations taking place across divergent fields of knowledge, Portuguese savants—many of whom based their theoretical and practical experiments on knowledge acquired during voyages overseas and to other lands, including Brazil—were no less important in this process. This was especially true in the area of *materia medica* and natural history, where knowledge acquired by Portuguese and Spanish agents in the Americas through the transplantation and acclimatization of new species, and their classification and description was a fundamental part of the transformations taking place in these fields throughout European culture since the sixteenth century. Moreover, much of this knowledge was acquired through contact with native populations of the New World.[15]

From this perspective, this essay seeks to show that medical and natural historical knowledge produced during the period of Dutch presence in Brazil should not be considered as a separate chapter in the larger history of how European intellectuals sought to understand and classify American nature. Nor was this scientific literature based on different epistemological principles than the Luso-Brazilian treatises of this period. Rather, New World knowledge was continually moving between the Iberian peninsula and other regions throughout Europe, and books served as crucial vehicles for this transmission. As we will see, Luso-Brazilian literature itself served a crucial role in the development of Dutch knowledge about the nature of the New World: both schools sought to create a paradisiacal vision of Brazilian nature through empiricist practices. As if in unison, these texts conjured up a land without sickness where Indians with robust bodies lived in an almost timeless splendor. In addition, in the eighteenth century, this literature came to be produced by a new class of natural historical practitioners: the barber-surgeons, who thus broke the monopoly over erudite knowledge enjoyed by doctors and naturalists. It was this group that became the most innovative contributors to the study of tropical medicine in the Americas because of their acute observations of the Brazilian world that surrounded them. Their knowledge was more popular and less academic in nature, but it was through this knowledge that tropical medicine emerged from the forests and hinterlands of colonial Brazil.[16]

Medicine in Dutch Brazil

An important source of information about Brazil was the literature produced during the Dutch invasion along the colony's northeastern coast. In 1578, a dynastic crisis in Portugal caused by the death of King Sebastian in northern Africa brought the Spanish and Portuguese crowns together, giving rise to the period known as the Union of the Two Crowns (1580–1640). Owing to disputes with Spain brought about by battles for their own autonomy, the United Provinces, who had traditionally been allies of the Portuguese in the refinement and commercialization of Brazilian sugar for European markets, chose to support an invasion of the sugar-producing region in northeastern Brazil. The first of various invasion attempts, organized this time by the Dutch West India Company, occurred in Bahia in 1624. The most important and longstanding invasion, however, took place in Pernambuco between 1630 and 1649.[17]

Many important studies in medicine and natural history in Brazil were carried out during the government of Count Johan Maurits of Nassau (1637–44), who served as the representative of the West India Company. Nassau brought with him a group of scholars who were charged with studying and illustrating Brazilian nature and geography, and the result of their work under Nassau's patronage was nothing less than spectacular. While Portuguese authorities attempted to prevent detailed information about the region from circulating freely within the public domain for fear of foreign invasion, nevertheless two significant works from this period were published under Count of Nassau's patronage. These were the *History of deeds recently practiced during eight years in Brazil and in other parts under the government of the illustrious João Maurício, Prince of Nassau*, by Gaspar Barleo [i.e. Barléu], and the *Natural History of Brazil*, published in Amsterdam in 1648, by Willem Piso, Georg Marggraf, and Johannes de Laet (see Figure 5.1). Later, in 1648, Willem Piso also published the *Natural and Medical History of West Indies*.[18]

To understand these authors' works and the links between the study of nature in Portugal and in Holland, it is necessary to look back some years and analyze the formal education of these savants. In 1593, the French naturalist Charles de L'Écluse, also known as Clusius, settled at the University of Leyden in Holland. Clusius was a great scholar and an important commentator on American flora. Having graduated from the University of Montpellier and after successful navigations to Portugal, Spain, and northern Africa, he devoted himself to the study of plants from the New World and their introduction into European botanical gardens. He was obsessed by the question of natural classification, and his plant garden, created in Leyden, served as a model for his European colleagues. Clusius was also responsible for translating into Latin the most important works of those Portuguese and Spanish naturalists who devoted themselves to the study of American nature. He translated the works of various authors whose texts examined the fauna and flora of both the East and the Americas, including Garcia d'Orta,[19] Cristóvão da Costa,[20] and Nicolas Monardes.[21] These treatises proposed new forms of classification and new uses of the *materia medica* that would serve as the basis for later studies. Clusius was primarily interested in translating these texts, and the frequent citations of these works by authors, such as Piso and Marggraf, demonstrate that Luso-Hispanic knowledge served as the basis for many investigations into the nature of the New World and that this knowledge circulated with regularity between European intellectuals.

In Leyden, Clusius was responsible for awakening an interest among his students in the natural diversity of the New World. The University of Leyden thus became an important center for the training of naturalist-doctors who were also interested in exotic plants. As such, when it came time for Nassau to recruit the group of scholars who would accompany him to Dutch Brazil and carry out the survey of local nature and geography, Willem Piso, a doctor who had graduated from Leyden, was the natural choice. The invasion of Brazil allowed Dutch scholars direct access to American nature, and the practical experience they accumulated reformulated many of the concepts about the subject that were previously in vogue.[22]

The *Natural History of Brazil* is remarkable for its role in the development of the natural sciences and of South American ethnography in particular.

Figure 5.1 *The Natural History of Brazil* is remarkable for its role in the development of the natural sciences and, in particular, for South American ethnography. Front page of Willem Piso, Georg Marggraf, and Johannes Laet, *Historia naturalis Brasiliae: Auspicio et beneficio Illustriss. I. Mauritii com. Nassau* (Leiden: Franciscum Hackium/ Amsterdam: Ludovicus Elzevirium, 1648). Reproduced courtesy of Osler Library of the History of Medicine, McGill University, Montreal, Quebec, Canada.

Doctor Willem Piso was responsible for many chapters about diseases, the disposition of air and water, and the topography of Brazil, in addition to many of the chapters that dealt with poisons and their antidotes and medicinal plants. Marggraf devoted himself to natural history and provided descriptions, classifications, and drawings of plants, fish, birds, quadrupeds, snakes, and insects. He also included some information about astronomy. Laet wrote an appendix about the Indians. The book, which was edited by Laet, also included some of his own notes. It provided an image of a virgin nature untouched by human hands and revealed the same bucolic and paradisiacal images that the earliest commentators had invented to describe Brazil. The *Natural History* compiled information collected by scholars through many incursions from the coast through the forest and into the hinterlands. It contained descriptions of the diverse exotic plants and animals that were constantly brought by the region's inhabitants for use in the zoological and botanical gardens and the cabinet of curiosities that Nassau had built in the new city under construction in the Captaincy of Pernambuco.[23]

The book was divided into three parts: one dedicated to medicine, subdivided into four chapters; another dealing with natural history and subdivided into six chapters; and an appendix about Brazilian Indians. The first chapter describes the benign quality of Brazilian air, the pureness and abundance of water, and the pleasant winds and mild climate. It emphasizes the diversity of fauna and flora described in detail in the chapters that follow, and shows how a salutary climate provided ideal conditions for a fertile nature. The second chapter treats endemic diseases. Piso noted that the diseases had very different origins and resulted from the combination of very different population groups—Indian, European, and African—that ended up generating a complex set of ailments. The fourth chapter, which deals with the poisons and antidotes, was the result of a thorough study of Indian knowledge on the subject. This knowledge was also shared by slaves who, quite often, used these skills either against their masters or against one another.[24]

The book also reveals the very limited knowledge base that Europeans had on the subject. Piso himself recognized that native Brazilians were extremely skilled in the application of nature, a practice he referred to as an art of poisons and counter-poisons. What does this acceptance tell us about the colonial project in Brazil? Does it suggest that this was in some ways a cooperative venture, rather than the exercise of power imposed onto Brazilian peoples by Europeans? When we examine Piso's writings, we find on the one hand that Piso seems to suggest that there are more similarities between European and indigenous medicine than differences. On the other hand, he consistently lambastes the beliefs inherent in indigenous rituals, including the cures of the witch doctors, which he calls barbarous. That being said, and despite these prejudices, Piso does not denigrate native knowledge about American nature because he realizes that they could be useful to European medicine. From Piso's perspective, he could see that native medicine was based on the use of medicinal simples, which required a profound knowledge of elements of the natural world. However, how was this knowledge to be appropriated? Piso reveals the methods he used for this purpose. First, he relied on his own experimentation; second on information from natives of the land who were not Indians (i.e., Portuguese and mestizos); and finally, on contact with the Indians themselves. Let us examine his procedure more closely.[25]

Dutch savants made various trips along the coast and into the interior of the Brazilian Northeast, where they were able to collect and sketch native specimens of local flora and fauna. Many of these objects were taken to the palace of Maurits of Nassau, the site of an important cabinet of curiosities, a zoological space, and a plant garden. These voyages were essential in carrying out the necessary observations of the natural world.[26]

Piso recounts as well that the local populations were already familiar with using native products as medicinal cures. He writes that during his journeys he was able to observe such customs, especially among the more rustic populations that lived along the banks of the São Francisco River. This region, known generically as the *sertão*, was located far from the coast and far from urban settlements along the littoral, which forced its inhabitants to live a largely rural life that owed much to contact with the indigenous populations. Nevertheless, Piso explains that he often had great difficulty finding out this information from the locals, as they claimed that their medical knowledge was secret. Apparently, what was truly at play were the rivalries between those populations of Portuguese origin and the Dutch, with whom the former refused to cooperate.

After encountering so much resistance to getting information out of the local populations, Piso explains that he had no alternative other than relying on information from native populations. However, even here the situation was not easy. More often than not, the Dutch were in contact with indigenous tribes that were closer to the coast, had long-standing relationships with the Portuguese, and had been used for manual labor in the sugar industry. In general, these populations spoke Tupi and were known as Brazilians by the Dutch, as they had been acculturated by and remained loyal to the Portuguese. The indigenous populations that inhabited the interior and were less used to dealing with Europeans were generically known as *tapuias*, and many of these tribes—especially those in Rio Grande do

Norte and Ceará—were allies of the Dutch. In both cases, Piso encountered difficulties in extracting information from them, and he was obliged to beg them for assistance, but rarely were his demands satisfied. In general, he had to rely on his own direct observation of the natives' methods, whenever possible. This took place, for example, when he himself fell ill or when he was able to observe the behavior of a witch doctor; for, according to Piso, the Indians did not hide the remedies they used in treating ill individuals to the extent the Portuguese did. The biggest problem for Piso was acquiring information about poisonous plants; in these circumstances, the natives were totally inflexible. Piso was able to extract information only from some "Indian sorcerers who had been condemned to death. I begged of them to provide me with some secrets before they died and, very much against their will, they did in the end reveal a few of their secrets to me."[27]

This first part of the book ends with a description of the properties and virtues of medicinal simples. Again, a large amount of the information was extracted from the Indians, but their systems of knowledge as a unit were discarded as mere superstitions. It is clear that in recognizing the Indians' information about nature and its uses, the traditional origin of knowledge was inverted: European scholars learned from the New World's rustic inhabitants. However, empiricism in the colonial context in fact became a way to extract *practical* knowledge from natives without embracing their heathen and superstitious beliefs about nature, magic, and their gods. By cementing this new knowledge in written language, Europeans converted it into an erudite framework. They adopted practical medical techniques from the indigenous populations, but they insisted on divorcing these techniques from the natives' superstitious accounts of why they functioned the way they did. This process was emblematic of the European colonists' appropriation of technical knowledge while separating these techniques from the native systems of thought in which they initially appeared.

The second part of the book, written by Marggraf, was a summary of Brazilian flora and fauna. This text can be considered one of the first natural histories of Brazil and was, until this date, the most complete text written about a particular region of South America. It is true that this text revised all previous classifications regarding the local *materia medica*, but it did so by using works that were Iberian in origin. Marggraf cites Garcia d'Orta, Clusius (*Histoire des Plantes)*, and Nicolas Monardes as the recognized authorities on American nature, as indeed they were throughout Europe at the time. In the book, plants and animals no longer appear as wonders of nature but rather are positioned to reflect a rationalized and natural order of the world. The proposed classification system was a decisive influence on Linnaeus's work in the following century, which would definitively bury the vision of nature inherited from antiquity, to give place to another, characterized by rationalization of the natural world. The elements of nature were grouped into six categories: plants, fish, birds, quadrupeds, snakes, and insects. This classificatory system, nonetheless, still retained vestiges of conceptions from classical antiquity. What enabled their classification was a synthesis of both their relationship with their environment and their morphology. As such, plants are above the ground and are immobile, fish move in water, birds fly in the air, quadrupeds move on the ground on four legs, snakes move on the ground, and insects fly as well, but they are not birds as they do not have feathers.[28]

Each one of the entries began with the presentation of the nomenclature of the simples to be described, followed by the name given by the Indians, then by the popular name in Portuguese, and ending with the version proposed in Latin. Each entry gave a detailed description of the element (vegetal or animal), its morphology and its utility in the medical subject, and was followed by a minute iconography. In the case of the vegetables, the picture of the plant was complemented by the picture of the flower, the fruit, and the seed. The images produced by Marggraf had the objective of describing the American world and revealing it to the Europeans, but not only that. This structure was the result of the combination of an investigative impetus with the spirit of observation, uniting science and art.

Though many analyses of the *Natural History* seem to point toward the progressive demystification of the world and an increasing control over nature, they tend to forget the extent to which Luso-Brazilian literature was held in the grip of a paradisiacal vision of the New World. The visions of Brazil that appeared in the book, whether in the images or the text, are reminiscent of the myth of the terrestrial paradise. Still, a vestige of the idea of sympathy formulated by Hyppocrates, the interaction of the climate, the environment, the stars, and the waters ascribed positive virtues to land and men. As Piso himself asked rhetorically, "[W]ould there be, in the name of immortal God, something that does not have any utility?"[29]

The Barber-Surgeons

The expulsion of the Dutch did not put an end to this process of appropriation by which an erudite European culture sought to understand American nature and attempted to use its products as a medical panacea according to similar theoretical bases and empirical understanding. During the eighteenth century, many books were written by Luso-Brazilians on the basis of direct observation of Brazilian territory. Nevertheless, two novelties emerged from this later literature. Contrary to what had occurred in the previous centuries, these eighteenth-century authors of medical treatises were not medically trained doctors but rather surgeons, or surgeon-barbers (see Figure 5.2). Second, the geographical center of this new production was not the sugar-producing coastal region of northeastern Brazil but rather the more central southeastern region of Brazil, around the territory of the recently discovered gold mines.[30]

Why did these surgeon-barbers, products of an educational background that was more practical than erudite, become the authors of the most important books on this subject? First of all, these practical men were able to ally acute observation of the cases they treated with their medical book learning and thereby produced a mix of popular and erudite knowledge. Second, as soon as they arrived in Brazil, they realized that years of apprenticeship in Portugal were not sufficient for the practice of the profession in the region: the diseases were not always the same and neither were

Figure 5.2 Medicine in Portugal and its overseas dominions was divided into two branches: a learned tradition practiced by doctors trained in universities and another, more practical tradition carried out by surgeons, midwives, barbers (who performed bloodletting and extracted teeth), pharmacists (who produced medicines), and "algebrists" or bonesetters (who treated broken bones and muscles). This image features a surgeon-barber, a midwife and a priest. *Ex-voto em louvor de Bom Jesus de Matosinhos a Rita Angélica da Costa–Acervo Museu do Diamante* (Diamantina). Reproduction courtesy of the Museu do Diamante e Bibiloteca António Torres.

the components of the climate, the air, and the living conditions that defined the illnesses and their treatments. Third, through an empirical approach, they were able to observe the many diseases that attacked the population and the arsenal of medications that they had. Fourth, contradicting the specialization of functions that characterized the practice of medicine in Portugal, these Brazilian surgeons made prognostics and cures, theorized about the diseases, and prescribed medications—all attributions restricted to doctors. Fifth, they also produced their own medicines and, in this case, they were able not only to benefit from traditional medications but very quickly to learn to use local herbs in their treatments, whose uses they often learned from the Indians and *mestizos*. Sixth, the chronic lack of doctors in the colony led them to write their books, which were primarily directed not to the elite and erudite but rather to all inhabitants of the region, although they intended them to be recognized as well by foreign doctors and especially those from Portugal. The popular format of these books, and the primary audience toward which these texts were directed, allowed their authors to write more freely than would otherwise have been possible: that is, in the form of practical advice for a home-grown and popular medicine.[31]

Regarding this more popular and practical medicine, characteristic of the eighteenth century, three surgeons stand out: Luís Gomes Ferreira, who wrote the *Erário Mineral (Mineral Treasury)* (1735);[32] João Cardoso de Miranda, who wrote the *Surgical and medical relation, which deals with and presents specially a new method to cure the scorbutic infection* (1741)[33] and the *Prodigious Lagoon discovered in the bushes of the mines of Sabará* (1749);[34] and José Antônio Mendes, author of *Government of Mineiros* (1770).[35] The little information that exists about the life of these three surgeons was almost all provided by the authors themselves in their books. They were all of Portuguese origin, had been apprenticed in Portugal and, like many other Portuguese, had come to Brazil in search of the many possibilities that the New World opened up for them. Some of them were attracted to Minas Gerais after the discovery of gold there in the final decade of the seventeenth century.

And why did Minas Gerais become the major theater of empirical medical literature during this period? In the eighteenth century, the discovery of gold in southeastern Brazil—and in the captaincy of Minas Gerais, in particular—brought to the region a significant population that combined a central urban nucleus and a rural surrounding (see Figure 5.3). As the century advanced, the number of slaves multiplied, and they came in the end to constitute the vast majority of the population, living and working under harsh conditions and lacking in medical assistance. However, doctors were few and charged exorbitant fees.[36]

As keen observers of the reality that surrounded them, the three surgeons collected knowledge necessary to correctly diagnose the diseases, provide adequate treatment, and to effect many cures. They understood that the specificity of the region's diseases required different treatments from the ones they knew and attempted therefore to add local herbs and products to the kingdom's pharmacopoeia. Many of them were already known and used by the region's dwellers. Sérgio Buarque de Holanda has shown that the knowledge of almost all of these products was transmitted to the

Figure 5.3 The Dutch invasion of Brazil (1624–49) was a period in which a great amount of new knowledge about Brazilian diseases was produced. In the eighteenth century, the discovery of gold in southeastern Brazil—and in the captaincy of Minas Gerais, in particular—produced a significant population whose illnesses were studied in eighteenth-century medical treatises. Map of Brazil with Minas Gerais and area of the Dutch invasion.

bandeirantes[37] from São Paulo by the Indians, and these *bandeirantes* were responsible for the exploration of the interior of Brazil.[38] Gomes Ferreira mentioned a powder known in Minas Gerais as *For all,* made of the thick bark of a tree, as having been introduced by an inlander *paulista*.[39] He also affirmed that the *paulistas* had learned from the Carijós Indians the use of firewater for the cure of colds and the incorporation of local roots, such as *butua* and *pacacoanha,* in other treatments.

Gomes Ferreira (see Figure 5.4) incorporated other medications as he "heard" that they were used successfully in the region, such as the application of *embaúba* for the treatment of hernias, and an emulsion of almonds, which in Minas was called *pevitada,* the liana known as *poalha, sapé,* or the *caraíva* tree. He suggested in one of his medical prescriptions that green apples, not found in America, could be replaced by green genipaps. He also exalted the virtues of the water smartweed, or *cataia, raiz-de-mil homens, poaia, erva orelha de onça,* and the jalap, or *batata de purga,* also known as *dos paulistas purga.* From these observations, we can perceive that some practical knowledge was formed, based on the daily experience of these men, who went around the region performing their curative practices.[40]

Gomes Ferreira warned that the "herbs, roots, mineral and animal things, that exist in Brazil and its backlands" are "very useful to public health" and that the local apothecary shops should develop medications incorporating them. José Cardoso de Miranda and his associate, Gomes Ferreira, prescribed *chá de picão,* "very well known in Minas," and Santa Maria's herb. They also added to their medical prescriptions many products from Africa or the East, such as coral red, myrrh, dates, cinnamon, musk, sandal and nutmeg.[41]

The three surgeons were particularly interested in those diseases that attacked the large populations of slaves in Minas Gerais. Slaves were these authors' main patients, and they were often sent by their masters to be cured of the diverse maladies that afflicted them. Many of these diseases were associated with the terrible feeding and living conditions the slaves had to endure. In the eighteenth century, Minas Gerais became the primary destination for a growing population that moved to Brazil. Slaves provided the bulk of the manual labor, both for the mining industry and for rural activities and basic services in the mining communities. In this way, over the course of the eighteenth century, the region held a gigantic concentration of slaves, estimated at close to 65 percent of the total population of the captaincy, which itself oscillated between 300,000 individuals at mid-century to 400,000 by century's end. The living conditions of these slaves were worse than imaginable. High mortality and low fertility rates meant that manual labor had to be constantly replenished through the Atlantic slave trade.

Figure 5.4 Contrary to what had occurred in the previous centuries, eighteenth-century authors of medical treatises were not medically trained doctors but rather surgeons, or surgeon-barbers. Luís Gomes Ferreira, who wrote the *Erário Mineral* (*Mineral Treasury*) in 1735, was one of the most famous. Reproduced courtesy of Junia Furtado.

As such, after being imprisoned under the harshest circumstances in the interior of Africa and during the long overseas journey, and subjected to the worst conditions in terms of food and health, slaves still had to endure many months of forced marches from Rio de Janeiro or Bahia toward the country's interior, where the mines were located.[42]

Under these conditions, slaves garnered high prices in Minas and were expensive investments that masters attempted to preserve by caring for their illnesses through surgeon-barbers, who were less costly than medical doctors. Despite this care, alimentary, labor, and living conditions were the worst imaginable for these slaves. A large number of the diseases that afflicted the slaves could be attributed to their work in the gold and diamond mines, that forced slaves to spend many hours in water or in the subsoil (aside from the many accidents that occurred because of landslides, soil avalanches, or floods). João Cardoso de Miranda was settled in Bahia at this time and was particularly concerned with the scurvy that killed or incapacitated a large number of the slaves that disembarked in Salvador. After the Middle Passage, around 2,000 slaves died annually in this single location.[43]

In Brazil, João Cardoso de Miranda became blind, which made practicing medicine difficult, so he became involved in commerce, such as the slave trade and the trade with Costa da Mina in Africa. For this reason, he became the first to discover the medicinal treatment for scurvy.[44] Miranda's medical prescription consisted of tea of fresh herbs followed by a diet reinforced with fresh food, such as chicken, lettuce, endives, wild chicory, and purslanes. This probably provided the physically debilitated patients with vitamin C, the absence of which was the cause of the malady. As Cardoso de Miranda was involved in the slave trade, his discovery combined his medical and commercial interests. In the single year of 1731, when he wrote to the King, he had cured more than 500 slaves attacked by scurvy, and in that year he had prevented many deaths associated with the disease.

Luís Gomes Ferreira was also interested in scurvy. As Cardoso de Miranda did not get the license to publish his discovery at once, Gomes Ferreira in 1735 transcribed in the *Erário Mineral* the medical prescription that the associate had developed. On the way back to the kingdom, when he passed by Bahia, Miranda asked him to be the carrier of a letter in which he asked that his invention be licensed. The reporting of Gomes Ferreira's medical prescription exemplifies the production of medical knowledge in Brazil, the contact that was established between the many men that practiced medicine in the colony, and the forms of circulation of practical knowledge that they developed. Cardoso de Miranda was able to publish his study on scurvy only in 1741, and it had profound implications.[45]

The comparison of these three surgeons' writings about scurvy also allows us to analyze the transformation that was occurring not only in the treatment of the disease but in the understanding that doctors had of this and other maladies that afflicted populations in the tropics.[46] Guided by practical experience, these doctors revealed ever more frequently the origin of various maladies and, despite disagreeing as to the causes, they agreed that practical experience was the best way to fight them. Though João Cardoso de Miranda affirmed that the cause of diseases was related to the slaves' spirits during the overseas trip and attributed their physical maladies to a disorder of moods, he not only realized that slaves were the majority of the disease's victims but that it was important to improve their living conditions. For Luís Gomes Ferreira, the disease came from Africa, and he blamed the slaves for its transmission: for this reason, it became known as the disease of Luanda. He no longer believed that heavenly bodies were the cause of disease, but he did realize that slaves were the group that was most commonly affected. Despite believing that the disease was contagious, he was able to perceive that better living conditions allowed for a more robust recovery. He was the first to promulgate the cure that João Cardoso de Miranda developed for this disease. Contrary to common practice, Cardoso de Miranda also suggested that the fact that his slaves were plump and well nourished improved their resistance and helped their recuperation from the illness. In the second half of the century, however, scurvy was already associated with types of food, as surgeon José Antônio Mendes observed. To him, to call it "disease of Luanda" was justified just because it spread in the ships from this region, but it was not a disease that originated in that land; it derived from the maltreatment to which the slaves were subjected during the middle passage, because since the most urgent cause of such complaints was the "gross, thick, and putrid food" that was given to slaves throughout the Americas.[47]

The art of medicine in the colony also required profound knowledge of the daily activities of the local inhabitants, as the communal ties characteristic of their way of living were clearly reflected in the prophylaxis of the diseases. The sick body forged ties of sociability. Alongside compassion and kinship, the doctor had to acquire a knowledge of the interior of houses and the details of patients' lives. In his medical prescriptions, Gomes Ferreira added old shoes, *enxofre de verrugas*, cistern water, human oil, feces from horses or healthy boys, and garden vegetables, among others. José Antônio Mendes prescribed breast milk, moldy white bread, powder of chimney rust, mare's milk, dove fat, blacksmith's filings—ingredients that needed to be begged, borrowed, or bargained for in the community.[48]

Among the prescribed medications, there were also some with mineral and animal origins. These medicines, based on sympathy and antipathy, were based on the notion that the various elements transmitted among themselves their qualities and defects. Consequently, there was a pragmatic preoccupation to isolate the diverse elements and find their specific uses, but they also shared the idea of a magical capacity to transmit their virtues. Because of this, it was enough that some objects were used by the patient for him to get cured. Gomes Ferreira suggests the use of bezoar stone or dogtooth for earache, and José Antônio Mendes recommends to tubercular patients that they use "vervain [sic] root" on the neck, "that is averse to these complaints and heal them."[49]

Water, for example, was the cause of many diseases, but it could also be the cure as well. João Cardoso de Miranda's text proclaimed the virtues of a prodigious lagoon in Minas Gerais. After arriving in Brazil, he became partially blind, and it was his eye disease that led him to Minas Gerais in 1749 in search of a treatment for his blindness in that lagoon, whose miraculous waters were becoming famous. Feeling much better after the baths, he decided to write a book about the prodigious waters of Lagoa Santa, as the place became known. The belief in the miraculous power of water was compatible with the medicine of that time, as the world for these writers was ruled and governed by God. However, Cardoso de Miranda's account was not limited to the magical and religious explanation: he also reported the studies done there by the Italian doctor Antonio Cialli, who "consented that the water had in itself the two most useful minerals that impregnate the water, as vitriol and iron do."[50]

Medical knowledge was impregnated with religious and magical practices and beliefs as well. God's time was the time of eternity and transcended by far terrestrial life. Because of this, José Antônio Mendes himself advised the doctors that, before assisting anyone, "you should soon demand that the patient confess and receive the sacrament, [...] the communion, in pursuit of God as the Father and Master of everything created, that this patient will overcome such complaints."[51] The treatments should be applied with faith, as many of the diseases were thought to originate from spells or disbelief. Likewise, the time of cure also belonged to God and depended on His will. Often the treatment was to be given while reciting a Hail Mary. This technique not only standardized and universalized the duration of the treatment: it elevated the patient's mind to heaven as well.

Conclusion

In eighteenth-century Brazil, Luís Gomes Ferreira, João Cardoso de Miranda, and José Antônio Mendes carried on the tradition of what came to be called "tropical medicine," since they were primarily concerned with knowing the specificity of local diseases and treatments. The most important feature of their work was the emphasis on observable practical experience as a central premise of learned knowledge. This allowed these authors to firmly incorporate plants, animals, and the natural remedies that they learned from Brazilian popular medicine. The use of these medicinal simples as a panacea-like cure derived from their contact with native populations. The three surgeons were able to valorize this common or vulgar knowledge because they were able to *see* its utility through their own empirical, practical experience. However, they injected this newfound knowledge into a cognitive system the structure of which was increasingly based on empiricism. In the end, they attempted to ground this knowledge in a language that was accessible to the common reader by constructing texts that were primarily based on easy recipes uing materials easily available to the average reader in Brazil. However, they also wanted recognition from doctors who were the ultimate arbiters in determining whether their contributions to the knowledge of *materia medica* would be incorporated into their more erudite frameworks.[52]

Knowledge of the treatments and natural circumstances of the New World circulated regularly between Europe and the Americas. As such, at the same time that books were being produced in America and becoming works of reference among European intellectuals, the inverse was also true, making it difficult to distinguish a single origin for the formulation of this knowledge. The example of this dual process can be observed in the formulation, circulation, and appropriation of a series of medicinal cures developed by the Portuguese doctor João Curvo Semedo, who published his medical prescriptions in a set of books that became extremely popular throughout Portugal's empire in the eighteenth century. In 1695, he published the *Medicinal Polyanthea*, which was followed by the *Sentinel of life against the hostilities of death* (1720) and the *Memorial of many simples*.[53] His books were based on medical prescriptions that he called "Curvian secrets," a set of compound elements—including bezoars, antidotes, and water—with a majority of these treatments made with ingredients from Brazil, Africa, and Asia. Many of these prescriptions can be found in the books of the three surgeon-barbers discussed in the last section of this chapter. Many of these remedies were learned from native populations but were also present in the treatises of Curvo Semedo, whose books were part of Gomes Ferreira and Mendes Miranda's libraries. If, on the one hand, this seems to confirm the hypothesis that knowledge was regularly exchanged between America and Europe, it also shows, on the other, that by incorporating this knowledge within an erudite framework, European intellectuals took advantage of the techniques and knowledge of the indigenous populations while at the same time isolating this knowledge from the indigenous context.

Notes

1 James Goodyear, Jr, "Agents of empire: Portuguese doctors in colonial Brazil and the idea of tropical disease" (Ph.D. dissertation, Johns Hopkins University, 1982).
2 Peter Coates, "The world beyond Europe," in *Nature: Western Attitudes Since Ancient Times* (Berkeley: University of California Press, 1998), 82–109; Antonio

Barrera, "Local herbs, global medicines: commerce, knowledge, and commodities in Spanish America," in Pamela H. Smith and Paula Findlen, eds, *Merchants and Marvels: Commerce, Science and Art in Early Modern Europe* (New York and London: Routledge, 2002), 163–81.

3   Lorraine Daston and Katharine Park, *Wonders and the Order of Nature, 1150–1750* (New York: Zone Books, 1998).

4   I am not resurrecting any kind of positivist narrative of knowledge, as empirical practices are always in relation with belief systems and require languages of representation that mediate them. See Peter Dear, "The meanings of experience," in Katharine Park and Lorraine Daston, eds, *The Cambridge History of Science*, vol. 3: *Early Modern Science* (Cambridge: Cambridge University Press, 2006), 106–31; Sharla M. Fett, "Sacred plants: the cross-cultural context for African American herbal medicine," in *Working Cures: Healing, Health and Power on a Southern Slave plantation* (Chapel Hill: University of North Carolina Press, 2002), 60–83 (ch. 3).

5   See for example Peter Dear, "Totius in verba: rhetoric and authority in the Royal Society," *Isis* 76 (1985), 145–61; Thomas daCosta Kaufmann, "Empiricism and community in early modern science and art: some comments on baths, plants and courts," in Anthony Grafton, and Nancy Siraisi, eds, *Natural Particulars: Nature and the Disciplines in Renaissance Europe* (Cambridge, Mass.: MIT Press, 1999), 401–17.

6   Anthony Grafton, *New Worlds, Ancient Texts: The Power of Tradition and the Shock of Discovery* (Cambridge, Mass.: Harvard University Press, 1992); Gianna Pomata and Nancy Siraisi, eds, *Historia: Empiricism and Erudition in Early Modern Europe* (Cambridge, Mass.: MIT Press, 2005); see more generally Steven Shapin, *The Scientific Revolution* (Chicago: Chicago University Press, 1996); Steven Shapin and Simon Schaffer, *Leviathan and the Air-Pump: Hobbes, Boyle, and the Experimental Life* (Princeton: Princeton University Press, 1985).

7   See Smith and Findlen, eds, *Merchants and Marvels*.

8   Claudia Swan, "From blowfish to flower still life paintings: classification and its images, circa 1600," in Smith and Findlen, eds, *Merchants and Marvels*, 109–36; see also Brian W. Ogilvie, *The Science of Describing: Natural History in Renaissance Europe* (Chicago: Chicago University Press, 2006); David J. Arnold, *The Problem of Nature: Environment, Culture and European Expansion* (Oxford and Cambridge, Mass.: Blackwell, 1996).

9   Júnia Furtado, "Science and scientists—Brazil/Portugal," in J. Michael Francis, ed., *Iberia and the Americas – Culture, Politics and History: A Multidisciplinary Encyclopedia* (Santa Barbara: ABC-Clio, 2005), 916–23; Mary Lindemann, *Medicine and Society in Early Modern Europe* (Cambridge: Cambridge University Press, 1999).

10 Irisalva Moita, *V Centenário do Hospital Real de Todos os Santos* (Lisbon: Casa da Moeda, 1992).

11 Roy Porter, "Medicine and the body," in *Flesh in the Age of Reason: The Modern Foundations of Body and Soul* (New York: Norton, 2003), 44–61; Katharine Park, "Natural particulars: medical epistemology, practice and the literature of healing springs," in Grafton and Siraisi, *Natural Particulars*, 347–67.

12 Sérgio Buarque de Holanda, *Visão do Paraíso*, 6th edn (São Paulo: Brasiliense, 1994), 35–130; Anthony Pagden, *European Encounters with the New World* (New Haven: Yale University Press, 1993); James Goodyear Jr, "Brazil: terrestrial paradise?" in "Agents of empire," 33–72; Eduardo Blázquez Mateos, *Viajes al paraíso: la representación de la naturaleza en el Renascimento* (Salamanca: Ediciones Universidad de Salamanca, 2003).

13 Licurgo de Castro Santos Filho, *História Geral da medicina brasileira* (São Paulo: Hucitec/Edusp, 1991). Eustáquio Duarte, ed., *Morão, Rosa e Pimenta: notícia dos três primeiros livros em vernáculo sobre a medicina no Brasil* (Pernambuco: Arquivo Público Estadual, 1956).

14 Francisco Moreno-Carvalho, "Medicina no Brasil Holandês: apontamentos sobre o tratado de Zacuto Lusitano," in Alberto Dines, Francisco Moreno-Carvalho, and Nachman Falbel, eds, *A fênix ou o eterno retorno: 460 anos da presença judaica em Pernambuco* (Brasília: Ministério da Cultura, 2001), 165–79.

15 Jorge Cañizares-Esguerra, "Changing European interpretations of the reliability of indigenous sources," in *How to Write the History of the New World: Histories, Epistemologies and Identities in the Eighteenth-century Atlantic World* (Stanford: Stanford University Press, 2001), 60–129.

16 Jorge Cañizares-Esguerra, "Iberian science in the Renaissance: ignored how much longer?" in *Perspectives on Science*, 12(1) (2004), 86–124; Júnia F. Furtado "Barbeiros, cirurgiões e médicos na Minas colonial," *Revista do Arquivo Público Mineiro*, Belo Horizonte, 41 (July–December 2005), 88–105.

17 Paulo Herkenhoff, ed., *O Brasil e os holandeses (1630–1654)* (Rio de Janeiro: Sextante, 1999).

18 Caspar Barlaeus, *Rerum per octennium in Brasilia, et alibi nuper gestarum* (Amsterdam: Ioannis Blaeu, 1647). Willem Piso, Georg Marggraf and Johannes Laet, *Historia naturalis Brasiliae. Auspicio et beneficio Illustriss. I. Mauritii Com Nassau* (Leiden: Franciscum Hackium/Amsterdam: Ludovicus Elzevirium, 1648). Gilberto Freyre, *Johan Maurits van Nassau-Siegen from a Brazilian Viewpoint* (The Hague: The Johan Maurits van Nassau Stichting, 1979), 237–46 (http://www.bvgf.fgf.org.br/portugues/obra/opusculos/johan_maurits.htm).

19 The pioneer in the incorporation of plants in the *Materia Medica*, which the maritime trips revealed to Europeans, was Garcia d'Órta (1501–68), a Portuguese doctor who lived in India. From his practice of medicine in the region, by observing local doctors and how they used the elements from nature around them, he wrote the *Colloquy of the simples and drugs and medical things from India*; Garcia d'Órta, *Colóquios dos simples e drogas da Índia* (Lisbon: Imprensa Nacional/Casa da Moeda, 1987). James Goodyear Jr, "The empirist: Garcia d'Orta," in "Agents of empire," 81–93.

20 Cristóvão da Costa (1538–94), surgeon from the Portuguese armada, born in Africa, lived in India and was deeply affected by Garcia d'Órta's book. He wrote in Spanish the *Treatise of Drugs and Medicine of the East Indies*, where he attempted to review and extend the observations done in the *Colloquy of the Simples and Drugs*. He added other simples, corrected some information and accompanied his notes with drawings of almost all plants described.

21 The Sevillian doctor Nicolas Monardes, despite having never left his hometown, wrote an important medicine book incorporating medicinal plants from the New

World: Nicolás Monardes, *Historia medicinal de las cosas que se traen de nuestras Indias Occidentales que sirven en Medicina* (Seville: Alonso Escrivano, 1574).

22 Lucile Allorge and Olivier Ikor, *La fabuleuse odyssée des plantes: les botanistes voyagers, les Jardins des Plantes, les herbiers* (Paris: J. C. Lattès, 2003), 91–3.

23 C. R. Boxer, *The Dutch in Brazil 1624–1654* (London: Oxford University Press, 1957).

24 Piso, Marggraf and Laet, *Historia naturalis Brasiliae*.

25 David Freedberg, "Ciência, comércio e arte," in Herkenhoff, ed., *O Brasil e os holandeses (1630–1654)*, 192–217.

26 Pedro Correia do Lago, "L'expédition de Jean-Maurice de Nassau," in Frans Post, *Le Brésil à la cour de Louis XIV* (Paris: Louvre, 2005).

27 Willem Piso, *História natural e médica da Índia Oriental* (Rio de Janeiro: Instituto Nacional do Livro, 1957), 46.

28 Lisbet Koerner, *Linnaeus: Nature and Nation* (Cambridge, Mass.: Harvard University Press, 2001).

29 Piso, *História natural et médica*, 45; Svetlana Alpers, *The Art of Describing: Dutch Art in the Seventeenth Century* (Chicago: Chicago University Press, 1984).

30 Francis Dutra, "The practice of medicine in early modern Portugal: the role and social status of the físico-mor and the surgião-mor," in Israel J. Katz, ed., *Libraries, History, Diplomacy and the Performing Arts: Essays in Honor of Carleton Sprague Smith* (New York: Pendragon Press/New York Public Library, 1991), 135–69; Júnia F. Furtado, "Barbeiros, cirurgiões e médicos na Minas colonial."

31 Márcia Moisés Ribeiro, *A ciência dos trópicos: a arte médica no Brasil do século XVIII* (São Paulo: Hucitec, 1997).

32 Júnia F. Furtado, ed., *Erário Mineral de Luís Gomes Ferreira*, 2 vols (Critical Edition) (Belo Horizonte: Fundação João Pinheiro, 2002); Charles Boxer, "A rare Luso-Brazilian medical treatise and its author: Luis Gomes Ferreira and his *Erário Mineral* of 1735 and 1755," *Indiana University Bookman* 10 (1969), 48–70; Charles Boxer, "A footnote to Luís Gomes Ferreira, *Erário Mineral*, 1735 and 1755," *Indiana University Bookman*, 11 (1973), 89–92.

33 João Cardoso de Miranda, *Relação cirurgica, e médica, na qual se trata, e declara especialmente hum novo methodo para curar a infecção escorbutica* (Lisbon, 1741).

34 João Cardoso de Miranda, *Prodigiosa Lagoa descoberta nas congonhas das minas do Sabará*, ed. Augusto da Silva Carvalho (Coimbra: Imprensa da Universidade, 1925).

35 José Antonio Medes, *Governo de Mineiros* (Lisbon: Oficina de Antonio Roiz Galhardo, 1770). "Mineiros" means a native or inhabitant of the Minas Gerais captaincy.

36 Júnia F. Furtado, "Gold," in Shepherd Krech III, J. R. McNeill, and Carolyn Merchant, eds, *Encyclopedia of World Environmental History* (London and New York: Routledge, 2003), vol. 2, 597–8.

37 Members of the expeditions to the hinterland were called bandeiras. Edward Goodman, "The Bandeirantes," in *The Explorers of South America* (New York: Macmillan, 1972), 102–15.

38 Sérgio Buarque de Holandia, "A botica da natureza," in *Caminhos e fronteiras* (São Paulo: Companhia das Letras, 1995), 74–89.

39 *Paulistas* were natives or inhabitants of the São Paulo captaincy, who discovered the gold mines in Minas Gerais.
40 Júnia F. Furtado, "Arte e segredo: o licenciado Luís Gomes Ferreira e seu caleidoscópio de imagens," in Júnia F. Furtado, ed., *Erário Mineral de Luís Gomes Ferreira*, 3–30.
41 Vera Regina Beltrão Marques, *Natureza em boiões: medicinas e boticários no Brasil setecentista* (Campinas: Editora da Unicamp, 1999).
42 Laird W. Bergad, *Slavery and the Demographic and Economic History of Minas Gerais, Brazil, 1720–1888* (Cambridge: Cambridge University Press, 1999).
43 Joseph C. Miller, "Floating tombs: the maritime trade of the Brazilians," in *Way of Death: Merchant Capitalism and the Angolan Slave Trade (1730–1830)* (Madison: University of Wisconsin Press, 1988), 314–78. Herbert Klein, "The middle passage," in *The Atlantic Slave Trade* (Cambridge: Cambridge University Press, 1999).
44 The discovery of the treatment for scurvy is most often attributed to James Lind (surgeon of the Royal Navy and Royal Naval Hospital) who, in 1747, spread news about how the disease could be treated through a diet including citrus fruits. See James Lind, *A Treatise of the Scurvy* (1753).
45 "Carta de João Cardoso de Miranda ao Físico-mor do Reino sobre um novo medicamento para o Mal de Luanda," in Júnia F. Furtado, ed., *Erário Mineral de Luís Gomes Ferreira*, vol. 2, 689–95.
46 Mirko D. Grmek, "Les grands fléaux," in *Histoire de la pensée médicale en Occident: De la Renaissance aux Lumières* (Paris: Éditions du Seuil, 1996), 253–78; Philip D. Curtin, "Europe and the Atlantic World: Africa and the disease exchange," in Jeremy Adelman, ed., *Colonial Legacies: The Problem of Persistence in Latin American History* (New York: Routledge, 1999), 15–27.
47 Mendes, *Governo de Mineiros*, 85–6.
48 Betânia Gonçalves Figueiredo, *A arte de curar* (Rio de Janeiro: Vício de Leitura, 2002).
49 Mendes, *Governo de Mineiros*, 114.
50 Miranda, *Prodigiosa Lagoa*, 9–11.
51 Mendes, *Governo de Mineiros*, 42, 43, 52.
52 de Holandia, "A botica da natureza," in *Caminhos e fronteiras*, 74–89.
53 João Curvo Semmedo, *Atalaia da vida contra as hostilidades da morte* (Lisbon: Oficina Ferreyrenciana, 1720).

CHAPTER 6

American Climate and the Civilization of Nature

JAN GOLINSKI

The climate of North America confronted European colonists with a series of problems in the seventeenth and eighteenth centuries. Old World settlers soon discovered the susceptibility of the continent to extremes of heat and cold, to humidity, snowstorms, lightning, tornadoes, and hurricanes. American weather seemed excessive and anomalous, unfamiliar in the magnitude of its effects and the threat they posed to settlers' livelihoods. The American climate also diverged from what was experienced at the same latitudes in Europe. In the classical tradition, a "climate" was supposed to correspond to a zone between parallels of latitude on the earth's surface; but the weather of the New World refused to conform to such expectations. Instead, it administered a series of sharp lessons to those who hoped to transplant European customs directly across the Atlantic. Citrus fruits could not in fact be grown in Maryland and Virginia, although they shared the latitude of Seville. A severe winter in Newfoundland sufficed to discourage those who had hoped to settle on the same parallel as London. Even the region optimistically designated "New England" sprang some surprises with its relatively short growing season and snowy winters. European settlers were soon made to realize that the American climate could not be predicted on grounds of latitude alone. Instead, they had to consider the physical environment as a whole, to study it methodically to learn how to survive in it.¹

Accordingly, descriptions of the American climate became parts of travelers' narratives, topographical surveys, and natural histories of the New World. In the early eighteenth century, systematic records were begun and instruments used to gather data about the American weather, in imitation of the methods being applied in European countries and in voyages of exploration overseas. British scientific institutions received reports of atmospheric measurements from travelers in India, the Caribbean, and the Arctic and on voyages in the Atlantic and Pacific oceans. The adoption of these methods in North America signified the continent's links to the Atlantic world and the global British empire. As a number of historians have noted, American Creole intellectuals participated in European discourse about nature in the New World, sharing in some respects the preoccupations of their colleagues across the Atlantic while also asserting their privileged knowledge of conditions on the ground.² Similarly, they contributed to a wide-ranging debate regarding the air's influence on human life and health. Medical writers, natural and moral philosophers, and historians were all discussing how climate would affect the development of civilization.³ The debate has sometimes been represented rather reductively as positing either climatic determinism or "moral" causes for the diversity of human cultures. In fact, almost all writers on the matter acknowledged that the forces of the physical environment interacted in a complex manner with human customs and institutions at different stages of development. The consequences were of pressing concern in places where Europeans had only recently settled and where the long-term effects of the climate were unknown. Hence, the observations of Creole naturalists and medical writers were closely attended to, both in Europe and in the colonies. The emerging picture of the climate in North America seemed to have fundamental implications for the development of settler society. This was the broader context in which studies of the American climate were interpreted in the eighteenth-century Atlantic community.

European settlers in North America saw the characteristics of the atmosphere as impinging on their prospects; they also believed they could tame it by extending settlements, cutting down forests, and bringing the wilderness under cultivation. In the second half of the eighteenth century, the notion that the climate was already being "civilized" by these means assumed particular prominence. Many Creole writers claimed that the clearing of forests, draining of marshes, and cultivation of land had already had a noticeable impact, moderating the extremes of weather in the period since Europeans first landed. Comparable effects were said to have been noticed in other colonial outposts, but they were emphasized with unparalleled regularity in North America.[4] Everyone seemed to agree that the American climate had been softened in recent decades, aligning itself more with the European norm. The transformation was universally ascribed to the effects of colonial settlement and agriculture on the landscape. Changes in the air, it was believed, had followed from these changes in the land.

In this chapter, I shall explore the roots of this conception of climatic change in eighteenth-century North America. I shall argue that it arose both from the familiarity of Creole intellectuals with Atlantic discourse concerning the effects of climate on civilization and from the specific situation in which the settler population found itself. American colonization was a project in which many Enlightenment hopes were invested; writers on both sides of the Atlantic expected that social progress would bring about the civilization of nature. Modern climatologists doubt that the climate had actually been transformed to the extent or with the rapidity that the settlers believed, but eighteenth-century observers had little doubt that it had.[5] Though it proved difficult to quantify the change with instrumental records, the widespread belief in its actuality testified to the settlers' confidence that they were molding their environment to suit their needs. Their conviction was consistent with the widespread assumption that nature could be civilized as a complement to the progress of human society.

A more localized reason for this belief among the settler population in North America was their awareness of the fate of the continent's native peoples. Beginning in the seventeenth century, settlers had observed a catastrophic decline in the native population, primarily due to epidemics of such diseases as smallpox.[6] The catastrophe gave particular urgency to consideration of the role of the American climate. European writers consistently claimed that the American atmosphere had a role in the natives' decline and that its long-term effects could be fatal for a population exposed to it. In response, Creole writers acknowledged the natives' susceptibility to certain diseases while defending their reputation for general vigor. They could not, however, dispose of the assumption that the climatic environment would eventually affect the health of their own population. They therefore sought reassurance in the belief that the climate was being changed. It was important to assert that, whatever its undesirable consequences in the past, its effects would become less fearful in the future. Americans of European descent thus came to regard their climate as something they were shaping by their own labor and enterprise—the virtues that supposedly differentiated them from the native population. They saw themselves as reaping the benefits of having cultivated the land, sparing themselves by their labor the natives' dreadful fate.

It was in this mode that the American climate came to be recruited as a national asset by the citizens of the new United States. During and after the revolutionary break with Britain, Americans often spoke of nature as an ally in the cause of independence and an attribute that would support the growth of their new country. Increasingly, they took pride in their climate as a nursery of liberty and a resource for the nation as it pursued its continental destiny. Earlier in the century, the British had already forged a sense of their national climate. As they charted its daily fluctuations between generally moderate bounds, they came to view the British weather as benignly temperate but bracingly varied. Such conditions were thought to aid the country's agriculture and commerce, reflecting the benevolence of providence to a nation destined for greatness and prosperity.[7] Though the British had associated their climate with their island identity as a maritime and trading power, Americans came to value theirs as the attribute of a nation that was hewing its habitat from the wilderness, slowly bringing the virgin continent under the sway of civilization.

Mapping the New World's Weather

This conception of the American climate emerged only gradually. The first British settlers in North America confronted an atmosphere that was poorly known and perhaps unfavorable to human welfare. Europeans venturing across the Atlantic quickly noticed that American locations were much colder in winter than the corresponding latitudes in Europe; in summer they could be hotter and considerably more humid. To determine the prospects for settlement, it was essential to find out what local conditions were like and how they varied with the seasons. Promoters of colonization were often not trusted when they touted the idyllic qualities of the New World climate. By the late seventeenth century, the Royal Society of London was welcoming reports that used methods of systematic recording to chart conditions on the other side of the Atlantic. The young physician Hans Sloane recorded a daily weather journal as part of his medical topography of the island of Jamaica, which he visited in 1688–89. It was published nearly twenty years later, as Sloane was beginning the social ascent that would take him to the presidencies of the Royal Society and the Royal College of Physicians. Already in the 1690s, letters by John Clayton about the natural history and climate of Virginia were published in the *Philosophical Transactions*. Striking a chord with the Hippocratic outlook of contemporary British physicians, Clayton noted how sudden changes in the Virginian weather affected the health of the inhabitants. Thomas Robie kept a weather journal (though without the benefit of instruments) from 1715 to 1722 at Harvard College in Massachusetts, sending it later to the Essex weather observer, William Derham, to share with the London virtuosi. James Jurin, secretary of the Royal Society in the 1720s, issued a general invitation for observers to submit weather journals to the society. Isaac Greenwood, a professor at Harvard, responded with proposals for compiling a "natural history of meteors" that appeared in the *Philosophical Transactions*. A few years later, Paul Dudley of Boston sent another weather journal to the Royal Society covering the years 1729 to 1733.[8]

These colonial observers looked to London for accreditation and publication of their research; they attached themselves to communication networks that were anchored in the metropolis. Factual observations flowed in from the peripheries of these networks, consolidated at a central point where (according to the Baconian model) interpretation was supposed to occur. Instruments flowed in the other direction, supplied by London makers for quantitative observation overseas. After a series of mishaps attending their transportation across the Atlantic, barometers and thermometers appeared in the colonies a few decades after they had been made available to consumers in London. Harvard received a portable barometer for its instrument collection in 1727, and it was used for observations by Greenwood. John Winthrop, Greenwood's successor in the Hollis chair of natural philosophy at Harvard, compiled a daily record of temperature and pressure in Massachusetts from 1743 to 1747, which he sent to the Royal Society.[9] John Lining, a Scottish physician from Lanarkshire who settled at Charleston, South Carolina in the 1730s was using instruments around the same time to take regular measurements of atmospheric pressure, temperature, rainfall, and humidity. The decade of the 1750s saw measuring apparatus being used systematically by Lining's colleague, Lionel Chalmers (also a Scottish emigrant to Charleston), and by the Yorkshire physician William Hillary on Barbados. Simultaneously, observers armed with instruments were compiling records in Maryland, Pennsylvania, and Quebec. Their numbers were swelled in the following decades as weather journals became a widespread pastime among a "calculating people" who were accustomed to quantifying many aspects of their daily lives.[10] By the beginning of the nineteenth century, the French visitor Constantin-François de Volney noted acutely that British and American meteorologists shared the same quantitative approach to their subject, when, "conformably to the national genius, [they] reduce every thing to direct and systematic calculations."[11]

The American climate was regarded as exotic because of its tendency to produce extreme conditions and its apparent fertility in atmospheric wonders. Reports of tornadoes, waterspouts, hurricanes, thunderstorms, and other prodigious meteors appeared frequently in metropolitan and colonial publications. In colonies where the Puritan inheritance remained influential, such phenomena continued to be regarded as divine portents well into the eighteenth century. Arguments about whether such events could be reduced to the workings of natural law or whether they had to be seen as supernatural or preternatural events surfaced periodically, as they did in Britain.[12] In 1719, Thomas Robie insisted that unusual meteors should be regarded not as portents but as entirely natural phenomena. Later, John Winthrop championed the naturalistic view, initially in connection with an earthquake that struck New England in 1755, then on the occasion of a comet in 1758, and later in descriptions of a series of fiery American meteors sent to the Royal Society in the 1760s.[13] Each description of an anomalous event tended to emphasize its uniqueness, paradoxically making it more difficult to assimilate to the regular order of nature. Descriptions of remarkable meteors played the same role as reports of American natural history, testifying to the exoticism of the New World even as they supposedly demonstrated that its natural phenomena could be reduced to scientific order. In a similar way, Benjamin Franklin's work on lightning from the 1740s, while aimed at showing the naturalistic basis of the phenomenon and lessening its dangers with lightning rods, also made everyone aware of the violent electricity of the American atmosphere. As the fame of Franklin's accomplishments spread, it came to be generally accepted that the air was more electrically charged in America than in Europe.[14] These wonders added to anxieties about the hazards of the American climate and to the urgency of the task of taming it.

American weather observers shared with others in the Atlantic world the conviction that the atmosphere posed specific dangers to human health. Their comments often focused on extreme conditions, as if any deviation from the temperate norm of the British Isles was hazardous. The *Philosophical Transactions* published accounts of winters at Hudson's Bay, when it was so cold as to freeze the mercury in the thermometer tube, and summers in Georgia, when 102° was recorded on the Fahrenheit scale. In the summer of 1752, Chalmers measured the heat in his kitchen in Charleston at 115°. He anticipated that his record of the occasion might "not displease the curious," as no register of such a hot season had previously been published.[15] Such extreme departures from average temperatures gave rise to serious worries about their effects on health. Chalmers fretted particularly about the physiological damage done by hot and humid summers, which were thought to thin the blood and relax the bodily fibers. Summer in Charleston, he noted, was a particularly dangerous time for fevers.[16] He cast an anxious eye over the apparently unhealthy aspects of the local milieu: the marshes with their mephitic stagnant water, the unwholesome fogs and dews, and the seasonal hazards of heat waves, tornadoes, and hurricanes. So long as forests continued to surround his town, Chalmers wrote, the stagnant air "in those close recesses ... renders them more proper for the habitations of wild beasts than of men."[17] However, though improvements of the surrounding landscape would be welcome, not all the attributes of civilization were. Chalmers feared that increased luxury and dissipation would weaken people's resistance to disease. He was particularly concerned about tea and coffee drinking that, he worried, "cannot fail in having ill consequences, in some constitutions, particularly during the relaxing heat of summer."[18]

Chalmers's concerns reflected the situation of a Scottish doctor working in a colonial climate that—in comparison with his homeland—would have seemed tropical. On the one hand, the heat and humidity of the South Carolina summers evoked fears of bodily torpor and lassitude, thought particularly conducive to fevers. On the other hand, the development of civilization in the colonies was believed to bring other pathogenic factors in its train. British writers, such as George Cheyne, had already developed the theme of the bad consequences of luxury and fashion for personal health.[19] Chalmers was reproducing this moralizing tendency in British medical discourse and reorienting it to the climatic situation of the colonies, where the dangers of lax behavior were thought to be heightened by an unfriendly tropical environment.

In eyeing the topographical hazards of marshes and forests, Chalmers was also responding to writers on tropical medicine who believed such features aggravated the risks from atmospheric heat and humidity. This became a perennial theme in writings by medical men who attended the British armed forces on overseas deployments. Army surgeon (and later president of the Royal Society) Sir John Pringle urged in 1752 that military camps should always be sited away from marshy ground to avoid the "putrid miasma" it emitted. Pringle led a movement to alter the environment surrounding human habitations to improve the healthiness of the air, a movement that also inspired the draining of swamps and clearing of forests in British overseas possessions.[20] James Lind, a naval surgeon who pioneered methods for treating scurvy on ships at sea, noted that stagnant water and marshes—even in England—produced vapors that were noxious to health. Such features were necessarily much more hazardous in the tropics and in hot locations in North America. Settlers and observers in many British territories shared the belief that clearing forests and marshes would improve the quality of the local air. Chalmers looked forward to the time when improvements in the vicinity of Charleston would allow refreshing breezes from the ocean to circulate more easily. Lind believed that even the air of equatorial Africa could be improved by cutting down vegetation. John Hunter, whose *Observations on the Diseases of the Army in Jamaica* (1788) painted a shocking picture of army losses to disease during the military campaigns of the period, wrote that "noxious exhalations from wet, low, and marshy grounds" had been shown unhealthy "by repeated experience and observation in all parts of the world."[21] He advocated clearing and draining the land near army camps as an urgent military priority.

Civilizing the American Climate

Medical writers, such as Chalmers, who worked in the southern colonies, looked to British possessions in the Caribbean for comparable experience of tropical climates. In Europe, however, the prevailing outlook saw America primarily as a cold continent. This was the view of many European writers since the sixteenth century, given renewed authority in the eighteenth century by the monumental work of the naturalist Georges Louis Leclerc, comte de Buffon. The Scottish historian William Robertson, whose *History of America* appeared in 1777, echoed Buffon and reflected the concerns of contemporary Scottish intellectuals with the question of climate and civilization.[22] Robertson asserted that, in the Western hemisphere as a whole, "cold predominates." Because of the northward extent of the continental landmass, he declared, frigid winds were conveyed from the Arctic to a remarkable distance. This accounted for "the extraordinary dominion of cold" across the whole continent. Robertson saw this as a reflection of America's primitiveness, of the absence of civilizing influences on its topography that might have moderated its climatic extremes. Europe, on the other hand, had benefited from centuries of civilization that had seen its land cultivated and its climate correspondingly improved. Many writers on both sides of the Atlantic shared the conviction that social development in Europe had moderated the brutal extremes of weather recorded by the ancients, whereas America had barely begun to experience a similar change. The reason, according to Robertson, was that the native peoples, "destitute of arts and industry," had taken no steps to improve their environment. Imagining the scene that confronted the first Europeans to arrive, the Scottish author evoked a landscape entirely lacking in cultivation, in which, "the air stagnates in the woods, putrid exhalations arise from the waters; the surface of the earth, loaded with rank vegetation, feels not the purifying influence of the sun; the malignity of the distempers natural to the climate increases, and new maladies no less noxious are engendered."[23]

This perspective denigrated the accomplishments of Native American peoples in agriculture and the development of settled civilization. Robertson ignored evidence of agricultural production among the indigenous peoples of North America, though he was forced to qualify his dogmatic generalizations when faced with the undeniable civilizations of the Aztecs and the Incas. However, his assertions, limited as they were by Eurocentric prejudice, reflected a widespread assumption among enlightened writers that amelioration of the climate would accompany improvement of the landscape, that both were aspects of the civilizing of nature. European settlers in the New World regarded the process as having begun with their ancestors' arrival. They welcomed signs that the American climate was becoming more moderate as indications that they were placing the stamp of improvement on the environment around them. It became widely accepted that the American weather was being brought into line with the temperate ideal of the Old World. The belief reassured the settler population that nature was succumbing to their efforts at civilization, that they were quickly catching up with the climatic progress achieved over centuries in Europe.

This idea was first put into literary circulation by Pehr Kalm, a Swedish follower of Linnaeus, whose account of his visit to North America was published in his native language in 1753 and translated into English in 1771. Kalm established that the belief that the climate was changing was already quite widespread among European settlers. Conversing with elderly members of the Swedish community in southern New Jersey, he was told that winters used to begin and end earlier and had previously yielded more snow than they now did. Old people agreed that the weather now varied more rapidly than it used to and that the characters of the seasons were no longer as distinct as they had been. When he traveled north to Quebec, Kalm was told the same thing: that the winters used to be much colder and more snowy and that crops used to ripen later in the fall. He was also told that the changes were connected with the clearing of the woods, which had allowed the sun's rays "more room to operate" to ripen the grain.[24] Kalm reported these beliefs without much comment, though he suggested his own meteorological observations were consistent with them. The question was examined more systematically by Hugh Williamson, a Philadelphia physician, in a paper delivered to the American Philosophical Society in August, 1770. Like Kalm, Williamson described the belief that the climate was changing as commonplace in Pennsylvania and the other colonies. He accounted for it as an effect of settlement, especially deforestation, which allowed more freedom for winds to flow and affected the rate at which the land heated and cooled. The result was more temperate weather, with milder winters and cooler summers. As the American climate moderated toward the European norm, Williamson anticipated, it would bring additional benefits of civilization: increased yields of crops and improvements in health. He believed a decline in the incidence of tropical fevers could already be observed in lands where woods had been cleared and marshes drained.[25]

Williamson's analysis was noted by European writers, including Buffon. The Pennsylvanian doctor's claims were consistent with the European view that America had originally been blighted with a climate that stunted the growth of living things. Nature was thought to have been less prolific of species in the New World and to have produced animals of inferior size and fierceness.[26] As Robertson put it, "[T]he principle of life seems to have been less active and vigorous there, than in the ancient continent."[27] However, the notion that the climate was changing demonstrated how human beings could mold nature to their wishes. Notwithstanding their subjection to natural forces, they could improve even the climate, modifying it in the direction of the temperate ideal. Robertson's compatriot James Dunbar, a professor of moral philosophy at Aberdeen, recognized the importance of America's changing climate for the debate about the role of the physical environment in the development of civilization. He hailed the fact that, "by opening the soil, by clearing the forests, by cutting out passages for the stagnant waters, the new hemisphere becomes auspicious, like the old, for the growth and population of mankind." This showed that climate did not set absolute limits on what human cultures could achieve, that the effects of natural forces varied "with the course of political events, and the general state of human improvement." According to Dunbar, the civilizing of the American climate showed that Montesquieu and David Hume could be reconciled—the former having argued for the role of physical causes, the latter for that of moral causes, in human development. The American case demonstrated that humans could master the natural forces that threatened to subdue them—that, at least "under certain limitations, soil and climate are subject to [man's] dominion."[28]

Though it nourished the speculations of philosophers and historians in Europe, the theme of the changing climate was also taken up as central to Americans' own hopes for their future, particularly in the years after the United States won its independence. Thomas Jefferson, in his *Notes on the State of Virginia* (1787), reported as a result of his own inquiries among the settler population that "both heats and colds are become much more moderate within the memory even of the middle aged."[29] Jefferson shared with European writers the conviction that the climatic environment shaped human life and that it could also be improved by cultivation and deforestation. He insisted that winter snowstorms were much less severe than they used to be and that refreshing winds from the coast "have advanced into the country very sensibly within the memory of people now living."[30] The climate was already showing signs of being tamed by civilization, giving much less reason to fear that it would inhibit the development of intellectual or moral life in Virginia. Decades later, toward the end of his life, Jefferson called for a national network of weather observers to compile observations over a long period "to show the effect of clearing and culture towards changes of climate."[31] William Currie, another Philadelphia physician, who published *An Historical Account of the Climates and Diseases of the United States* in 1792, compared the American situation to that in the ancient civilization of China, where by extensive alterations of the landscape, "the air, in very unfavourable situations, has been rendered exceedingly wholesome." England, Italy, and Germany had improved their climates within historic times, according to Currie. The same could be expected in the United States: "When in the course of time, this continent becomes populated, cleared, cultivated, improved, and the moisture of the soil exhausted, … the bleak winds will become more mild, and the Winters less cold."[32]

By the last decade of the century, a general consensus had clearly formed among Americans that improvements in the landscape had already reduced the severity of winter cold and moderated other climatic extremes. Samuel Williams reported, in his *Natural and Civil History of Vermont* (1794), that the transformation was "so rapid and constant, that it is the subject of common observation and experience. It has been observed in every part of the United States; but is most of all sensible and apparent in a new country, which is suddenly changed from a state of vast uncultivated wilderness, to that of numerous settlements, and extensive improvements."[33] Throughout the country, according to Williams, winters had become shorter, summers less intensely hot, and the weather in general subject to more rapid variations. When Volney visited in the late 1790s, he reported that these changes were recognized by everyone and "have been represented to me not as gradual and progressive, but as rapid and almost sudden, in proportion to the extent to which the land is cleared."[34] Americans were anticipating that the progress of civilization would continue to modify their climate as it had that of Europe in the previous centuries.

The perception of climatic transformation was widely shared among the population of the new United States; it seems to have become a topic of experimental investigation after having surfaced in popular consciousness.

Learned authors debated its magnitude and discussed whether it was altogether a good thing. Benjamin Rush, the most famous Philadelphia physician of the era and professor of chemistry at the University of Pennsylvania, addressed the question in a paper published in 1789. He accepted that "accounts which have been handed down to us by our ancestors" gave reason to believe that the climate had changed.[35] However, the question was tricky to specify empirically because of the paucity of exact records from the early stages of colonization. Rush suspected that memories of the elderly, the source of Kalm's information, were unreliable, perhaps because people's perceptions of heat and cold altered as they aged. He concluded that there was no decisive evidence that winters had been colder before 1740 than after, but he agreed that the seasons had tended to merge into one another and the weather had become more variable in recent years. He accepted Williamson's assertion that clearing of forests and cultivation of the land had had a significant effect. A few years later, David Ramsay made another assault on the question in his *Sketch of the Soil, Climate, Weather, and Diseases of South Carolina* (1796). Ramsay drew on the long tradition of weather observations in Charleston, comparing contemporary records with those compiled by Chalmers in the middle of the century. He concluded that both maximum and minimum temperatures had moderated over the period but that it was too soon to say whether this represented a long-term trend. He nonetheless reasserted the basic article of enlightenment faith, that improvements in the natural environment would bring permanent benefits in terms of climate and health: "The advantages resulting to the temperature of the air, and to the healthiness, as well as to the appearance of any country, from the art of man, inhabiting and cultivating it, are inconceivably great. We may, therefore, indulge the hope that our [climate] is progressively meliorating from permanent and encreasing causes."[36]

Williams, who was a member of the American Philosophical Society in Philadelphia and of the Palatine Meteorological Society in Germany, adopted a different experimental approach to the question in his *History of Vermont*. He measured soil temperatures in uncut woods and in open fields and concluded that deforestation had measurably warmed the soil. He also measured the rate of evaporation of water from leaves, trying to estimate how much atmospheric humidity was reduced by removing forests. He understood that discoveries by Joseph Priestley and others had shown the importance of vegetation in restoring the air's suitability for respiration. His final conclusion was finely balanced, reflecting the consensus that change had happened, while entering some reservations about the overall benefits of cutting down trees.[37] Williams's caution reflected the beginnings of a realization that clearing forests might not be altogether a good thing. As Richard Grove has shown, the dangers of deforestation were beginning to be realized in the late eighteenth century by some observers of tropical islands in the Caribbean and elsewhere.[38] Priestley's findings called into question the equation of "improvement" with the removal of trees, as they suggested that this could reduce the local air quality. The question remained a controversial one, entangled with the politics of relations between local settlers and landowners on the one hand and colonial administrators and improvement societies on the other. In North America, also, the question of the forests was politically contentious, the harvesting of New England timber for the Royal Navy being resented by settlers as an aspect of a repressive colonial regime. A movement to defend the forests emerged only rather slowly after independence; but it was foreshadowed in some of the commentary on climatic change around the turn of the nineteenth century. Rush and Ramsay, notwithstanding their general endorsement of landscape improvement, both thought that clearing the woods had increased the incidence of certain kinds of fevers.[39] Especially with the resurgence of yellow fever, in both southern and northern states in the 1790s, medical writers were given reason to qualify their optimism that changes in the landscape were improving the health of the population. Jeremy Belknap, in his *History of New-Hampshire* (1791–92), voiced a more custodial attitude to the American forests, emphasizing their value for absorbing noxious vapors and producing salubrious air. He argued that New Hampshire owed the good health of its population to its rugged environment, including its forests. People lived longer there than in European cities, benefiting from the "balsamic quality" of the air released by the trees.[40]

Despite emerging differences over the health benefits of forests, American commentators were united in taking pride in their national climate. Belknap joined a chorus of denunciations of Buffon for his assertion that the New World climate had caused the degeneration of animal species transplanted from Europe. "[N]otwithstanding the dreams of European philosophers," Belknap thundered, "America can best be described by those who have for a long time resided in it."[41] Jefferson had already led the defense of American climatic conditions against the French naturalist's condemnation of them on these grounds. He had confronted Buffon personally on the matter during a visit to Paris and tried to clinch his point by sending a specimen of an American moose, a mammal that was undeniably larger than any European equivalent. In addition, Belknap took Buffon to task for his ignorance of skunks, bears, raccoons, and wildcats, all of which had been found in America to be larger than they were reputed to be in Paris.[42] American naturalists presented themselves as advocates for

New World nature in these transatlantic debates, reasserting a privileged knowledge of their own environment to which they had already laid claim in the colonial period.

Climate and American Destiny

The transatlantic dispute about the effects of the American climate was particularly fraught when it touched on the native peoples of the New World. Buffon followed earlier Spanish writers in asserting that the same cold and humid conditions that had stunted the growth of American wildlife had also limited the vigor of the continent's peoples. He thought that European settlers could be spared the fate of degenerating to the natives' level only by altering the climate in which they lived. Robertson agreed that nature in the New World had checked the growth of "the more noble animals" while encouraging odious reptiles and insects, "the offspring of heat, moisture, and corruption."[43] He cited the abbé Cornelius de Pauw's view that, "under the influence of an unkindly climate, which checks and enervates the principle of life, man never attained in America the perfection which belongs to his nature, but remained an animal of an inferior order."[44] Afflicted with bodily weakness and a disposition to melancholy, the native peoples lacked the force of mind to take charge of their environment, even to the extent of domesticating animals or plowing the soil.[45] Robertson did, however, concede that the natives' backwardness was not entirely to be ascribed to climatic causes, and he acknowledged that European settlers were now taming the American wilderness. The latter development would permit them to escape the debilitating effects of the original climate and even offered the prospect of improving the character of the natives.[46]

Though Buffon and Robertson exempted European settlers from the American climate's negative effects, their criticisms drew a vigorous response from American writers. Jefferson rejected the claim that the New World had produced human beings who lacked vitality and social sentiment. Native Americans were not, according to the Virginian, deficient by nature; they were simply at an early stage of social development. He rejected, in other words, the assertion of a line of commentators going back to the Spanish conquistadors that American natives were naturally suited to serve as slaves for Europeans.[47] However, as recent scholars have noted, Jefferson's sympathy for native people was not extended to the Africans enslaved on his own estate and throughout the South. He held that Africans were intellectually inferior to Europeans "by nature" and hence were likely to remain in the condition of slaves.[48] Apparently it was more important to defend the reputation of Native Americans than that of Africans, presumably because the former were products of the continent and its climate with which the new nation had cast its lot. As Charles A. Miller has put it: "Jefferson identified the human nature of America with its natural history, thus establishing a bond with the Indians that was inconceivable with the Africans."[49]

Samuel Stanhope Smith, professor of moral philosophy and later president at the College of New Jersey, whose *Essay on the Causes of the Variety of Complexion* (1787) was published in the same year as Jefferson's work on Virginia, shared his view that Native Americans were more amenable to having their physical characteristics changed by civilization than were Africans. In general, Smith held that racial attributes had a climatic origin, but he qualified this view when considering whether people of European descent might ever decline to the condition of the natives under the influence of the American climate. He declared that the "arts and conveniences of civilization," including cultivation of the landscape, would tend to correct the effects of climate, which were most severely damaging in "a savage state of society."[50] Even if European-Americans were to "sink into a state of savagism, perhaps the resemblance might not, in every point, be complete; because [they] would receive the impressions of the climate on the ground of features formed in Europe, and in a high state of civilization; [whereas the natives] have received them on the ground of features formed in a very different region of the globe, and in a much ruder state of society."[51] Smith was situating himself in a tradition of European Creole writers who assigned themselves a distinct racial identity from the American natives. Like earlier Spanish writers, he differentiated the settler population from the natives, seeing the latter as inherently more susceptible to the forces of climate.[52] It is notable, however, that he did not entirely discount atmospheric influences on the settler population's physical condition. Rather, he claimed that such influences would be diminished by the biological inheritance of European physique and modified by the advance of civilization. Members of the settler community were thus offered the reassurance that their bodily constitution and their work to civilize the environment would together save them from climatic degeneration.

Williams also engaged with the question of the natives in his *History of Vermont*. He criticized Buffon as an armchair philosopher ignorant of the realities of American natural history: "No such animal was ever seen in America, as the Indian M. de Buffon described in Paris."[53] He preferred to take Rousseau as his guide, seeing Native Americans more in the light of the proverbial noble savages. It was not true that they lacked sensibility, energy, or sexual drive: "Nature is the same in the Indian, as it is in the European."[54] What was true was that virtues and vices were less developed in the "savage" state. Although Williams saw the moral sense as natural to man, he believed its dictates became less simple and straightforward as civilization developed. Nonetheless, he thought primitive people possessed the basic virtues of love of country and fierce independence, virtues that Williams clearly felt all Americans would be proud to embrace. If Buffon had missed this, it must be because he had been misled by "the debilitating and degrading effects, which luxury, intemperance, and excess, are constantly producing in the populous cities of Europe."[55] Thus, Williams portrayed European commentators as tainted by the corruption surrounding them. He neatly reversed the charge of "degeneration," throwing it back at the decadent societies of the Old World. In contrast to the corrupting effects of European luxury and indulgence, the simple virtues of New World natives were said to derive from their closeness to American nature. They were therefore virtues that all inhabitants of the new republic might hope to share.

Jefferson and Williams were not the only citizens of the new republic motivated to defend the reputation of Native Americans against European criticism. As Antonello Gerbi has shown in his magisterial survey of the disputes surrounding New World nature, other American writers, including Benjamin Franklin and Thomas Paine, worked to restore the image of the natives as models of nobility for the new United States.[56] Their project also connected with the debate over climate, as Native Americans were identified metonymically with the environment in which they lived. Buffon, de Pauw, Robertson, and others were resented for their suggestion that nature in the New World had a debilitating effect on the human beings who lived there. The claim was rejected by Americans as factually incorrect, but the underlying premise—that climate had an effect on the characteristics of a people—could not be categorically disposed of. Americans asserted instead that their rugged environment nurtured sentiments of freedom and self-reliance; and they also clung to the belief that they were civilizing it by cultivating the landscape. It was under this dual aspect of an entity that was both natural and in the course of becoming civilized that the climate was recruited as an asset for the newborn nation.

Foreign visitors often remarked on the new nation's self-investment in its climate. Volney, who fled to the United States in 1795 as a refugee from the revolutionary regime in France and left disillusioned three years later, wrote that the Americans took a quite irrational pride in their weather. As far as he could see, it was altogether less desirable than that of Mediterranean countries, but Americans stubbornly defended its qualities. Volney thought this could only be ascribed to self-interest and the simple fact that people got habituated to the conditions in which they lived. Their judgment was distorted by imbibing "a physical and moral atmosphere, which we breathe without perceiving it."[57] Americans themselves spoke of their newfound liberty as a "climate," or an "atmosphere," or as a result of the workings of nature on their moral constitution. Williams wrote that, though European monarchs delayed the progress of reform, "nature was establishing a system of freedom in America."[58] Currie compared the ideal climate that Americans could expect to enjoy when the landscape was improved with the political liberty they were already experiencing, having "already, in a great measure, regained the native dignity of our species."[59] Belknap, who claimed the New Hampshire air was already highly salubrious, dubbed his state "a nursery of stern heroism."[60] Americans of European descent were beginning to identify themselves with the effects of their climate, seeing it as the source of their virtues, especially the vigorous defense of liberty that had won them independence.

There was no inconsistency between this belief and the assertion that the climate had been transformed since European settlement began. If climate shaped the characteristics of a people, it could also be changed by industrious cultivation of the landscape. It was the willingness to change their environment in this way that differentiated the settlers from the natives, at least in the eyes of the former. From the earliest stages of colonization, they had appropriated the land according to European notions of property rights that assigned exclusive possession and a consequent responsibility for improvement.[61] The natives, conversely, whatever their inspirational virtues as freedom-loving people, were regarded as having failed to take charge of their natural surroundings or to civilize the world around them. The European pattern of landholding and cultivation was thought to have had benefits for the climate that were already noticeable. The settlers thereby reassured themselves that they had no reason to fear the natives' appalling fate, their catastrophic decline from epidemic diseases. Believing that they were moderating the extremes and anomalies of the American weather and molding it to fit the needs of civilization, the settlers allayed their anxieties that climatic factors had played a part in the disastrous collapse of the native population.

Although it was articulated in arguments against European philosophers and historians and was developed in the context of the war of independence from Britain, this attitude was rooted in enlightened ways of thinking that were common to Europe and colonial America. Americans understood their climate as integral to their destiny because they saw it as one of the means by which nature exerted its pervasive influence on human life in all its manifestations. The philosophers and historians who debated the role of physical and moral causes in history believed that nature was a shaping presence even in highly civilized societies. However, they did not believe human beings needed to remain passive objects of natural forces. Rather, active intervention in the natural environment was an aspect of human nature itself. People—by their nature—acted on their physical surroundings. The same assumption was made in the works of medical writers who advocated reshaping the landscape around settlements. Civilized people could take action to reform their atmosphere, for example by removing the sources of unhealthy air in marshes and forests. In doing this, they were not acting against nature, but giving expression to an element in the natural constitution of humanity itself. The settler community in America shared this basic outlook with the European writers whom they often criticized for their factual ignorance of American realities. In terms of fundamental principles, Smith and Jefferson, for example, agreed with Robertson and Dunbar. On both sides of the Atlantic, enlightened intellectuals believed that human beings should actively intervene in nature rather than passively accepting the fate that they were dealt. They expected people to assert their prerogative to remold their natural surroundings, for example, taking charge of their circumstances to alter the way the air affected them. This understanding of the human relationship to the climatic environment was a characteristic product of the European Enlightenment, and it was in America that it was adopted with the greatest enthusiasm.

Notes

1. Karen Ordahl Kupperman, "The Puzzle of American Climate in the Early Colonial Period," *American Historical Review* 87 (1982), 1262–89; idem, "Fear of Hot Climates in the Anglo-American Colonial Experience," *William and Mary Quarterly*, 3rd ser., 41 (1984), 213–40; William B. Meyer, *Americans and their Weather* (Oxford: Oxford University Press, 2000), 17–42.
2. See, for example: Carla Mulford, "New Science and the Question of Identity in Eighteenth-Century British America," in Mulford and David S. Shields, eds, *Finding Colonial Americas: Essays Honoring J. A. Leo Lemay* (Newark: University of Delaware Press, 2001), 79–103; Joyce E. Chaplin, "Nature and Nation: Natural History in Context," in Sue Ann Prince, ed., *Stuffing Birds, Pressing Plants, Shaping Knowledge: Natural History in North America, 1730–1860* (Philadelphia: American Philosophical Society, 2003), 75–95.
3. See, especially: Clarence J. Glacken, *Traces on the Rhodian Shore: Nature and Culture in Western Thought from Ancient Times to the End of the Eighteenth Century* (Berkeley: University of California Press, 1967), 501–705. Also relevant are: Christopher J. Berry, "'Climate' in the Eighteenth Century: James Dunbar and the Scottish Case," *Texas Studies in Literature and Language* 16 (1974): 281–92; David Wallace Carrithers, "The Enlightenment Science of Society," in Christopher Fox, Roy Porter, and Robert Wokler, eds, *Inventing Human Science: Eighteenth-Century Domains* (Berkeley: University of California Press, 1995),

232–70; Roxann Wheeler, *The Complexion of Race: Categories of Difference in Eighteenth-Century British Culture* (Philadelphia: University of Pennsylvania Press, 2000); and Mark Harrison, *Climates and Constitutions: Health, Race, Environment, and British Imperialism in India, 1600–1850* (Oxford: Oxford University Press, 1999).

4   James Rodger Fleming, *Historical Perspectives on Climate Change* (New York: Oxford University Press, 1998), 21–32.

5   Kenneth Thompson, "The Question of Climatic Stability in America before 1900," *Climatic Change* 3 (1981), 227–41.

6   For the general situation, see: Alfred W. Crosby, Jr., *The Columbian Exchange: Biological and Cultural Consequences of 1492* (Westport, CT: Greenwood Press, 1972), 35–63. On the response of North American settlers, see: Joyce E. Chaplin, *Subject Matter: Technology, the Body, and Science on the Anglo-American Frontier, 1500-1676* (Cambridge, Mass.: Harvard University Press, 2001), esp. 157–98.

7   See, for example: John Campbell, *A Political Survey of Britain* (2 vols, London: Richardson and Urquhart, et al., 1774), 1: 55, who concluded that "we cannot but acknowledge the singular Bounty of Providence" in bestowing upon Britain its temperate climate.

8   Raymond Phineas Stearns, *Science in the British Colonies of America* (Urbana: University of Illinois Press, 1970), 188–9, 434, 448–50, 468; James H. Cassedy, "Meteorology and Medicine in Colonial America: Beginnings of the Experimental Approach," *Journal of the History of Medicine* 24 (1969): 193–204.

9   Cassedy, "Meteorology and Medicine," 197–9; Stearns, *Science in the British Colonies*, 646–7.

10 Patricia Cline Cohen, *A Calculating People: The Spread of Numeracy in Early America* (New York: Routledge, 1999), esp. 81–115.

11 Constantin François de Volney, *View of the Climate and Soil of the United States of America* (English trans., London: for J. Johnson, 1804), 134.

12 Peter Eisenstadt, "The Weather and Weather Forecasting in Colonial America" (New York University Ph.D. dissertation, 1990), 41–55, 146–51; David D. Hall, *Worlds of Wonder, Days of Judgment: Popular Religious Belief in Early New England* (Cambridge, Mass.: Harvard University Press, 1990), 79–80, 106–9, 221–2.

13 Hall, *Worlds of Wonder*, 108; Stearns, *Science in the British Colonies*, 648–52.

14 James Delbourgo, *A Most Amazing Scene of Wonders: Electricity and Enlightenment in Early America* (Cambridge, Mass.: Harvard University Press, 2006), ch. 2.

15 Lionel Chalmers, *An Account of the Weather and Diseases of South Carolina* (2 vols in 1, London: Edward and Charles Dilley, 1776), 18.

16 Chalmers, *Account of Weather and Diseases*, 47–56.

17 Chalmers, *Account of Weather and Diseases*, 10.

18 Chalmers, *Account of Weather and Diseases*, 35.

19 See, especially: George Cheyne, *The English Malady, or a Treatise of Nervous Diseases of all Kinds* (London: G. Strahan, 1733). Modern studies include: Anita Guerrini, *Obesity and Depression in the Enlightenment: The Life and Times of*

George Cheyne (Norman: University of Oklahoma Press, 2000); G. J. Barker-Benfield, *The Culture of Sensibility: Sex and Society in Eighteenth-Century Britain* (Chicago: University of Chicago Press, 1992), 6–15; Roy Porter, *Flesh in the Age of Reason* (New York: W. W. Norton, 2003), 237–40.

20 Dorothea Wade Singer, "Sir John Pringle and his Circle," *Annals of Science* 6 (1948–50): 127–80, 229–61.

21 John Hunter, *Observations on the Diseases of the Army in Jamaica* (London: for G. Nichol, 1788), 15.

22 On Robertson and his precursors, see: Antonello Gerbi, *The Dispute of the New World: The History of a Polemic*, trans. Jeremy Moyle (Pittsburgh: University of Pittsburgh Press, 1973), especially 158–60, 165–9; Jorge Cañizares-Esguerra, *How to Write the History of the New World: Histories, Epistemologies, and Identities in the Eighteenth-Century Atlantic World* (Stanford: Stanford University Press, 2001), 38–55; Stewart J. Brown, ed., *William Robertson and the Expansion of Empire* (Cambridge: Cambridge University Press, 1997); Jeffrey Smitten, "Impartiality in Robertson's *History of America*," *Eighteenth-Century Studies* 19 (1985), 56–77. On the prior debate in the Spanish-speaking world about the climate and its influence on New World natives, see: Jorge Cañizares-Esguerra, "New World, New Stars: Patriotic Astrology and the Invention of Indian and Creole Bodies in Colonial Spanish America, 1600–1650," *American Historical Review* 104 (1999), 33–68.

23 William Robertson, *The History of America* (London: W. Strahan, T. Cadell, and J. Balfour, 1777), 252, 254, 257, 258.

24 Adolph B. Benson, ed., *Peter Kalm's Travels in North America* (2 vols, New York: Wilson-Erickson, 1937), 1: 271, 275–7; 2: 509, 513.

25 Hugh Williamson, "An Attempt to Account for the Change of Climate, which has been Observed in the Middle Colonies in North-America," *Transactions of the American Philosophical Society* 1 (1771), 272–80.

26 Glacken, *Traces on the Rhodian Shore*, 587–91; Gerbi, *Dispute of the New World*, 3–34.

27 Robertson, *History of America*, 259.

28 James Dunbar, *Essays on the History of Mankind in Rude and Cultivated Ages* (2nd edn, London: W. Strahan, T. Cadell, and J. Balfour, 1781), 297, 354, 356.

29 Jefferson, *Notes on the State of Virginia* [1787], in *The Portable Thomas Jefferson*, ed. Merrill D. Peterson (New York: Viking Press, 1975), 119.

30 Jefferson, *Notes on Virginia*, 116.

31 Charles A. Miller, *Jefferson and Nature: An Interpretation* (Baltimore: Johns Hopkins University Press, 1988), 42.

32 William Currie, *An Historical Account of the Climates and Diseases of the United States of America* (Philadelphia: T. Dobson, 1792), 82, 86. On the supposed changes in the European climate since ancient times, see also: Williamson, "Attempt to Account for the Change of Climate," 340–1; Theodore S. Feldman, "The Ancient Climate in the Eighteenth and Early Nineteenth Century," in *Science and Nature: Essays in the History of the Environmental Sciences*, ed. Michael Shortland (Chalfont St. Giles: British Society for the History of Science, 1993), 23–40.

33 Samuel Williams, *The Natural and Civil History of Vermont* (Walpole, NH: Isaiah Thomas and David Carlisle, Jr, 1794), 57.
34 Volney, *View of Climate and Soil*, 269.
35 Benjamin Rush, "An Account of the Climate of Pennsylvania and Its Influence upon the Human Body," in Rush, *Medical Inquiries and Observations* (Philadelphia: Prichard and Hall, 1789), 57–88, quote on 61.
36 David Ramsay, *A Sketch of the Soil, Climate, Weather, and Diseases of South Carolina* (Charleston: W. P. Young, 1796), 8. See also: Robert Croom Aldredge, *Weather Observers and Observations at Charleston, South Carolina, 1670–1871* (Charleston: no publisher, 1940).
37 Williams, *History of Vermont*, 57–65.
38 Richard H. Grove, *Green Imperialism: Colonial Expansion, Tropical Island Edens, and the Origins of Environmentalism, 1600–1860* (Cambridge: Cambridge University Press, 1995), esp. 264–308.
39 Rush, "Climate of Pennsylvania," 83; Ramsay, *Sketch of Soil*, 21.
40 Jeremy Belknap, *The History of New-Hampshire, Containing a Geographical Description of the State,* 2nd edn, 3 vols (Dover, NH: for O. Crosby and J. Varney, by J. Mann and J. K. Remick, 1812), 3: 172.
41 Belknap, *History of New-Hampshire*, 3: 172.
42 Gerbi, *Dispute of the New World*, 252–68; Belknap, *History of New-Hampshire*, 3: 109–12.
43 Robertson, *History of America*, 261.
44 Robertson, *History of America*, 287.
45 Robertson, *History of America*, 257–61, 270–72, 287–92.
46 Robertson, *History of America*, 292–96.
47 Gerbi, *Dispute of the New World*, 67–76.
48 Miller, *Jefferson and Nature*, 63–76.
49 Miller, *Jefferson and Nature*, 75. See also: Anthony F. C. Wallace, *Jefferson and the Indians: The Tragic Fate of the First Americans* (Cambridge, Mass.: Harvard University Press, 1999).
50 Samuel Stanhope Smith, *An Essay on the Causes of the Variety of Complexion and Figure in the Human Species*, ed. Winthrop D. Jordan (Cambridge, Mass.: Harvard University Press, 1965), 93.
51 Smith, *Essay on Complexion*, footnote 45. On Smith, see also: Winthrop D. Jordan, "Introduction," in Smith, *Essay on Complexion*, vii–liii; David N. Livingstone, "Geographical Inquiry, Rational Religion, and Moral Philosophy: Enlightenment Discourses on the Human Condition," in Livingstone and Charles W. J. Withers, eds, *Geography and Enlightenment* (Chicago: University of Chicago Press, 1999), 93–119, especially 103–11.
52 On Spanish writers, see: Cañizares-Esguerra, "New World, New Stars." On consciousness of racial identity in North America, see: Joyce E. Chaplin, "Natural Philosophy and Early Racial Idiom in North America: Comparing English and Indian Bodies," *William and Mary Quarterly*, 3rd ser., 54 (1997), 229–52; Chaplin, *Subject Matter*; Eve Kornfeld, "Encountering 'the Other': American Intellectuals and Indians in the 1790s," *William and Mary Quarterly*, 3rd ser., 52 (1995), 287–314; Michael Zuckerman, "Identity in British America: Unease in Eden," in Nicholas Canny and Anthony Pagden, eds, *Colonial Identity in*

the Atlantic World, 1500–1800 (Princeton: Princeton University Press, 1987), 115–57.
53 Williams, *History of Vermont*, 183.
54 Williams, *History of Vermont*, 158.
55 Williams, *History of Vermont*, 184.
56 Gerbi, *Dispute of the New World*, 240–52.
57 Volney, *View of Climate and Soil*, 330–1. Compare the remark of the English visitor Thomas Hamilton in 1833: "on the subject of climate … there is no topic on which Americans are more jealously sensitive. It delights them to believe that theirs is in all respects a favoured land" (quoted in Thompson, "Question of Climatic Stability," 233).
58 Williams, *History of Vermont*, vii.
59 Currie, *Historical Account of Climates*, 89 footnote.
60 Belknap, *History of New-Hampshire*, 3: 196.
61 On the ecological effects of the colonists' belief that land could be appropriated as private property, see: William Cronon, *Changes in the Land: Indians, Colonists, and the Ecology of New England* (New York: Hill and Wang, 1983); Carolyn Merchant, *Ecological Revolutions: Nature, Gender, and Science in New England* (Chapel Hill: University of North Carolina Press, 1989).

SECTION III

Itineraries of Collection

CHAPTER 7

Empiricism in the Spanish Atlantic World

ANTONIO BARRERA-OSORIO

In 1518, a royal official in Santo Domingo, Alonso de Zauzo (1466–1527), wrote a letter to Charles V about the "marvelous multiplication of livestock. Cows give birth generally to two calves, but many times to three."[1] He mentioned that in Hispaniola there were already hogs, sheep, and mares. Zauzo suggested sending merino sheep and proposed a mechanism for doing so: "[T]hose coming, should be accustomed to eat... grains and things that can be brought in a ship so they would not die."[2] Wheat was already adapted in some areas of the island. He also mentioned some "marvelous" native medicines, and attempts to adapt Old World spices, such as pepper, in the New World.[3]

A year later, in what seems to be an unrelated event, the cosmographer Diego Ribeiro was appointed (1519) cosmographer for making instruments and charts at the House of Trade. Ribeiro also participated in the junta of Badajoz (1524) to discuss the matter of the demarcation line with Portugal (Treaty of Tordesillas) and in the organization of Joffrey de Loaisa's expedition to the East, departing from La Coruña.[4] His activities included the making of charts and instruments and the design of a bilge pump.[5] Zauzo's interests in the wonders of the Hispaniola and Ribeiro's activities as cosmographer and instrument maker in Seville point to the confluence of artisans, royal officials, and scholars around the Atlantic World (Seville and Hispaniola, for instance) and their attempts to describe and control natural entities.

My thesis is simple: the Atlantic World fostered the creation and development of institutions and institutionalized empirical practices for the study and organization of the New World. This empiricism was articulated in reports based on personal experience, testing practices, and collaborative practices that put together partial reports into general views of natural things. Natural things from the New World and artificial things (instruments) to control the New World shaped and transformed attitudes toward the natural world and its environment. In the imperial and commercial context of the Atlantic World, natural things and instruments together with the artisans, royal officials, and scholars who manipulated them produced institutions and practices that encouraged empirical approaches to nature. This institutionalization resulted from the need to organize empirical information about the New World, but in turn this institutionalization produced tensions between traditional understandings about the relationship between theory and practice. The Spanish case illustrates this process, a process that I suggest fostered Europe-wide transformations of attitudes toward natural entities and the role of experience in the production of knowledge.

Empiricism, for the purposes of this argument, means an epistemological practice based primarily on observations and personal experience and the articulation of these observations and personal experiences in reports. The empirical practices that emerged in the context of the Spanish American empire, I argue, constituted an alternative practice to the Aristotelian, Neo-Platonic, magical, and alchemical practices of the sixteenth century.[6] The empirical practices I describe in this chapter emerged in a commercial and imperial context, which means first, that both practitioners and audiences had a practical and utilitarian aim in mind. Royal officials needed to understand the nature of the resources in the New World. Thus, when reports of a blue dye in Guatemala circulated in Spain, for instance, the crown asked for reports about it, tests, and samples. There was no mention at all of Aristotelian principles or of hidden relations between these dyes and the planets. Similarly, entrepreneurs wanted to exploit resources, and for this they also needed to understand them and their uses, as I discuss below. A key element in the development of these practices is that they were institutionalized in new sites of knowledge production outside traditional sites of knowledge production, such as the universities.[7] These new sites were the House of Trade and the Council of Indies and the viceroyal court system in the New World.

As I mentioned above, this empiricism constituted an alternative to traditional European practices that, in general, articulated empirical practices within academic and classical explanatory frames, such as Aristotelian Scholasticism and Neo-Platonism and Hermeticism. I argue that the empirical practices I describe in this chapter, articulated in reports, were framed within practical and utilitarian principles that pushed aside concerns about causes and principles (Aristotle) and hidden relations (Hermeticism and magic). These were reports that describe specific events or objects based on observations and personal experience.[8] The Spanish crown, within the House of Trade, for instance, sought to solve problems concerning observations, such as bias and incompleteness in reports, by requesting multiple reports and evaluating those reports by committees (juntas of experts), by requesting tests to be performed before experts, samples to be sent to experts, and by organizing expeditions to collect both samples and information about natural things in the New World.[9] I simply argue that an alternative set of empirical practices in the sixteenth and seventeenth centuries emerged from the commercial and colonial empires of the New World in such sites as the House of Trade and the Council of Indies.[10]

The story of Spanish activities in the Atlantic World helps to explain the emergence of empirical practices in the sixteenth and seventeenth centuries. Why did sixteenth-century Europeans rely increasingly on empirical information to understand the world around them? Edgar Zilsel argues that "the rise of the methods of the manual workers to the ranks of academically trained scholars at the end of the sixteenth century is the decisive event in the genesis of science."[11] In what context did these artisans and scholars come together? Several historians have examined the relationship between artisans and scholars in royal courts, or even in city neighborhoods.[12] I explore the relations between artisans, merchants, royal officials and the resulting emergence of empirical practices in centers of the Atlantic World (Seville, Santo Domingo, and Mexico) during the sixteenth century. These empirical practices are reflected, for instance, in reports sent to the House of Trade beginning in the sixteenth century about natural things from the New World, in debates about experience and theory that took place at the House of Trade, and in the several patents and contracts granted to entrepreneurs to exploit natural resources in the New World. What these practices, reports, debates had in common was an emphasis on empirical approaches and collaborative practices to gain a better understanding of the New World.[13]

I am interested in explaining how experience moved to the center of epistemological concerns in sixteenth- and seventeenth-century Europe. I am also interested in exploring the consequences of institutionalizing those practices at the House of Trade. Peter Dear characterizes the new science of the seventeenth century in these terms:

> In characteristically seventeenth-century philosophical discourse, by contrast [to scholasticism], experience increasingly took the form of statements describing specific events. These are exemplified most famously by the research reports found in countless contributions to the *Philosophical Transactions* by the early Fellows of the Royal Society, but they did not appear there *de novo*. Natural philosophers and, especially, mathematical scientists increasingly used reports of singular events, explicitly or implicitly located in a specific time and place, as a way of constructing scientifically meaningful experiential statements.[14]

Why did natural philosophers and mathematical scientists increasingly use reports of singular events? At least in the case of the Spanish-Atlantic World, those reports constituted the most efficient way of understanding the New World since the sixteenth century. They came, for instance, from people searching for commodities in the New World and from pilots who brought navigational information back to the House of Trade. The Spaniards sent many reports about the New World to the House of Trade, the Council of Indies, the court and vice-royal courts. Eventually, those institutions created mechanisms to collect and generate that kind of reports (about singular events and based on experience) for the understanding of the New World. Those institutions also created tensions between those who supported an emphasis on the collection of empirical information and those who preferred to subordinate it to theoretical frameworks.

In this chapter I discuss: first, the practices employed for searching after commodities in the New World; and, second, the establishment of a veritable chamber of knowledge within the House of Trade for collecting information about the New World, and how the House fostered debates about the relationship between theory and practice.

Transforming the New World

The humanist Juan Pérez de Oliva reports that Columbus, in his second voyage (1493), took "artisans necessary for the building and maintenance of cities. His ships were loaded with all the seeds of herbs and plants, and animals that we [Spaniards] use most frequently, for their multiplication in those strange lands, so those lands would become attractive to our mariners." Columbus, Perez concluded, left Spain "to mix the world, and give those strange lands the shape of ours."[15]

This transformation of nature implied experimentation and the transfer of technology to the New World and knowledge about the natural products specific to the Indies. It was a project inspired in a Christian ideology that assumed the unity of the natural world and the transformation of nature for human purposes. What is particular to, and significant in, the Spanish American empire was the royal support of the experts engaged in the transformation of the New World. The activities of peasants, merchants, and artisans—by themselves practical and empirical—gained in status by the support of the crown. In the search for commodities and the ecological transformation of the New World, Spaniards made single natural entities the focus of empirical observations and reports.[16]

Personal experience became a knowledge-producing tool when artisans, entrepreneurs, and royal officials in distant lands began transforming the land and searching for commodities. They sent information based on their own experience about new natural entities (in many cases lacking references in the classical traditions) to commercial partners and royal officials in Europe. These partners and officials made their political and commercial decisions on the basis of those reports.

In fifty years, the Spaniards populated and transformed the New World from Florida to Chile with European people and viruses, horses, cows, pigs, sheep, silkworms, dogs, sugar cane, wheat, wine grapes, vegetables, oranges, lemons, fig trees, pomegranates, quince trees, banana trees, mulberry, melon, and cucumbers.[17] The adventurer Cieza de León, who traveled in Tierra Firme around 1536, tells us that in Panama there were already oranges, citrons, figs, and cows but not wheat or barley. Flour at that time was coming from Peru and Spain.[18]

The spread of animals was astonishing. Recall Alonso de Zauzo's letter to Charles I in 1518. Zauzo mentioned that in Hispaniola there were already hogs, sheep, and mares everywhere.[19] Hispaniola became the source of livestock for the Indies and the new center of activities for the Spaniards in the New World. The introduction of plants and trees, instead, was more difficult. Adapting plants and tress to the New World required more experimentation. Animals multiply by themselves, but plants need cultivation. More testing was necessary to adapt plants successfully. From 1494 to 1517, the crown and royal officials in the Caribbean tried many different methods for cultivating grains and trees. In 1514, a royal provision to San Juan Island ordered that its residents plant apples, pears, pomegranates, quince, peaches, walnuts, chestnuts, "and whatever grows well."[20] The crown granted tax exemptions and financial incentives to laborers to go, stay, and work in the new lands and experiment in the planting of familiar crops. Different varieties of cereals, for instance, were tried in different parts of the Caribbean throughout the century.[21] The crown, royal officials, and ordinary immigrants shared a similar desire to transform the new lands to make them economically and ecologically inhabitable for Europeans.

The transformation of nature was not limited to adapting European plants and animals. In 1534, Charles V ordered the governor of *Tierra Firme* (the area covering the coasts of present day Venezuela, Colombia, and Panamá) to take "expert people" to the Chagre river (in present Panamá)—navigable from the Atlantic ocean—and determine "what form and order could be given to open the land [between the Chagre river and the Pacific ocean], so that opening that land, the said river be connected to the South sea, allowing in this way navigation" from the "North sea" to the "South sea." This oceanic channel would increase commerce and provide, thus, a "great service" to the crown, Perú, and Tierra Firme.[22] By October the king obtained an answer from Pascual de Andagoya, governor of Tierra Firme: "I do not believe that there is a prince in the world that, with all his power, would be able to do it."[23] Although the idea of a canal through Tierra Firme could not be implemented at the time, Spaniards studied nature to reshape the natural world to fit their own designs.

The crown, royal officials, and ordinary immigrants shared a similar concern to transform the ecology of the new lands to make them economically and ecologically inhabitable for Europeans. From these activities eventually emerged a practical and experimental culture. The introduction of commodities to the New World illustrates this point. The cases studied here share some elements: private initiative, royal support, transference of knowledge (from experts), and finally an emphasis on empirical evidence and experimentation. The cultural contacts that developed in conjunction with the circulation of knowledge between experts from different geographical areas was a constant feature of the American enterprise during the sixteenth century.

From 1518 onward, the crown had promoted the cultivation of mulberry trees (for the production of silk), clove, ginger, pepper, and such dyes as pastel and madder in the New World. In 1535, the crown—after previously unsuccessful attempts—granted to two Germans a license to cultivate pastel and saffron in New Spain, "Micer Enrique" (or "Enrique Ynger") and "Alberto Cuon." Little is known about Alberto Cuon, but Micer Enrique was Heinrich Ehinger, brother of Ambrose and Ulrich and partner of the German bankers, the Welsers.[24] In the contract with the crown, the Germans obliged themselves to take to New Spain, at their own expense, "masters, laborers, seeds, and tools." In return, the crown granted them the monopoly in the cultivation and trading of pastel and saffron, "all the lands and people necessary" for the cultivation and manufacture of these products, and economic and financial privileges, such as tax exemption and a license to take 200 slaves to the New World.[25]

The saffron project failed because *tuzas* or *totzans*, a rodent, ate the roots of the saffron.[26] The pastel project fared better, at least in the beginning. By 1537, the Germans had brought five "masters of pastel-making" from Toulouse, France to Seville. Royal officials at the House of Trade refused the masters a license to pass, following secret instructions from Charles V to obstruct foreigners traveling to the New World.[27] The bankers protested, and the crown had to order the officials to let them pass to the New World "despite being Frenchmen."[28] Production began the following year. In 1538, Alberto Cuon brought a sample of pastel from New Spain to Segovia, a famous textile center, "to test it" there. The queen ordered the magistrate of Segovia "to find out the most skillful and honest people [among dyers and other officials] of the city to test [the pastel], under oath, before your presence, and to reveal their opinions and findings, and to provide this declaration to Cuon."[29] In New Spain, some pastel was sold for testing in 1539.

The introduction of pastel into the New World required the transference of expertise from France to New Spain and the testing of the product in different areas (Segovia and Mexico).[30] Similarly, the production of a medicine, the Santo Domingo balsam, in the 1520s and 1530s had required the transfer of knowledge from the Taino Catalina de Ayanbex and her family to the Spaniards and the testing of the product in different areas (Santo Domingo and Spain).[31] In the exploitation of natural species in the New World, the crown supported private initiatives and articulated state policies from those initiatives (i.e., the search for new commodities). In the alliance of private and state commercial interests, natural entities became the object of empirical observations, tests, and reports—practices that took place outside the traditional venues of guilds, universities, and humanist circles.

These empirical, testing, and reporting practices became part of the American enterprise. The pastel enterprise soon failed because it could not compete with the quality of the pastel from Toulouse. As a result, the crown became interested in the search for native dyes in the New World. In the late 1550s, Marco de Ayala, a resident of the town of Valladolid, Yucatán, presented a report on Campeche wood, a dark blue dye, from Yucatán to viceroy don Luis de Velasco. Ayala tested this dye before the viceroy.

The viceroyal court became the place for performing "tests" in the New World. Or, in the case of technological innovations, it was the place for legitimizing those experiences.[32] Don Luis de Velasco granted Ayala a license to exploit Campeche wood in Yucatán. In 1563, the king renewed the original license for ten years, despite Ayala's failure to exploit the product in the previous years.[33] The crown was already collecting information on another blue dye for, simultaneously with Ayala's report, the entrepreneur, Pedro de Ledesma, had sent a report on indigo (a blue dye) to the king. Ledesma explained that the method used by the indigenous people to process indigo was costly. He requested royal support to find a cheaper way to exploit New Spain's indigo.[34]

By 1564, Ayala had begun to sell his dye.[35] In 1565 the crown dispatched a royal decree to the governor of Yucatán requesting samples of Campeche wood, indigo, and cochineal (an organic red dye) "because We want to know the qualities of this wood and seeds and understand their effects and uses in these kingdoms (…) and to test [*el ensaye*] them as it is appropriate, We order you to send a reasonable amount of campeche wood, indigo, and cochineal to Our officials of the *Casa de la Contratación* in the first ships leaving those provinces."[36]

By 1576, the cultivation and manufacture of Campeche wood was already an important activity in Yucatán and Campeche. Meanwhile, Pedro de Ledesma had established a partnership with the Marqués del Valle to exploit his indigo. The partnership, however, was dissolved before 1572, and indigo could be grown by anyone from that time on. By 1577, there were already forty-eight indigo mills in Yucatán.[37]

The crown saw in the production of these dyes a mechanism to stabilize the prices of the textile industry in Spain and to diminish Spain's dependence on other countries. Thus, the crown requested viceroy don Martin Enríquez to send a report about the habitat of Campeche wood and indigo, the methods for using them, plans for maintaining stable production, and the cost of production. In this way, the study of nature and the welfare of the kingdoms became the double face of economic enterprises in this period.[38]

Similarly, the developments of mining technologies, for instance, were subject to similar empirical and reporting practices. In the 1540s, the German miner John Tetzel tested copper samples from mines in Cuba, but he was unable to find the method for smelting it. He took samples of it to experts in Spain and the Holy Roman Empire and found the method. He wrote a report to the Spanish crown and obtained a contract for the exploitation of the copper mines in Cuba.[39]

Equally, the crown established empirical methods for the making of charts at the House of Trade. Central to the project of making charts and training pilots was the gathering and organization of empirical information about the New World. By 1508, the crown had already established, informally, a practice of gathering empirical information that would eventually require some sort of organization. For the making of charts, the crown ordered House officials to organize a meeting with the "most expert pilots present at the time [of the meeting], and in the presence of Amerigo Vespucci, our chief pilot, to make a portolan of all the lands and islands of our Indies." This portolan would be called the *padrón general*, royal portolan, and from it all particular charts should be copied.[40] Pilots were ordered to bring information "about new lands or islands or ports or anything worthy of being noted in the royal portolan" and report it to the chief pilot (see Figure 7.1).[41] The reporting activity was similar to the one discussed above. The collaborative dimension in organizing and collecting information at the House became one of the characteristics of this institution. The crown would eventually design at least two strategies to bring information from several sources: meetings of experts, as the one just mentioned, and questionnaires. The institutionalization of information-gathering practices began with the chief pilot.

These cases, and many others, illustrate the approach to nature that emerged from the commercial and imperial interests in the New World.

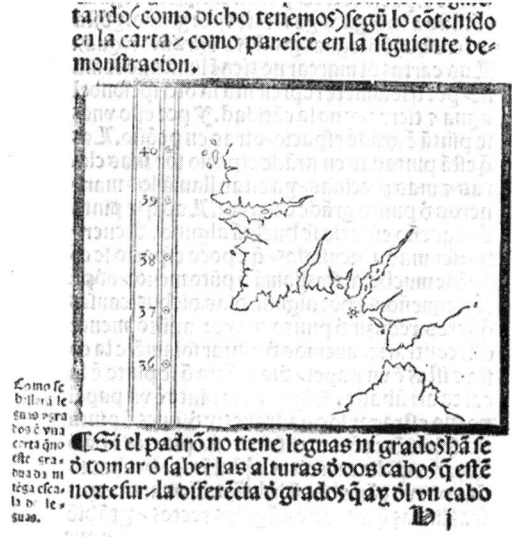

Figure 7.1 The result of information-gathering practices at the House of Trade was the elaboration of the Padrón General, from which pilots made their charts. Chart in Martín Cortés, *Breve Compendio de la Sphera* (Toledo, 1551). Reproduced courtesy of the John Carter Brown Library at Brown University.

Reports and tests became standard practices at the court and viceroyal courts and at the governing bodies of the American empire—the House of Trade and Council of the Indies—to determine the usefulness, value, or efficiency of natural products (i.e., dyes, medicines), technologies (i.e., mining methods), and instruments (i.e., mills and fishing rakes).[42] These empirical practices were an efficient method to study and deal with the natural world—as artisans and farmers have known for centuries. The breaking of the barriers between artisans and other social groups during the commercial expansion of the sixteenth and seventeenth centuries made possible the institutionalization of empirical practices as knowledge-producing tools.

In Spain, as would be the case later in England and Holland, nature's commodities were tested at the courts, hospitals, and gardens; nature's curiosities were collected, studied, and described. By the early seventeenth century, there was already an international network of scholars studying American nature through Spanish books and institutions. For instance, Carolus Clusius not only traveled to Spain and visited gardens filled with American plants and curiosities but established communications with natural historians in Spain.[43]

The long-distance control of the New World was not only based on the mapping and control of the American lands but on the mapping of the ocean. The mechanisms for collecting and disseminating knowledge about the new lands and the ocean were institutionalized at the House of Trade over a period of fifty years. The establishment of this chamber of knowledge at the House facilitated contacts and negotiations between pilots and experts and scholars (cosmographers). These agents were engaged not only in collecting and organizing information but in their own transformations as imperial agents. The political and economic needs of the Spanish empire to extract resources from the New World created a context in which the production of knowledge served to transform human and material resources into agents and tools of imperial domination. In the next section, I discuss the establishment of a chamber of knowledge at the House of Trade. With all the reports and information arriving from the New World discussed above, the crown decided to establish the office of the chief pilot for the purpose of "bringing together practice and theory" (Figure 7.2).[44]

The establishment of this office had two important consequences for our purposes. First, the relationship between practice and theory was far from clear in the context of the Atlantic World. Important groups (pilots) engaged in the American enterprise supported empirical approaches over theoretical ones, which would cause tensions between them and the cosmographers of the House. Second, the decision to emphasize

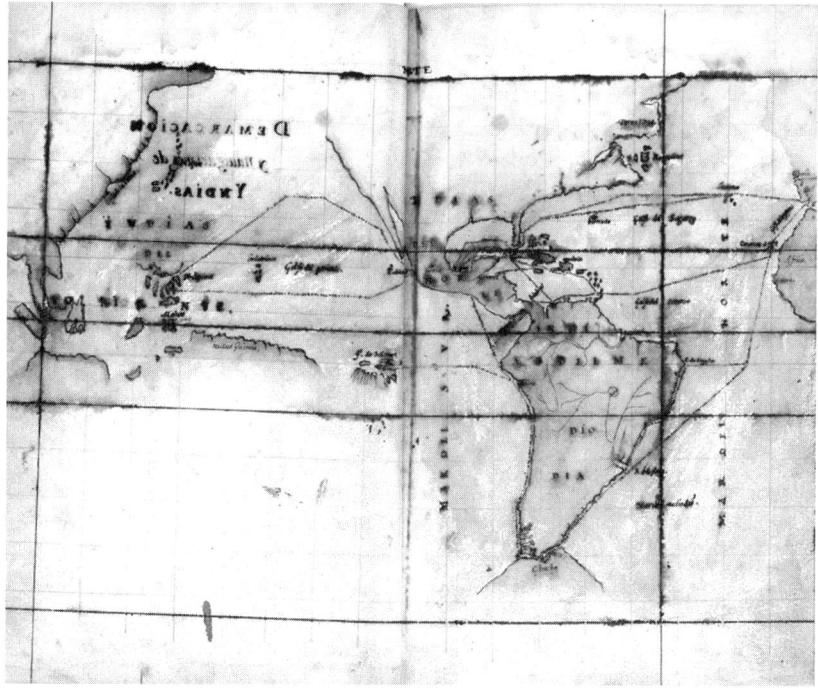

Figure 7.2 This map was the result of information gathered at the Council of the Indies. It is an "empty" map with only some geographical characteristics because that was the kind of information gathered at the Spanish centers of information. Map of the New World in Juan López de Velasco, *Descripción y división de las Yndias* manuscript, one map (manuscript, c. 1575). Reproduced courtesy of the John Carter Brown Library at Brown University.

theoretical approaches over empirical ones would cause changes in the structure of the House (for training pilots in cosmographical methods) and foster continuous tensions between pilots and cosmographers. These tensions forced both groups to continuously revise the relationship between practice and theory.

A Chamber of Knowledge

In 1598, the Englishman Richard Hakluyt described the navigational activities of the House of Trade, the *Casa de la Contratación*, in Seville for his English readers:

> [The] late Emperour Charles the fift, considering the rawnesse of his Sea-men, and the manifolde shipwracks which they susteyned in

passing and repassing betweene Spaine and the West Indies, with an high reach and great foresight, established not onely a Pilote Major, for the examination of such as sought to take charge of ships in that voyage, but also founded a notable Lecture of the Art of Navigation, which is read to this day in the Contractation house at Sivil. The readers of which Lecture have not only carefully taught and instructed the Spanish Mariners by word of mouth, but also have published sundry exact and worthy treatises concerning Marine causes, for the direction and incouragement of posteritie. The learned works of three of which readers, namely of Alonso de Chavez, of Hieronymo de Chavez, and of Roderigo Zamorano came long ago very happily to my hands, together with the straight and severe examining of all such Masters as desire to take charge for the West Indies.[45]

Although the office of the chief pilot was established by Ferdinand the Catholic (Ferdinand V of Castile, r. 1479–1516) in 1508 and not by Charles V, Hakluyt's description captures well the teaching and training activities of the "Contractation house at Sivil." However, the office of the chief pilot was established not only for the training of pilots but for the making of charts and, eventually, navigational instruments.

These activities of the House evolved into a veritable *chamber of knowledge*. By chamber of knowledge I mean the offices and practices developed and institutionalized within the House for collecting and disseminating information about the New World, for training pilots in the new navigational techniques, and for hiring cosmographers and pilots for research and teaching activities.[46] The historian Clarence Haring calls it a "Hydrographic Bureau and School of Navigation, the earliest and most important in the history of modern Europe."[47] This chamber of knowledge predated the gathering activities of such scientific academies as the Royal Society of London for Improving Natural Knowledge (1660).

The House's scientific activities belonged to a long medieval cartographical and astronomical tradition, which had already proved significant in the exploration and expansion of Portuguese trading routes in and around Africa and Asia. What was original in the case of the House was the way knowledge and information were collected, reproduced, and disseminated among networks of pilots, artisans, cosmographers, and royal officials.

The navigational activities of the House emerged around groups of artisans, such as pilots, who contributed their skills and knowledge and who negotiated with university-trained experts to produce artifacts, such as navigational instruments and charts and information about the New World.[48] The royal support for the House's scientific activities promoted the status of pilots and cosmographers and their knowledge: their activities were central for the expansion and control of the new empire and for the development of trade and commerce.[49]

Commerce and empire depended on the knowledge of pilots. The pilots' lack of knowledge in the use of navigational instruments and in interpreting cosmographical charts caused "many risks and harm to the crown and royal treasury and to the merchants working in the New World."[50] Navigational knowledge and the training of experts in this knowledge became key factors in the establishment of long-distance empires. The crown established the office of the chief pilot "to bring together practice and theory" so pilots could learn how to navigate to the New World.[51] And with these trained pilots, the empire would expand and commerce would prosper.

This chamber of knowledge emerged slowly from the interaction between crown officials, pilots, and cosmographers in the exploration of the New World.[52] In this world of exploration, knowledge became a central element in imperial politics. The gradual creation of a chamber of knowledge within the House of Trade mirrored the establishment of the Spanish kingdoms in America. By 1550, the Spanish had already established towns and cities from Mexico to Chile and had already transplanted Old World products to the New World. This process relied on force and violence and on luck and determination. However, it also relied on the production of knowledge. By 1550, the chamber of knowledge consisted of a group of professionals, a set of institutionalized practices for the gathering, dissemination, and classification of knowledge, the training and certification of laymen in those practices, and their eventual hiring by the House of Trade.

Joining Theory and Practice

After the short reign of Philip I (1506), Ferdinand returned to Spain and reorganized the administration of the Indies. In addition to appointing the bishop Juan Rodríguez Fonseca and the royal secretary Lope de Conchillos to oversee the general administration of the Indies,[53] Ferdinand appointed Amerigo Vespucci (1454–1512) as the first chief pilot of the House on March 22, 1508.[54]

The office of the chief pilot was established after Ferdinand called to court the pilots Juan Díaz de Solís (1470–1516), Vicente Yáñez Pinzón, Juan de la Cosa (d. 1510), and Amerigo Vespucci to organize new discoveries in the Indies for these "had been neglected during his absence" (1506), during the reign of Philip I. In this meeting, Amerigo Vespucci "was chosen to remain at Seville to draw sea-charts, with the title of *piloto mayor* [chief pilot]."[55] Thus America is named after the first professional who was officially engaged in charting routes of access to the New World.[56]

The chief pilot's activities emerged as a response to the accidents and delays in trans-Atlantic voyages. Sometimes, ships had to wait in harbor for lack of good pilots, or bad pilots steered ships into wrong ports, in one instance landing in France.[57] The situation did not improve after the establishment of the chief pilot. In 1512, the king stated the situation thus:

> It has come to our knowledge and it is known from experience that pilots are neither as expert nor as trained as they should be to steer and govern the ships under their command in their voyages to the Indies, islands and *Tierra Firme* of the Ocean Sea. The pilots' lack of knowledge to steer and govern [their ships], to use the quadrant, astrolabe, and to take the altitude has caused and still causes many mistakes and faults in their navigation. This causes a great disservice to us and great harm to the merchants of the Indies.[58]

Thus imperial administration, concerned with increasing the commerce with the New World and finding new sources of income for the royal treasury, decided to appoint an expert on navigational techniques to make standard navigational charts to cross the Atlantic.

The problem was precisely how to define standard navigational charts. The House of Trade's chamber of knowledge was a response to the need to organize information arriving from the New World. The chief pilot was supposed to bring together practice (empirical reports) and theory (cosmography), but the pilots were more interested in practice than in theory. Within the House two groups emerged, one emphasizing a practical, empirically based approach to making charts and the other emphasizing a theoretical approach to their making.

The establishment of a standard set of practices for making charts at the House was affected by a royal administrative decision regarding the dispute about chart making between Pedro de Medina and Diego de Gutiérrez. In 1538, Pedro de Medina received a royal license to make sea charts and navigational instruments. Two years later, Diego Gutiérrez (cosmographer of the instruments) and Sebastian Cabot (royal pilot) denounced Medina's charts as false. Medina took the case before the Council of the Indies.[59]

Originally, Cabot and Gutiérrez dominated the market of charts and instruments in Seville; Gutiérrez made the instruments and charts, and Cabot approved them. Diego Gutiérrez (1485/1488–1554) had been in Seville ever since the time of the Magellan expedition; perhaps he began working with Nuño García de Torreño, who was then the instrument maker. Gutiérrez received a royal license to make charts and instruments in 1534. His son Sancho received a license in 1539, and it seems that two other sons, Diego and Luis, were also involved in the manufacture of instruments.

Over the years, the Gutiérrez family established close connections with the chief pilot, Sebastian Cabot, insuring a virtual control of the instrument-making business in Seville.[60]

When Medina arrived, he began selling his own instruments and charts; he made charts different from those made by Gutiérrez and approved by the chief pilot, Cabot. At issue was the authenticity of Gutiérrez and Medina's charts. Gutiérrez, as it turned out, was making the charts according to the traditional ways of navigation using two systems of gradations (latitudes) in the same chart to correct for the magnetic alteration of the compass in the Atlantic. Medina, instead, was following the portolan chart, which used the same system of grades and distances for all regions.

In 1545, following the statutes of 1536, the Council decided in favor of Medina and required Gutiérrez to make his charts according to the portolan chart. After learning of this decision, the House wrote to the Council and explained that if the intention were to change the methods of navigation, the pilots would have to learn how to navigate with the new charts (that is, those of Medina). The support for Medina's charts implied a more instrument-based and mathematical approach to the natural world. It consequently implied a change in methods of navigation and retraining of the pilots. The efforts to instruct pilots would now need more drastic measures. Joining theory and practice was much more than collecting reports from the New World: it was about restructuring epistemological practices and identities in the Atlantic World. The empirical activities of the House promoted the establishment of a chair of cosmography. This means that the empirical activities of the House, as they were integrated into theoretical frameworks, required the retraining of pilots into those theoretical frameworks.

Instructing Pilots

In 1552, the crown established the chair of cosmography for the instruction of pilots. Jerónimo de Chaves was the first officer appointed to this position. Originally the chief pilot had to teach the theory of navigation to those sailors who wished to learn it. As it was a voluntary learning activity, the number of pilots with theoretical knowledge and expertise in the use of instruments did not increase as rapidly as the voyages to the New World increased in the early sixteenth century. This situation would generate some tensions between pilots and royal officials throughout the sixteenth century. The administrative solution to these tensions would shape, in turn, the teaching and practice of navigation during this period.

Cosmographers and mathematicians had already been offering their services to the crown for instructing pilots in the art of navigation. In 1525, for instance, fray Juan Caro—a Spanish Dominican, working for the Portuguese crown in Cochin, India—wrote to his relatives in Seville and to Charles V offering his services and requesting a paying position for teaching Spanish pilots the art of navigation in Seville. Caro, who had traveled to India "to learn...[the] secrets [of navigation], so that with this knowledge he could receive honors" in Spain, "wanted very much to live [again] in...[Castile] because...[he] was a native from there."[61]

Caro represents the entrepreneur who sought to advance his position by learning "secrets" to obtain a job in the institutions created by the crown for imperial practices. Caro, however, was ordered to his monastery in Portugal. There the Portuguese king ordered his arrest—his knowledge of the Moluccas made him dangerous for Portugal's interests. Caro was exiled to Sofala in Africa where he died. It should be noted that Caro's proposal for teaching pilots came before the crown had established a separate lecture of cosmography from the teaching duties of the chief pilot. At this time, the teaching of pilots still belonged to the duties of the chief pilot.

In 1528, during the absence of the chief pilot Sebastian Cabot, Alonso de Chaves taught the theory of navigation and the use of the astrolabe, quadrant, and navigational charts at Hernando Columbus's house. These lectures were the result of Columbus's and Chaves's initiative.

Columbus's participation in this initiative was important because he was well known and respected at the court. Although he was not an official member of the House, he had already assisted the crown in various cosmographical activities. In 1517, for example, he had begun a "cosmography and description of all Spain"[62] by collecting, through messengers, "all the peculiarities and memorable things that there are [in Spain]."[63] In 1523, for reasons that are still a mystery, Charles V eliminated the project.[64] Still the crown requested Columbus's services again to join the Junta of Badajoz (1524) to assist in determining the exact line of demarcation between the Portuguese and Spanish Empires.[65] Later, in 1526, the crown ordered Columbus to make a master portolan chart, a mappamundi, and a globe representing the new lands.[66] In 1528, when Chaves was already instructing pilots at Columbus's house, Columbus recommended that the Council hire Chaves for teaching pilots at the House. This recommendation constituted the first attempt to separate the instruction of pilots from the office of the chief pilot.

Although the Council did not follow Columbus's recommendation, it thanked Chaves and asked him to continue the instruction of pilots.[67] Officially the chief pilot continued with the instruction of pilots. As mentioned above, the creation of a separate office for the teaching of pilots was an unintended consequence of the royal decision regarding the dispute between Pedro de Medina and Diego de Gutiérrez and the realization that pilots needed special training. The result of this trial was an administrative decision in favor of instrument-based methods of navigation.

This litigious debate between cosmographers and instrument makers demonstrated to the royal officials the need to establish formal mechanisms to train pilots in the new methods of navigation. However, establishing the lectureship in cosmography came only after Sebastian Cabot ceased to be the chief pilot in 1549 (Diego Sánchez Colchero was appointed temporarily as chief pilot from 1550 to 1552). In 1552, the crown appointed Alonso de Chaves as chief pilot, issued new statutes for the House, and established the chair of cosmography with the appointment of Alonso's son, Jerónimo de Chaves.[68]

Jerónimo de Chaves was born in Seville in 1523 and attended the university there.[69] He obtained a bachelor's degree in mathematics and perhaps studied medicine. In 1545, he published a Spanish translation of Sacrobosco's *Tractatus de Sphaera Mundi*.[70] He intended to make Sacrobosco's popular cosmography available among practical astronomers ignorant of Latin. In 1548, he published a chronology and almanac.[71] This work proved very popular and went through some eighteen editions by 1588.[72] Chaves was also a collector of natural curiosities, books, maps, and instruments. Jerónimo de Chaves's appointment in 1552 to the newly created chair of cosmography for instructing pilots in "the art of navigation and cosmography" acknowledged his growing status within the community of cosmographers at Seville. Pilots could no longer take their examinations "without listening for a year to that science (or most of it) to gain competence."[73]

Jerónimo de Chaves received instructions from the king specifying the duties of his professorship. These instructions came together with the statutes of 1552. According to the new rules, it was no longer voluntary for pilots to study cosmography and the new navigational methods. The program of instruction consisted of a year of lectures covering Sacrobosco's treatise of the sphere and the magnetic variations of the compass, using the Spanish translation of the treatise by Jerónimo de Chaves; the use of regiments; the use and construction of the astrolabe, compass, quadrant, and cross-staff; and the use of diurnal and nocturnal clocks. In addition, pilots had to memorize the lunar calendar to know the time of tides. Classes took place from Monday to Friday at 3 to 4 o'clock from October to March and at 5 to 6 from April to September.[74] A year proved to be a long period for the pilots to be away from their jobs at sea, and Jerónimo de Chaves requested that the Council of Indies reduce the lectureship of cosmography to three months.

With the addition of the chair of cosmography and the statutes of 1552, the structure of the House's chamber of knowledge was completed. The statutes of 1552 codified the following practices: examination of pilots had to take place only at the House before royal cosmographers and regular pilots, and certification was granted by a majority of votes from those present at the examination (statute number 128). Examinees needed to have six years of practice in navigation and knowledge in the theory of navigation (number 129). Furthermore, the chief pilot could neither instruct nor sell instruments and charts to prospective examinees so that no conflict of interest would arise (numbers 130 and 131). Finally, the chief pilot, cosmographers, and pilots were entrusted with the crucial task of emending the royal sea chart together (number 134).[75] In essence, the statutes of 1552 institutionalized methods for gathering, organizing, and disseminating information appropriate to the development of the imperial state.

The statutes established different offices for different purposes, rules for training and certifying pilots, procedures for hiring professionals, and methods for research. As the American enterprise grew, the activities of the House became more specialized.[76] First in stature and authority was the chief pilot, as a supervisor of pilots and instruments; next came the cosmographer of instruments, himself a product of the needs of the House to improve and design instruments for navigation. Finally came the cosmographer-lecturer who instructed pilots in the art of navigation and cosmography. The House's chamber of knowledge became the place to normalize, institutionalize, and distribute knowledge and practices according to the political interest of the crown and merchants.

Zauzo's letter to Charles I in 1518, Ribeiros's appointment for the making of instruments, the establishment of the chief pilot "to bring together practice and theory," the disputes within the house about the making of charts—all point to the emergence of an empirical culture in the Atlantic World: a culture characterized by reports on single entities, such as the cochineal dye, based on empirical and testing practices. It was also a culture characterized by collaborative activities: the making of charts was possible by the reports from pilots and the knowledge of the cosmographers. This culture questioned the traditional relationship between theory and practice, as seen in the debates within the House of Trade, and created the context for new institutions and agents for the control of the New World and its resources. The Scientific Revolution received its first quantitative impulse in the Atlantic World, for in the Atlantic World experience became the tool of knowledge production.

Acknowledgments

For their helpful comments and questions, I owe thanks, *mil gracias*, to the participants in the conference "Atlantic Knowledges" (Clark Library, UCLA, Los Angeles, 2005) and, in particular, to James Delbourgo and Nicholas Dew, the organizers of the conference and editors of this volume.

Notes

1. Letter from Zauzo to Charles V in Marcos Jiménez de la Espada, *Relaciones Geográficas de Indias: Perú* (Madrid: Biblioteca de Autores Españoles, 1965), vol. I, p. 11ff; Alfred Crosby, *The Columbian Exchange: Biological and Cultural Consequences of 1492* (Westport: Greenwood Press, 1972), p. 76.
2. Zauzo in Jiménez de la Espada, *Relaciones*, vol. I, p. 12.
3. See Paula De Vos, "The Science of Spices: Empiricism, Economic Botany, and Entrepreneurialism in the Spanish Empire." Paper presented at the Society for Spanish and Portuguese Historical Studies, Charleston, 2005.
4. See Thomas F. Glick, José M. López Piñero, Víctor Navarro Brotóns, and Eugenio Portela Marco, *Diccionario histórico de la ciencia moderna en España* (Barcelona: Ediciones Península, 1983), pp. 225ff.; see also Germán Latorre, "Diego Ribero: Cosmógrafo y Cartógrafo de la Casa de la Contratación de Sevilla," *Boletín del Centro de Estudios Americanistas* (1918), pp. 27–31.
5. Carta de Hernando Colón, Madrid, August 21, 1528. Seville, Archivo General de Indias (henceforth AGI), Indiferente, 421, L. 13, F. 295r/v. See also Cortesão, *Cartografia*, vol. 2, pp. 130ff.
6. For a similar argument in the Portuguese case, see Rejer Hooykaas, "Science in Manueline Style," in *Obras Completas de D. João de Castro*, ed. Armando Cortesão and Luís de Albuquerque, pp. 231–426 (Coimbra: Academia Internacional da Cultura Portuguesa, 1968), p. 323. The Portuguese, however, did not develop the same type of institutionalized empirical practices as the Spanish because the Portuguese operated, mainly, in the Old World (Africa and Asia).
7. The study of printed material, such as books for understanding knowledge-producing practices in Spain, yields a more orthodox view than the one I proposed in this chapter because books were written usually by learned scholars for learned scholars. I use mostly archival material, material written by royal official, entrepreneurs, soldiers, and merchants for other merchants, royal officials, and entrepreneurs. Based on archival material, I argue only that empirical practices—reports based on personal experiences and testing practices—emerged in the context of imperial and commercial interests in the New World for the purpose of exploiting, managing, and reorganizing resources in the New World. For the persistence of classical traditions in sixteenth-century Spain, see the excellent work of Ralph Bauer, *The Cultural Geography of Colonial American Literatures:*

Empire, Travel, Modernity (Cambridge: Cambridge University Press, 2003); Anthony Pagden, *The Fall of the Natural Man* (Cambridge: Cambridge University Press, 1990).

8 Peter Dear mentions that this type of report became increasingly common in seventeenth-century England. In sixteenth-century Spain, they became increasingly common at the House of Trade and Council of Indies as I show in my work. See Peter Dear, *Discipline and Experience: The Mathematical Way in the Scientific Revolution* (Chicago: University of Chicago Press, 1995), p. 25.

9 The English would face similar problems about the bias on observational report, for this case see Steven Shapin, "The House of Experiment in Seventeenth-Century England," *Isis* 79 (1988), 373–404.

10 Empirical practices were not unique to the Spanish empire, of course. Physicians in Italy during the fifteenth century, for instance, had been developing a more empirical approach to their own discipline, as Katherine Park has shown in her work. The difference between that case and the Spanish American context is the magnitude of the American enterprise. For instance, in a great attempt to collect information from the New World the Spanish crown sent printed questionnaires, including questions about natural history, to almost every official in the New World from the Caribbean Islands and New Spain to New Granada and Peru in 1577 and, again, in 1584. The collection of answers to the questionnaires, held at the Council of Indies, is known as the *Relaciones de Indias*, and they fill volumes: at least ten for the case of New Spain and four for the case of Perú. In addition to this questionnaire, the cosmographer López de Valasco elaborated instructions to determine the longitude of significant places in the New World based on the moon eclipses of 1577, 1578, 1581, 1582, and 1584. As part of this overall project for determining longitudes and altitudes, the crown sent yet another expedition to the New World: the expedition of the cosmographer Jaime Juan, in 1583, to map the altitudes of New Spain towns (he was sent before the eclipse of 1584). There were also the well-known Dr. Francisco Hernández's expeditions to New Spain, Peru, and the Philippines in 1571. (He stayed six years in New Spain and returned to Spain in 1577.) As I said, the difference between those cases of empirical approaches to nature in Europe and the Spanish American enterprise is, in part, the magnitude of the American enterprise. On the Italian physicians, see Katherine Park, "Natural Particulars: Medical Epistemology, Practice, and the Literature of Healing Springs," in *Natural Particulars: Nature and the Disciplines in Renaissance Europe*, ed. Anthony Grafton and Nancy Siraisi (Cambridge, Mass.: MIT Press, 1999), pp. 348ff. On the Relaciones, see Howard F. Cline, "The Relaciones Geográficas of the Spanish Indies, 1577–1648," in *HandBook of Middle American Indians*, ed. Howard F. Cline, pp. 183–232 (Austin: University of Texas Press, 1972).

11 E. Zilsel, "The Sociological Roots of Science," in *The Social Origins of Modern Science*, ed. Diederick Raven *et al.* (Dordrecht, Boston and London: Kluwer Academic Publishers, 2000), p. 17.

12 Deborah E. Harkness, "'Strange' Ideas and 'English' Knowledge: Natural Science Exchange in Elizabethan London," in *Merchants and Marvels: Commerce, Science, and Art in Early Modern Europe*, ed. Pamela H. Smith and Paula Findlen (New

York: Routledge, 2002), pp. 137–55; Pamela H. Smith, *The Business of Alchemy: Science and Culture in the Holy Empire* (Princeton: Princeton University Press, 1994); Eric H. Ash, *Power, Knowledge, and Expertise in Elizabethan England* (Baltimore: Johns Hopkins University Press, 2004); Pamela H. Smith, *The Body of the Artisan: Art and Experience in the Scientific Revolution* (Chicago: University of Chicago Press, 2004).

13 I explore the issues discussed in this article further in Antonio Barrera-Osorio, *Experiencing Nature: The Spanish American Empire and the Early Scientific Revolution* (Austin: University of Texas Press, 2006).
14 Dear, *Discipline and Experience*, p. 25.
15 H. Pérez de Oliva, *Historia de la Invención de las Yndias* (Bogotá: Instituto Caro y Cuervo, 1965), p. 53; Crosby, *Columbian Exchange*, p. 67.
16 William Cronon argues that English colonists saw the natural things in the New World as a collection of single entities (as commodities). I think that a consequence of this understanding of nature was the production of reports about single entities or events. See Cronon, *Changes in the Land: Indians, Colonists, and the Ecology of the New England* (New York: Hill and Wang, 1983), pp. 20ff.
17 Río Moreno, *Los Inicios*, passim; Crosby, *Columbian Exchange*, ch. 3.
18 Pedro Cieza de León, *La crónica del Perú* (Madrid: Historia 16, 1985), p. 74.
19 Crosby, *Columbian Exchange*, pp. 75ff.
20 Royal provision from King don Ferdinand to the residents of San Juan island. September 27, 1514. AGI, Indiferente General 419, L. 5, ff. 248v–9v.
21 Río Moreno, *Inicios*, pp. 49–69.
22 Real cédula al governador del Tierra Firme. February 20, 1534. Toledo in *Colección de documentos y manuscriptos compilados por Fernández de Navarrete*, 32 vols (Nendeln: Kraus-Thomson Organization Ltd, 1971), vol. 13, part 2, p. 1101. I cite this document from now on as *Colección Navarrete*.
23 Carta de Pascual de Andagoya. October 22, 1534. Tierra Firme in *Colección Navarrete*, vol. 13, part 2, p. 1106.
24 On the names of these two Germans see Jean-Pierre Berthe, "El cultivo del 'pastel' en Nueva España," *Historia Mexicana* 9 (1960), pp. 340–67, p. 346. He suggests that Alberto Cuon could be Albrecht Cohen or Kuhn.
25 Asiento otorgado a Enrique Ynguer and Alberto Cuon, March, 1527. "Belpuche." AGI, Contaduría 672, N. 5; also see Berthe's analysis in "El cultivo," pp. 343–7.
26 Asiento otorgado a Enrique Ynguer and Alberto Cuon, March, 1527. "Belpuche." AGI, Contaduría 672, N. 5. The Spanish name *tuza* comes from Nahutl *totzan*, see Francisco Santamaría, *Diccionario de Mejicanismos* (Mexico: Ed. Porrúa, 1978), p. 1097.
27 Charles had given authorization to Genoveses and Germans to travel to the New World, but secretly sent instruction to the *Casa* officials to obstruct the movement of foreigners to the New World. See Carande, *Carlos V*, vol. 1, p. 266.
28 Real cédula a los oficiales de la Casa de la Contratación, December 30, 1537. Valladolid. AGI, Indiferente 1962, L. 5, ff. 309v–10r. I found a license to pass to New Spain granted to three of them (their names are in Spanish): Juan Barta, Domingo de San Pablo, and Bartolo de Rigazo "native of Toulouse and masters of making pastel," see AGI, Pasajeros, L. 2, E. 3784.

29 Berthe, "El cultivo," p. 355, cuadro 4. Juan Ximenez and Gonzalo Gómez brought some pastel *"para ensayes."*
30 In the case of the exploitation of pastel in México, nineteen indigenous towns with their jurisdictions were involved in the cultivation and manufacturing of pastel. See the *relaciones de los pueblos del pastel* from April, 1545 in AGI, Contaduría 672, N. 5; see also Berthes's analysis of these documents in his article "El cultivo," pp. 349ff.
31 Antonio Barrera, "Local Herbs, Global Medicines: Commerce, Knowledge, and Commodities in Spanish America" in *Merchants and Marvels*, ed. Smith and Findlen; see also the work of Daniela Bleichmar on medicines: Daniela Bleichmar, "Books, Bodies, and Fields: Sixteenth-century Transatlantic Encounters with New World Materia Medica," in *Colonial Botany: Science, Commerce, and Politics in the Early Modern World*, ed. Londa L. Schiebinger and Claudia Swan (Philadelphia: University of Pennsylvania Press, 2004), pp. 83–99.
32 See Merced otorgada por don Luis de Velasco a Francisco Mirantes por cuatro años para el uso de un ingenio de moler metales. December 15, 1550. Mexico. Library of Congres, Manuscripts, Kraus collection, item 140, ff. 10v–11r. Merced otoragada por don Luis de Velasco a Castañon de Aguero por cuatro años para el beneficio de metales. February 29, 1551. Mexico. Library of Congress, Manuscripts, Kraus collection, item 140, ff. 33r–v. On this topic, see my *Experiencing Nature*, ch. 3.
33 Real cédula a Marco de Ayala. November 1, 1562. Madrid. AGI, México, 2999, L. 2, ff. 6r–7r. See also, the Relación de méritos y servicios de Marco de Ayala. September 15, 1561. Mérida de Yucatán. AGI, Patronato 64, R. 7.
34 Carta de Pedro de Ledesma al Rey. May 22, 1563. México. AGI, México, 168. By the end of the sixteenth century there was indigo coming not only from New Spain but from Honduras, the Caribbean and Tierra Firme, see Huguette and Pierre Chaunu, *Séville et l'Atlantique*, 8 vols (1504–1650) (Paris: Librairie Armand Colin, 1955), vol. 6/2, pp. 988–93. By the early seventeenth century, indigo was already exported to Netherlands, see N. W. Posthumus, *Inquiry into the History of Prices in Holland*, 2 vols (Leiden: E. J. Brill, 1946), v. 1, pp. 415–18.
35 Carta de Juan Rodriguez de Noriega sobre el tinte del palo de Yucatán. April 28, 1564. N/P. AGI, Indiferente 1093, R. 11, N. 239.
36 Real cédula al gobernador de la provincia de Yucatán. June 25, 1565. El Escorial. AGI, México, 2999, L. 2, ff. 34r–34v.
37 François Chevalier, *Land and Society in Colonial Mexico* (Berkeley: University of California Press, 1972), pp. 73–4.
38 Carta real al virrey de Nueva España, don Martín Enriques. December 3, 1576. Madrid. AGI, México 109; carta real al gobernador de Yucatán, December 3, 1576. Madrid. AGI, México, 2999, L. 3, ff. 29v–31v.
39 Asiento con el Príncipe, Felipe. January 11, 1546. Madrid. AGI, Santo Domingo 99, R. 6, N. 22. On mining, see Peter Bakewell, "Mining," in *Colonial Spanish America*, ed. Leslie Bethell (Cambridge: Cambridge University Press, 1987), pp. 203–49; González Loscertales, Vicente Montaud, and Inés Roldan de Montaud, "La minería del cobre en Cuba, su organización, problemas administrativos y repercusiones sociales (1828–1849)," *Revista de Indias* 40

(1980), pp. 255-99; Peter Bakewell, "Technological Change in Potosi: the Silver Boom of the 1570s," in *Jahrbuch für Geschichte von Staat, Wirtschaft und Gesellschaft lateinamerikas*, vol. 14 (1977), pp. 57-77.

40 Real Provisión a Américo Vespuccio concediéndole facultad para examinar pilots. August 8, 1508. Valladolid. AGI, Indiferente 1961, l. 1, ff. 65v-67.

41 Real Provisión a Américo Vespuccio concediéndole facultad para examinar pilots. August 8, 1508. Valladolid. AGI, Indiferente 1961, l. 1, ff. 65v-67.

42 On instruments, see Isabel Arenas Frutos, "Inventos sobre Tecnología Submarina para la América Colonial," in *Separata de la Asociación de Historiadores Latinoamericanistas Europeos* (1992), pp. 421-34; Nicolás García Tapia and José A. García Diego, eds, *Vida y técnica en el renacimiento: manuscrito que escribió, en el siglo XVI, Francisco Lobato, vecino de Medina del Campo* (Valladolid: Universidad de Valladolid Secretariado de Publicaciones, 1987); Nicolás García Tapia, *Patentes de invención españolas en el siglo de oro* (Madrid: Registro de la Propidad Industrial, 1990); Pablo Emilio Pérez Mallaína Bueno, "Los Inventos Llevados de España a las Indias en la Segunda Mitad del Siglo XVI," in *Separata de Cuadernos de Investigación Histórica N°7, Fundación Universitaria Española* (1983), pp. 35-54; Manuel Luengo Muñoz, "Inventos para acrecentar la obtención de perlas en América, durante el siglo XVI," in *Anuario de Estudios Americanos* 9 (1952), pp. 51-72.

43 On Clusius, see Luis Ramón Laca, "The Spanish and American Plants in Clusius's Correspondence," in *Polish Botanical Studies: Guidebook Series 20* (1998), pp. 135-60; José López Piñero and José Pardo Tomás, "The Contribution of Hernández to European Botany and Materia Médica," in *Searching for the Secrets of Nature: The Life and Works of Dr. Francisco Hernández*, ed. Simon Varey et al. (Stanford: Stanford University Press, 2000), pp. 127ff; Enrique Alvarez-López, "Las plantas de América en la botánica europea del siglo XVI," *Revista de Indias*, no. 20 (1945), pp. 121-288.

44 Provisión real a Americo Vespuccio. August 6, 1508. Valladolid. AGI, Indiferente 1961, L. 1, F. 65v-67.

45 Richard Hakluyt, *Voyages* (London, 1973), vol. 1, pp. 16ff.

46 Paula de Vos, "Spain in the 'New World' of Europe: The *Casa de la Contratación* and the Organization of Navigation and Cartography in the Sixteenth Century" (MA Thesis, University of California, Berkeley, 1993).

47 Clarence Henry Haring, *Trade and Navigation between Spain and the Indies: in the Time of the Hapsburgs* (Cambridge, Mass.: Harvard University Press 1918), p. 35. On the *Casa's* Chamber of Knowledge, see Haring's discussion in chapter 2 of his *Trade*; see also J. Pulido-Rubio, *El Piloto Mayor de la Casa de la Contratación de Sevilla: Pilotos Mayores, Catedráticos de Cosmografía y Cosmógrafos* (Sevilla: Escuela de Estudios Hispano-Americanos de Sevilla, 1950): this is still the only book about the chief pilots and cosmographers of the *Casa*.

48 There is a long historiography on the relations between artisans and the development of scientific activities in the sixteenth and seventeenth centuries: see Paolo Rossi, *Philosophy, Technology, and the Arts in the Early Modern Era* (New York: Harper and Row, 1970); J. A. Bennett, "The challenge of practical mathematics," in Paolo L. Rossi, Stephen Pumfrey, and Maurice Slawinski,

eds, *Science, Culture and Popular Belief in Renaissance Europe* (Manchester: Manchester University Press, 1991); J. A. Bennett, "The Mechanics' Philosophy and the Mechanical Philosophy," *History of Science* 24 (1986), pp. 1–28; Paula Findlen, *Possessing Nature: Museums, Collecting, and Scientific Culture in Early Modern Italy* (Berkeley: University of California Press, 1994); Smith, *Business of Alchemy*; and more recently, Pamela O. Long, *Openness, Secrecy, Authorship: Technical Arts and the Culture of Knowledge from Antiquity to the Renaissance* (Baltimore: Johns Hopkins University Press, 2001).

49 On the development of these centers of knowledge during the sixteenth century, see John Law, "On the Methods of Long-distance Control: Vessels, Navigation and the Portuguese Route to India," *Sociological Review Monograph* 32 (1986), pp. 234–63; Steven J. Harris, "Confession-building, Long-distance Networks, and the Organization of Jesuit Science," *Early Science and Medicine* 1 (1996), pp. 287–318; and also his article, "Long-distance Corporations, Big Sciences, and the Geography of Knowledge," *Configurations* 6 (1998), pp. 269–304.

50 Provisión real a Americo Vespuccio. August 6, 1508. Valladolid. AGI, Indiferente 1961, L. 1, F. 65v–67.

51 Provisión real a Americo Vespuccio. August 6, 1508. Valladolid. AGI, Indiferente 1961, L. 1, F. 65v–67.

52 On the notion of common people, see Peter Burke, *Popular Culture in Early Modern Europe* (New York: Harper and Row, 1978); Rossi *et al.* eds, *Science, Culture and Popular Belief*; and William A. Christian, *Local Religion in Sixteenth-Century Spain* (Princeton: Princeton University Press, 1989).

53 This was the first governmental body in charge of the general administration of the Indies. With Charles, this governmental body was replaced by a group of royal officials that would constitute, in 1524, the Council of Indies; see, Schäfer, *El Consejo*, vol. 1, pp. 24–55; Ernst Schäfer, "El Origen del Consejo de Indias," *Investigación y Progreso* VII (1933), pp. 141–5.

54 Amerigo Vespucci was appointed chief pilot on March 22, 1508, see Cédula real a los oficiales de la Casa de la Contratación. March 22, 1508. Burgos. AGI, Contratación, 5784, l. 1, f. 4v.

55 Herrera, *Historia General*, década I, libro 7, p. 177; this translation is by John Stevens, see Antonio de Herrera, *The General History of the Vast Continent and Islands of America, Commonly call'd, The West-Indies, From the First Discovery thereof: With the Best Accounts the People could give of their Antiquities. Collected from the Original Relations sent to the Kings of Spain. By Antonio de Herrera, Historiographer of His Catholic Majesty. Translated into English by Capt. John Stevens* (London, 1725), vol. 1, pp. 321–2.

56 On the name "America," see Dietrich Briesemeister, "La imagen de América en la Alemania que conoció Hernando Colón," *Hernando Colón y su época* (Sevilla, 1991), pp. 27–46, p. 31.

57 Letter of King Ferdinand to Nicolás de Ovando July 13, 1508 in José Toribio Medina, *El Veneciano Sebastián Caboto al servicio de España y especialmente de su proyectado viaje a las Molucas por el estrecho de Magallanes y al reconocimiento de la costa del continente hasta la gobernación de Pedrarias Dávila* (Santiago de Chile, 1908), vol. I, p. 26, note 2.

58 Provisión real a Juan Díaz de Solís y Juan Vespucci. July 24, 1512. Burgos. AGI. Contratación, 5784, l. 1, ff. 20r–1r: "Por quanto a nuestra noticia es venydo y por esperiencia se ha visto que por no ser los pilotos tan espertos ni tan instrutos como seria menester para regir y governar los navios que llevan a cargo en los viajes que hazen para las yndias y slas y tierra firme del mar Oceano y por defecto dellos por no saber de que maña se han de regir y governar ny por donde han de tomar el quadrante y el astrolabio y el altura ny saber la quarta della les han acaecido y en cada dia acaecen muchos yerros y defectos en las navegaciones que hazen de lo qual a nos se … mucho deservicio y a los tratantes en las dichas yndias mucho daño y de cada dia se espera recibyr mayor [daño] sy no lo mandamos probeer y remediar."

59 I am following Ursula Lamb, "Science by Litigation: A Cosmographic Feud," *Terrae Incognitae* 1 (1969), pp. 40–57; see also her article "The Spanish Cosmographic Juntas of the Sixteenth Century," *Terrae Incognitae* 6 (1974), pp. 51–64. The documents to follow the case are in AGI, Patronato 259, R. 16; Justicia 1146, N. 3, R. 2; Indiferente 1207, N. 61; Indiferente 1963, l. 7, ff. 19v–20; Indiferente 1093, R. 4, N. 68. Cabot used his experience in Spain later in England, see Alison Sandman and Eric H. Ash, "Trading Expertise: Sebastian Cabot between Spain and England," in *Renaissance Quarterly* 53 (2004), pp. 813–46.

60 Lamb, "Science by Litigation;" Karrow, *Mapmakers*, p. 285; Medina, *Caboto*, vol. 1, p. 355; see also Alison Sandman, "An Apologia for the Pilots' Charts: Politics, Projections and Pilots' Reports in Early Modern Spain," *Imago Mundi* 56 (2004), pp. 7–16, and "Mirroring the World: Sea Charts, Navigations, and Territorial Claims in Sixteenth-Century Spain," in *Merchants and Marvels*, ed. Smith and Findlen, pp. 83–108.

61 Carta de fray Juan Caro a su cuñado, Dr. Porras, en Sevilla. December 19, 1525. Conchin, India in *Colección Navarrete*, vol. 16, pp. 278–80; carta de fray Juan Caro al Emperador, December 29, 1526. Cochin, India in *Colección Navarrete*, vol. 16, pp. 280–4. Caro's significance is related to the Moluccas dispute rather than to the evolution of the teaching practices at the Casa; however, his proposal is significant because it shows, once again, the degree of participation of "commoners" in proposing institutional novelties to the crown. I was unable to find the date of Caro's death. The Portuguese crown was so concerned about the knowledge of this Spanish friar that he was finally exiled to Sofala, Africa to avoid the requirements from the Spanish side for his liberation. See Cortesão, *Cartografia*, vol. 2, pp. 20ff and 416ff.

62 Cédula real a Hernando Colón. June 13, 1523. Valladolid in Rodolfo del Castillo-Quartiellerz, *Documento inédito del siglo XVI referente a D. Fernando Colón* (Madrid: Real Academia de la Historia, 1898), pp. 8ff.

63 Juan Pérez, "Memoria de las obras y libros de Hernando Colón," in Tomás Marín-Martínez, *"Memoria de las obras y libros de Hernando Colón" del bachiller Juan Pérez* (Madrid: Consejo Superior de Investigaciones Científicas de Sevilla, 1970), p. 47.

64 Cédula real a Hernando Colón. June 13, 1523. Valladolid in Castillo-Quartiellerz, *Documento inédito*, pp. 8ff.

65 Antonio Rumeu de Armas, *Hernando Colón, historiador del descubrimiento de América* (Madrid: Instituto de Cultura Hispanica, 1973), pp. 76ff. See the reports of the Junta in *Colección Navarrete*, vol. 16, pp. 745ff; Herrera, *Historia General*, década III, libro 6, chap. VI, pp. 183–4.

66 Real cédula a Hernando Colón para que termine la carta de navegación que se le ordenó en cédula real de octubre 6 de 1526 de Granada. May 5, 1535. Madrid. AGI, Indiferente 1961, l. 3, ff. 276r–276v.

67 Real provisión a Alonso de Chavez, August 21, 1528. Madrid. AGI, Indiferente 421, l. 13, f. 295v.

68 The statutes of 1552 are in *Colección Navarrete*, vol. 3, pp. 105–263.

69 On Jerónimo de Chaves's life, I am following Karrow, *Mapmakers*, pp. 116–17.

70 Jerónimo de Chaves, *Tratado de la Sphera que compuso el doctor Ioannes de Sacrobusto con muchas adiciones. Agora nuevamente traduzido de Latin en lengua Castellana por el bachiller Hierónymo de Chaves: el qual añidio muchas figuras, tablas, y claras demostraciones: juntamente con unos breves scholios, necessarios a mayor illucidation, ornato y perfection del dicho tratado* (Seville, 1545).

71 Jerónimo de Chaves, *Chronographia o Repertorio de los tiepos [sic], el mas copioso y preciso que hasta agora ha salido a luz; en el qual se tocan y declaran materias muy provechosas de philosophia, astrologia, cosmographia y medicina...* (Seville, 1548).

72 Karrow, *Mapmakers*, pp. 116–17; Antonio Palau-y-Dulcet, *Manual del Librero Hispano-Americano* (Barcelona: Librería Anticuaria de A. Palau, 1948–77), No. 7450.

73 Nombramiento de Jerónimo de Chaves. December 4, 1552. Monzón. AGI, Contratación 5784, l. 1, f. 95–95v.

74 Nombramiento de Jerónimo de Chaves. December 4, 1552. Monzón. AGI, Contratación, 5784, l. 1, f. 95–95v.

75 *Colección Navarrete*, vol. 3, pp. 197ff.

76 In general, the lecturer of the chair of cosmography also made instruments. See, Pulido, *El Piloto*, p. 25.

CHAPTER 8

Fruitless Botany

Joseph de Jussieu's South American Odyssey

NEIL SAFIER

It is only by traveling, by passing through forests and fields, that one can conduct research that is useful to botany.

—Joseph de Jussieu

How were the practical labors that produced natural knowledge in eighteenth-century transatlantic botanical exploration to be organized and carried out? After a decade of wandering through the highland scrub and tropical forests of equatorial South America, Joseph de Jussieu freely offered the following reflection: "Live first, then philosophize."[1] Roughly a century earlier, Francis Bacon had set forth the proposition that *in situ* description was to be followed by the *metropolitan* ordering and interpretation of natural data. Jussieu, however, appears at first glance to be proclaiming precisely the opposite: that to wear the garland of natural philosopher, one must have "lived" the experience of nature in the field. However, was this in fact a reversal of the natural historical regime as practiced in the metropolitan center, a philosophical challenge to naturalists who compiled and systematized data garnered from the four corners of the globe in their European salons and laboratories? By citing a

Latin aphorism in a letter sent to his brother Antoine at Paris's prestigious Jardin du Roi, Joseph de Jussieu seems to have been making a claim both about the privileged perspective of the botanical agent—capable of observing and understanding the particulars of the natural world precisely because of his location in the field—and the pragmatic challenges of surviving the difficult conditions in a laboratory far from books and microscopes. Ultimately, he appears to have been passing judgment on the qualifications of the metropolitan natural philosopher as well. The European philosopher's ability to describe and systematize knowledge was subordinated, in Jussieu's eyes, to the eyewitness observations and lived experience of the traveler, who could collect healthy specimens on site and profit from the myriad advantages of encountering the world's floral cornucopia in its native habitat.

When Joseph de Jussieu composed this letter in April of 1748, he had just completed an excursion to the mythic Land of Cinnamon, or "Canelos," on the eastern slopes of the Andean cordillera. At that very moment, Jussieu stood at the crossroads between two separate but interconnected theaters of nature: to the East, the mountains began their sharp descent toward the vast expanse of the Amazon River basin; to the West, the volcanic peak of Chimborazo stood majestically against the spiny Andean cordillera curving gently from north to south. It was at the intersection of these two worlds, in the hill town of Elén, that Jussieu reflected on his itinerant condition and informed his brother of a philosophical shift toward the natural spaces he had traversed and observed during the previous thirteen years. Frustrated by his lack of funds and discouraged by the obstacles preventing his return to France, Joseph cited a classical proverb to signal a *pragmatic* move away from the act of "philosophizing" toward more practical activities that would provide remuneration and sustenance for a scientific career whose royal sponsorship had run dry.[2] "I feel an obligation to abandon the curious for the useful and necessary," Jussieu went on to write in this letter, "[and] have been content to maintain my honor and well-being so that I may return to France without waiting for providence [to assist me]..."[3] For Jussieu, self-reliance and pragmatism appear to have become significantly more important than "reasoning" about the moral and physical worlds through the study of botany. And the circumstances he encountered in South America seem to have demanded that scientific curiosity be suspended when more practical matters were at stake.

However, the adage Joseph cited to his brother may have somewhat overstated the case. For what had Jussieu been doing all these years if not combining pragmatic and philosophical pursuits? The stage for his forays in and around the South American landscape provided ample opportunities for learning "philosophically" by acting "practically." In a region of volcanoes, glaciers, and curative trees, opportunities to observe nature in tandem with local populations abounded, and each of these settings served as a kind of portable laboratory for investigating American nature. Joseph had figured out how to extract "philosophical" information from the very practical situations in which he found himself, whether observing the volcano Cotopaxi's belching fumes of sulphurous gases, trying to prevent the deaths caused by a smallpox epidemic in Quito, or witnessing the peculiar local cures for intestinal maladies proffered by those whom he characterized as ignorant medical "charlatans." What had stood in his way of "philosophizing," in fact, were not by and large the opportunities afforded by practical situations (although this was sometimes the case) but instead the mechanisms for conveying this knowledge to a place where it could be codified, systematized, and distributed to a broader audience. As Londa Schiebinger has written in her study of plants and empires, the eighteenth century saw "a confluence of circumstances [that] allowed the cultivation of certain types of knowledge over others," and she includes economic, strategic, political, and gender considerations as potential sites at which, through ignorance or other unacknowledged mechanisms, the transmission of information and ideas could be obstructed.[4] For Jussieu, the pathway from the Andes across the Atlantic became a vector not for *ignored* knowledge but rather *thwarted* knowledge: a knowledge that was cut off from broader networks either by obligations imposed by (Spanish) colonial powers, climatic and meteorological complications, financial considerations, (English) maritime bellicosity, or Jussieu's own changing psychological disposition. In the case of Jussieu, each of these categories or obstacles impeded the transmission of knowledge, thereby inhibiting its circulation. This is not to say that the knowledge acquired from Jussieu's explorations was entirely lost, nor that every scientific pursuit in an Atlantic context necessarily followed this model. However, the battered paths on which his observations traveled and the inadequate mechanisms for recovering much of this knowledge in the metropolitan center speak to a modality of science that is firmly indebted to the particular conditions of the early modern Atlantic world, whose geographic, social, and racial dynamics seem to have affected natural knowledge in dramatically inhibitory and catalytic ways. Conditions at sea, on land, and in all of the many portable laboratories where investigations and inquiries into the natural world could take place diminished as often as they augmented the knowledge that was being produced at the fringes of the Atlantic.

Joseph de Jussieu's three-and-a-half decades of botanical wanderings are a dramatic example of what I consider to be an unexamined aspect of science in the early modern Atlantic: its failures. The history of Joseph's travails in South America was an anomalous strand in the success story that was his family's rise to botanical fame. However, the story of his travels and ephemeral encounters with South American nature nevertheless shows how an individual largely outside the orbits of formal European institutions (yet with an enviable degree of access to their inner sancta) construed natural knowledge on both sides of an Atlantic divide.[5] In Jussieu, we have the record of an extraordinary traversal of Atlantic spaces and communities, including La Rochelle, Saint Domingue, and Panama and several extra-Atlantic spaces as well: from Quito to Chuquisaca and from Lima to Potosi. However, Jussieu's disposition toward transatlantic travel developed in his later years into a foreboding, even terrifying sense of doom. Ultimately, his experiences reflect both the real and the symbolic difficulties of transatlantic circulation and the dangers this circuit posed both practically and philosophically for those who sought to interrogate nature in the early modern period.

This chapter will examine several specific moments in Jussieu's South American itinerary as benchmarks to judge the shifting identities of an extra-European medical, botanical, and philosophical career. His letters reveal a host of intimate interactions with local populations, plants, and myths that show Jussieu not only as a failed, melancholic botanizer but as a keen observer of the knowledge produced by multiple actors on the Atlantic stage. Joseph's letters also reveal the extent to which an approach that privileges institutional frameworks, national allegiances, and an overly constricting nexus between botany, science, and the state fails to account for a range of actors, physical factors, and sociocultural issues that had an equal if not more significant impact on the ultimate fate of his botanizing. Spanish viceroys, Creole savants, peripatetic Jesuits, indigenous guides, and African "charges" all took part in the physical, social, and intellectual processes by which knowledge emerged from Jussieu's pen. Despite these social interdependencies, however, Jussieu by and large believed himself to be operating independently from more formal institutional frameworks. This fiercely coveted freedom to carry out botanical activities without his brothers (and others) meddling in his activities may have been one of the prime factors in the erratic and ultimately unsuccessful itineraries his knowledge took as it made its way from the Andean highlands to the salons of the Seine.

Global Networks, Local Knowledge, Unreliable Shipping and Handling

Jussieu's initial contact with non-European botany took place in 1735 and 1736 on the islands and in the coastal cities of the greater Caribbean. For decades prior to his arrival, however, French mariners, merchants, and naturalists had already been plying the waters between France's Atlantic provinces and the ports of the Caribbean. The French interest in extra-European specimens took root in the early part of the seventeenth century as royal institutions with a keen interest in non-European plants came to play a central role in affairs of state, including the broad colonial ambitions of Louis XIII and his son, Louis XIV. The establishment of botanical gardens in the university centers of Nantes (1550) and Montpellier (1593) provided new spaces to house non-European species, and the medical faculties of French universities likewise came to take a much more serious interest in objects and exotica arriving from overseas. The opening of the Royal Garden in Paris in 1640 followed soon thereafter, spurred by Richelieu's sponsorship of several exploratory botanical missions to Guadeloupe and Martinique. When Guy-Crescent Fagon became director of Paris's Jardin du Roi in 1699, advancing French botanical knowledge of the Caribbean and South America was a central part of his agenda. During the nearly two decades of his tenure, Fagon sent two naturalist-missionaries to explore these two connected regions and to bring back any seeds or specimens that might eventually be transplanted into the garden. Charles Plumier (1646–1704) and Louis Feuillée (1660–1732) combed the southern hemisphere looking primarily for medicinal plants, but in the process they brought back an abundance of useful observations. Feuillée later published an important treatise entitled *Journal des observations physiques, mathématiques, et botaniques* (1714) while Plumier's *Nova plantarum americanarum genera* (1703) and his *Traité des fougères de l'Amérique* (1705) provided rich illustrations and ample descriptions of the botanical bounty of the Americas.[6] Despite the dramatic diminution of the size of French colonial holdings in South America by the early decades of the eighteenth century, a burgeoning interest in potentially lucrative natural specimens was replacing French territorial aspirations (at least in the southern hemisphere). And it was during this period that a scientific expedition to measure the shape of the Earth, sponsored by the French Academy of Sciences and impelled by Newton's hypothesis that the terrestrial sphere was flattened at its poles, gave a young Jussieu the chance to study in a part of the world that had been officially off limits to non-Iberian naturalists: the Andean highlands near Quito.[7]

Heir to a growing French interest in the Americas and the Caribbean as a theater of botanical exploration, Jussieu lost little time in sending shipments back to his brothers Bernard and Antoine at the Jardin du Roi. Along with the letters describing his activities in Petit Goave, Saint Domingue, Portobelo, and Panama, he enclosed a detailed list entitled "Description des Plantes observées dans le voyage du Perou année 1735–36–37" that provided descriptions of plant species primarily from "Porto Bello," a port on the coast of present-day Panama, and the city of Panama itself.[8] In addition to providing names (in Latin) of these trees and plants, Jussieu also included the author of the book from which each name and description derived. The texts of Charles Plumier, Carolus Clusius, André Thevet, Marggraf and Piso, and Louis Feuillée comprised at least a portion of Jussieu's bibliographic repertoire; it is likely that he had access to many of these texts *in situ*.[9]

Jussieu's excitement at being given the opportunity to compare his bookish knowledge with plant species consulted on the spot is evident in these initial letters. "The utility of the various botanical excursions I've taken is not insignificant," he wrote to his brothers, "since aside from the collection of seeds that I'm sending along to you, I have [also] become more familiar with the Botany of America."[10] He was quick to show the limits of Plumier's work by pointing out the many plants this "tireless botanist" had either failed to observe or had simply not sent back to France. His optimism and enthusiasm grew with each foray into the Caribbean landscape, his knowledge expanding with each new specimen encountered. He described the company's combination of peripatetic study and writing as an ideal mixture through which to bring a new, more complete library of American nature into being: "Along the route, we collect many instructive reports; all of this will make a body of works that are both substantial and interesting."[11]

In stark contrast to these academic emissaries sent from Europe with passports and special privileges, approximately twenty "nègres" accompanying the expedition had been sent much earlier from Africa under dramatically different circumstances and with radically dissimilar privileges. These persons inhabited the lowest rung of the social hierarchy in the French colony of Martinique and were obliged at the behest of the Governor to haul the "oversized bags" and large chests. Mules could not transport these objects because of their awkward shape.[12] The labor of these nègres, when it was described, was portrayed in terms of the benefits that could accrue to the expedition, but their efforts always appeared as disembodied and fleeting. During his visit to Martinique, which was the company's first stop after the thirty-seven-day Atlantic traversal from La Rochelle, Jussieu ascended a "naked mountain" and collected a variety of seeds that he later sent to his brothers. This excursion, which took him four hours, was greatly facilitated by the "very narrow paths that some blacks [nègres] had taken care to widen."[13] Though the labor of these groups was a critical factor in the successful completion of Jussieu's botanical activities, the presence of the nègres, presumably peoples of African descent, appeared in the text as little more than anecdote.

Yet later, these nègres came to play a role as conduits for the transmission of medical knowledge. Jussieu regularly demonstrated concern for the impoverished, the ill, and the otherwise downtrodden, especially as he became more deeply rooted socially in Quito, Chuquisaca, and Lima. During one of his first stops with the company in Panama, he commented in a letter to his brothers that the surgeon Seniergues had requested additional medicines from a Parisian apothecary named Bardon for "the sick who rely on us to cure them [*des malades qui se confient à nos soins*]." In particular, he discussed certain pills, useful for treating ulcers and smallpox (among other maladies), with which this apothecary had originally provisioned their company, but whose supply would soon be entirely used up.[14] These elaborate steps taken by Jussieu to ensure that the infirm local populations receive their medicines shows that the company's interest—or Jussieu's, at the very least—extended beyond merely profiting from plants and the knowledge gleaned about them; indeed, in this case, they did not seem to support the physical and bacteriological misfortunes of their local informers. Rather, both here and elsewhere in his correspondence, Jussieu seems to have taken particular pride in exercising his medical skills without regard to the socioeconomic status of his patients. (He was later to write that "if I had been more avaricious, I could have amassed more than 40,000 piasters.") Indeed, Jussieu inveighed against the Creole physicians in Saint Domingue for precisely this reason, decrying the arrogance and hubris of the "surgeons and doctors of this country" for not acknowledging the role that the illiterate local population—the so-called "gens sans lettres"—could play in advancing botanical and medical knowledge: "If the surgeons and doctors of this country were better instructed, and were not embarrassed to watch the illiterate locals [*gens sans lettres*] using medicinal simples to cure blacks [*nègres*] of diseases they regard as incurable, there would be a tremendous benefit to [learning] the virtues of plants, and making them known to the whole world..."[15]

Who precisely these "gens sans lettres" are and what their status was within the colonial society of Saint Domingue remains unclear. However, the reference to their using their untutored powers to cure the nègres suggests that Joseph quite likely witnessed a scene in which individuals not formally trained in the medicinal arts attempted to cure individuals of African descent. This local knowledge, embedded within the illiterate or semiliterate classes of Caribbean society, was in turn transmitted through the bodies of African or Afro-Caribbean slaves. Despite his apparent lack of interest in the broader affairs of these African or Afro-Caribbean laborers, Jussieu clearly learns from them when it comes to their hybrid knowledge within Caribbean society. The fact that this knowledge transfer appeared in one of the first letters Joseph wrote to his brothers speaks to the ubiquity of this embedded and embodied local knowledge. It also speaks to the close attention paid to the transmission of this knowledge by Jussieu, even if we have no explicit evidence of how this information may have percolated into his own writings. This early intimate interaction between Jussieu and local populations would only increase over the course of his South American sojourn as his medical skills came to be appreciated and exploited in the communities he visited.

Indeed, by 1745, the Spanish authorities of the Audiencia of Quito had discovered Joseph's medical aptitudes and had demanded that his talents be put to profitable use among the local populations. This request was presumably to stave off yet another demographic collapse in the population of able-bodied workers in the agricultural and mineral industries of the colony. Why would Jussieu agree to take on these additional duties? His subordinate position vis-à-vis the local authorities may be one explanation. However, this does not explain why he was willing to perform such a significant service for so little reward. Financial problems plagued the young botanist, as they did the entire expedition, and monetary problems were especially acute for Jussieu after the departure of the rest of the expedition. For the most part, Jussieu was required to seek sponsorship from the upper echelons of Quito society to maintain his commitment to assisting the lower classes. Though Jussieu's broad medical knowledge may have been useful to the local population in Quito, his enlistment in resolving their medical woes did not always yield the best results for the Jardin du Roi in Paris. In a letter from Quito, Joseph indicated quite clearly that because so much of his time had been dedicated to the resolution of these local problems, he had had little time to pursue his own studies and observations. This is not to say that Jussieu did not make many useful observations during the period of the Quito epidemic. When Cotopaxi erupted in November of 1744, Jussieu provided his brothers with a thorough description of the negative health effects associated with the increased presence of volcanic ash on the respiratory systems of local populations. He also described the "bicho de culo," a disease supposedly unique to Brazil (described by Piso in the *Historia Naturalis Brasiliae*) in which the intestines lose their shape and function.[16] Jussieu enumerated the various ingredients used to carry out intestinal "washings," the preferred treatment among local healers (boiled olive broth was particularly recommended), concluding that most of the local health problems were caused simply by bad food and too great a quantity of pepper in their diets.[17] He again took the opportunity to rail against the "charlatans and inferior doctors" who insisted on employing "herbs," suppositories, "washings," and the like to resolve these potentially life-threatening maladies. These peripheral events offered Jussieu the possibility of communicating local knowledge back to his brothers in France, even if there is no direct evidence that his medical conclusions were incorporated in any meaningful way once they arrived back in Europe. Jussieu's particular attention to regional circumstances and the medical condition of local residents stemmed from his deeper engagement with these populations, a relationship fostered by the social conditions in which he lived. His experience merged European training and a commitment to *in situ* analysis with a sensibility to local medical circumstances that could only be achieved by someone who considered himself something more than just a transient observer.

Despite the occasionally colorful descriptions of local maladies, however, the majority of Joseph's letters are notable for their lack of detailed analysis and description. Rather than providing extensive reports on his activities, he seemed instead to summarize many of his observations, always emphasizing his fear that these descriptions would not arrive at their appointed destination. Rather than going into great detail in his letters, he suggested waiting for his return to Europe so that he could provide a more complete account of his activities in person: "The impossibility or at least the lack of certainty [of knowing whether these packets of seeds will arrive] makes me question the wisdom of reporting on my botanizing and entering into too great a detail. I will wait until we meet in person to satisfy your curiosity."[18]

This cautionary refrain about the uncertainty of his shipments' arrival is repeated consistently throughout his letters; he even wrote at one point that "he grew weary of writing because of this danger," the perils of the seas shaping the content and tenor of his epistolary correspondence.[19]

In addition to this cautionary attitude vis-à-vis the safety of his shipments, Jussieu's other activities also directly affected the manner by which his botanical knowledge was withheld from transatlantic circulation. In 1739, a popular uprising in Cuenca against several members of the expedition (in which his friend and colleague Seniergues was killed because of his involvement with the daughter of one of the local élites) led to delays in writing up the botanical results of one of his recent expeditions. The young Joseph had traveled to Loja, the region's largest concentration of Cinchona trees, and he was undoubtedly eager to describe to his brothers "the most remarkable things that I have had occasion to observe in this province of Peru." Unfortunately for Jussieu, however, Seniergues's wounds were so severe that he was forced to attend to him constantly as "guard, apothecary, surgeon, and medical doctor." Meanwhile, his companion La Condamine had already visited Loja in 1737 (using an earlier study carried out by Jussieu back in Europe "as a guide") and would go on to publish an academic report based on his own observations of the Cinchona tree.[20] Jussieu's account, far more scientifically rigorous than his colleague's "mémoire," would remain in manuscript form and would not be published until 1936. Joseph's many roles and especially his medical prowess drained him of the time to communicate the results of his official capacity as company botanist to his two siblings. "I do not have enough free time… that would be required for a detailed assessment of the observations I have made," he wrote in this same letter describing Seniergues's deteriorating condition, "[but] I will take the first opportunity that presents itself to offer you greater detail [on my observations]. In the meantime, I take pleasure in sending you a sample of seeds."[21] If even with a single patient he did not appear to have enough time for a systematic assessment of his previous travels, it comes as no surprise that during the outbreak of smallpox in Quito, this would become an even greater problem. In short, his constant attention to ill patients did not "allow [him] the time to relax enough to write," nor did he have enough time to undress himself—so he claimed—before going to sleep at night.[22]

Aside from not having enough time to write down the results of his botanical observations, Jussieu also based his refusal to go into great detail on the grounds that his letters were not likely to arrive. In retrospect, it seems that the unreliability of shipments across the Atlantic during this period justified Jussieu's concern, as a large percentage of his letters did not, in fact, arrive at their final destination. Joseph expressed this insecurity explicitly, especially the fear that his brothers' correspondence may fall "into the hands of the English."[23] In one letter, Joseph complained of the very few letters he had received in the most recent post from France. The single letter that he had been able to read detailed a shipment that his brothers had asked an anatomy professor in Cadiz to send on to Peru. Its failure to arrive, however, not surprisingly put Joseph into an inconsolable state: "My disgrace was such that by receiving that [single] letter, I felt all the more acutely the loss of the [other] letters that I was certain never more to see."[24] In a different letter, one that informed his brothers of his desire to return to Europe by way of the "River of the Amazons" so that he could make "an ample collection of natural curiosities," he hastened to add that he was certain that many of his seed shipments had been lost, "the war [of Austrian Succession] having interrupted the free commerce" of goods on the open seas. Indeed, Bourdaz, a French agent at Port-Sainte-Marie near Gibraltar, described a dramatic scene in which an entire cache of letters was thrown overboard off the coast of Spain by a group of marauding Englishmen, who had overtaken the French frigate *Harmione* on its way back from Lima.[25]

However, though we must add the precariousness of the high seas to the list of challenges faced by Joseph, the conditions in Quito alone were often enough to damage the specimens that had been readied to make their transatlantic traverse. Precipitation was an especially significant culprit, particularly in the rainforests where the humidity made it difficult to collect and preserve seeds and saplings "in their state of perfection." However, rain was by no means the sole threat:

> At present, I have no seeds ready to send. The rats have eaten those that I had collected, and the humidity does not allow me to keep the herbs and dried plants for very long without their getting moldy. It rains here almost every day, and often all day long, so that even those specimens that are safe from the rats do not manage to escape from the humidity, which quickly causes them to decay.[26]

In addition to the delays owing to his medical responsibilities, the uncertainty of the maritime postal system, and the climatic conditions along his route, Joseph was often simply too overwhelmed by the task of enumerating the character and "properties" of each specimen to spend his time in this way. Even so, he still maintained a strong desire to visit the myth-shrouded "país de Canelos." Jussieu acknowledged the potential risks of such a path but wrote stoically that "[he] must do something in the service of botany." It would take him nine years from this first mention of his interest in Canelos (thirteen from his initial arrival) to bring this desire to completion. In 1743, he once again wrote that he had "resolved to penetrate the province of Canelos, where there is much to observe in the realm of Botany."[27] Unwilling to allow the fresh specimens he had recently amassed to be "corrupted" by the rigors of a transatlantic voyage, he wished instead to wait until he himself could carry these "conserved curiosities" back to France. No one knew, least of all Jussieu, that it would take another twenty-eight years before his battered body would arrive back in Paris, with neither his spirit intact nor the "fresh specimens" from the Andes to show for his efforts at botanical collection and observation these many years.

Canelos: Botanizing in the Land of El Dorado

The Andean region of "Canelos," also known as the mythic Land of Cinnamon, was not originally meant to be on Joseph de Jussieu's homeward itinerary. Instead, Jussieu had planned to pass through the province of Quixos east of Quito to arrive at the "River of the Amazons," eventually descending the river to Pará, from Pará on to Lisbon, and from Lisbon back to France. To Joseph, the road through Quixos had several merits to commend it, including the presence of Jesuit missionaries who frequently followed the same route on their way from Quito back to Rome and who had offered their services to the now middle-aged Jussieu for his safe and comfortable passage through the jungle. Most importantly, the Amazon River would provide the greatest opportunity to make botanical observations along the way (facilitated in part by the presence of these very Jesuits) and to collect far more information than would have been possible in the dry conditions of the Cordillera: "I hope that this renowned but little known country [*ce pais si célèbre et si peu connu*] will provide me with the opportunity to collect an ample assortment of natural curiosities," he explained to his brothers, acknowledging that his financial circumstances and regular service as a doctor in Quito during nine years had not given him the flexibility to visit "all the places to which my curiosity would have [otherwise] taken me."[28]

However, like so much else in Jussieu's career, events did not unfold as he had initially desired or foreseen. While making final preparations for his overland journey toward the Amazon, and hence home, he was ordered to collect the earth-measuring expedition's scientific instruments prior to his departure from South America. These orders came from Count Maurepas, Minister of the French Navy from 1723 to 1749 and a key sponsor of the Quito expedition. At that moment, the materials were in the possession of Louis Godin, who resided in Lima where he held the chair of mathematics at San Marcos University. Maurepas requested that Jussieu journey south to Lima instead of heading east, and Joseph complied by picking up the trail that led from Quito to Guayaquil and from Guayaquil down to Lima. Rather than delaying him for a couple of extra months, this detour at the behest of the French state indirectly caused the extension of his South American sojourn by more than two decades.

The muddy road he took from Quito to Guayaquil, washed out at the time by torrential rains, had a more immediate effect on Jussieu's itinerary, allowing him at last to reconnoiter a region that lay at the center of his geographical and botanical curiosity: Canelos, the Land of Cinnamon. It was early in the sixteenth century that Europeans learned of the presence of a specimen resembling Indian cinnamon (*Cinnamomum zeylanicum*) in the eastern foothills of the recently conquered Andes.[29] Through the writings of other early chroniclers and conquistadors around the time of the Spanish arrival in South and Central America, rumors soon spread across the land of an "American cinnamon" [*canela de América*] and a "land of cinnamon" [*país de la canela*], a site of mythical and geographical speculation where gold, spices, and other natural treasures abounded. Canelos was associated with the fable of an indigenous warrior-chieftain named "El Dorado" who, it was said, each day dipped himself in gold powder to demonstrate his wealth and impress his subjects. Most famously, the "País de la Canela" was sought after by Gonzalo Pizarro and his lieutenant, Francisco de Orellana, during the important *entrada* of 1542, when Orellana and his crew accomplished the first European exploration of the Amazon River. "I am determined to undertake this expedition toward El Dorado and the Land of Cinnamon," wrote Pizarro, and his words were echoed by those who followed him. Even after the Land of Cinnamon had been described in print, the hidden land of gold and cinnamon managed to intrigue and elude Spanish explorers well into the sixteenth and seventeenth centuries. As late as 1749, it fascinated Joseph de Jussieu as well, who wrote that "my present aim is to enter… the province of Canelos in order to observe the Cinnamon tree."[30] The province was known in that period as "savage [*inculta*], full of impenetrable forests and rivers." However, the fragrant scent of Pizarro's cinnamon could still be detected in the fervent imagination of eighteenth-century botanists such as Jussieu.[31]

From some of his earliest letters, Joseph expressed the hope that he would visit this fabled Land of Cinnamon and collect enough plants "to populate the royal garden with the riches of America," echoing the mythic notion of a land of botanical bounty that would provide lucrative products for the Crown as well.[32] However, after visiting the nearby mines of Zaruma in 1739—a journey from which he returned with a fever "that nearly cost me my life"—Jussieu also recognized that such an expedition would challenge him physically and psychologically. "I will have many trials and difficulties to overcome," wrote Jussieu, "but these things are not important." Emphasizing the sacrifices he was willing to make in service to his brothers and the Crown, he continued: "I must do something in the name of botany."[33] Intrepid, earnest, and eager to achieve what his brothers expected of him, he set his sights on a perilous adventure to a fable-laden land, all the while awaiting "the happy moment that the company would begin the return journey to France."[34]

In letters to his brothers some eight years later, Joseph sought to link his descent into Canelos along the Bobanaza River with the earliest reports associated with this fabled land, boastfully asserting that his "curiosity" and "desire" had finally brought him to the heart of this historic province: "I went to the very place where the cinnamon grows that is the namesake of this province."[35] Linking himself implicitly to Pizarro and Orellana by returning rhetorically to the famed species of this region, he declared his own authority as much through etymology as through observation: "There, I recognized the characteristics of the Cinnamon of Quixos, which is the way that those who inhabit the land where the cinnamon is harvested refer to it." Describing the "Canelle de Quixos" as a "very high Tree with a good strong wood… [and] a very pleasant odor," he went on to affirm that his own observations conformed to Plumier's account of the "flowers and fruits" of this genre. In the end, he concluded confidently that the "Borbonia" described by Plumier "can be referred to as the Cinnamon tree," finishing off his account of the "Borbonia peruviana" with a long and legitimating Latin coda. Jussieu's taxonomic conquest of Canelos was complete.[36]

Joseph's account of his journey to Canelos had overtones of a heroic quest to discover the true nature of a mythic American specimen. However, the report of his activities in Canelos ultimately presented the detour as a failure, at least from a botanical perspective. Jussieu's move from empirical description to heroic narration may in fact have been a strategy used to garner sympathy among his readers (i.e., his brothers), a strategy not uncommon in some eighteenth- and early-nineteenth-century accounts.[37] However, rather than presenting his tale as a triumph, he recounted the many grave dangers he faced to assuage the expectations of a distant audience. After much travail, he finally managed to make it into the heart of the Land of Cinnamon, but he was unable to carry away the specimens of this sought-after plant because of "continuous rains that made me lose the success of my work, causing the plants and seeds that I had collected with every possible precaution to rot."[38] Despite having picked the "best time of the year to take this trip," Jussieu was unable to compensate for the *in situ* circumstances that brought havoc to his well-laid plans. He was able to save some of the other specimens brought from Quito—"other specimens … that were able to resist the moisture"—but he refused to provide much detail about their physical characteristics. He explained to his brothers that he did not include detailed descriptions because he would "never finish if I had to account for the properties and particularities of each individual item." Instead, he put off the onerous task until the "interview" that he hoped would take place before the year's end (he would not arrive back in France for two decades). He consoled his brothers with the simple observation that "if the plants survive, you will have the greatest pleasure of all in recognizing their characteristics and admiring their singularities."[39]

The expedition to the region of Canelos, then, seems to conform to the pattern we saw previously. Rather than providing a detailed account of his botanical exploits, he summarized the key points of his journey, highlighting the misfortunes he had endured. Substituting empirical observations with heroic tales of rushing rivers, giant boulders, and impenetrable forests, Joseph portrayed his excursion into Canelos as a series of hardships from which he was able to emerge in the end but without the typical fruits of botanical exploration:

> I had to pass on foot through dense forests and rough paths [*des bois et sentiers fort rudes*], ascend high mountains with my bare hands, [and] cross fast and dangerous rivers with no more assistance than what one could hastily improvise in a village at the entrance to the province, and which was supposed to last until we reached the village of Canelos [as well as] the duration of our return trip.[40]

If judged according to the idealized standards set by Bacon in his *Novum Organon*, Jussieu's journey through Canelos lacks a crucial component: the retrieval of intellectual data. According to Mary Terrall, the heroic mode of narration in a Baconian vein was designed to "[give] the discoverer the power to reveal knowledge to his contemporaries."[41] However, Jussieu's account appears as a counter-narrative to the heroic text, a terrestrial equivalent of the Portuguese shipwreck narrative, a genre that emerged in print during this same period.[42] Rather than *augmenting* the portrayal of new raw materials with tales of swashbuckling natural adventures, Joseph portrayed the overall project as misguided and ultimately unachievable, excusing himself for *not* having brought back such curiosities by scattering his text with heroic moments that compensated for the poor harvest he ultimately collected. The attempt to conquer Canelos through taxonomy collapsed in response to the end result he achieved, underlining the frustrations inherent in transporting viable specimens from the heart of Amazonia to a metropolitan place where they could be assessed and described.

Canelos functions as a turning point not only in Jussieu's regional botanical itinerary but in his broader economic situation as well. The move from engaged "philosophical" explorer to a more pragmatic orientation seems to have taken root in the wake of his expedition to Canelos. Immediately afterward he wrote that "I have financed myself for the past eight years, and I would have had a difficult time making a living had it not been for my medical skills. [My stay in] Quito [and] the journey to Canelos cost me significant sums, [and] the voyage to Lima and my return to France could not be achieved without excessive expenditures."[43]

In subsequent letters, Joseph showed increasing signs of fatigue, frustration, and apathy about matters toward which he had previously shown great enthusiasm. Aside from emotional strain at being so long separated from his family, financial considerations forced him to take up medicine on a more permanent basis, despite a "natural repugnance" for the profession. Further debilitating ailments were to become even more significant barriers to his physical and psychological disposition.

The "idealistic" botanist whom the surgeon Seniergues had difficulty convincing that "one should always prefer a Quadroon to a plant and a Dubloon to a seashell" was transformed by the material challenges of his journey.[44] Although he would still in future years seek out opportunities to botanize in the diverse locales through which he passed, his fears and frustrations began to get the best of him. In one letter, there is a premonition of this shift in behavior, not caused by any innate or even explicit psychological event but rather by immediate, quite real events that seemed to cause him genuine trepidation: "I travel with the apprehension of falling into the hands of the English."[45] The frequent appearance of these fears in Jussieu's account lends credence to Johannes Fabian's description of (central African) exploration, in which the nonrational emotions and behaviors of European travelers, what Fabian calls "ecstasis," formed a constitutive element in the production of knowledge. Compensating for his fears, Jussieu devised a complicated itinerary by land so that he would avoid maritime conflict: "This is what I will do if the seas are still infected by English corsairs, since the war is still on." Hoping to ensure a more speedy arrival back in France, Joseph unwittingly delayed his return. He allowed his fear, trepidation, and possibly even paranoia about the effects of his journey to transform the eventual products it was to provide. By giving in to his sentiments, he secured his place within the pantheon of botanical failure and wrote himself out of the history of Atlantic science.

Conclusion

Successful transatlantic circuits are the preeminent paradigm by which we have long understood the features of the early modern Atlantic world. This is why Atlantic history has so frequently been written with reference to particular, self-affirming Atlantic worlds, whether British, French, Iberian, or Dutch. I have sought to demonstrate through Joseph de Jussieu's South American odyssey how thwarted knowledge—knowledge that was retrieved from dubious sources, precariously expedited, or diverted from its desired destination—circulated, if less predictably, within Atlantic networks across imperial lines. As I have shown, there were various layers and spheres within which this thwarted knowledge was observed, retrieved, packaged—and lost. In one instance, knowledge was collected from unlikely sources—from the volcanic gases of Cotopaxi and the laboring bodies of African slaves to the dying bodies of local Quiteños—and as such did not cohere to any clear program of Atlantic knowledge production within an imperial frame. In another case, an unplanned detour to a region replete with mythic overtones led to the observation of a natural specimen of significant interest, but the meteorological and climatic conditions there blunted the transcription and eventual impact of these observations. Each of these examples sheds light on the complications that could arise in situations where natural knowledge was produced. The successful circulation of information from one point in the Atlantic to another was often dependent on circumstances that could just as easily go awry as go right.

Joseph de Jussieu has historically been portrayed as a scientific failure, and historians since Condorcet have gone to great lengths to place his largely unsung achievements on an equal rung with explorers whose South American sojourns provided professional and scientific recognition later in their careers, including Charles-Marie de la Condamine, Antonio de Ulloa, and Alexander von Humboldt.[46] I would argue, however, that there is no need to vindicate or excuse the challenges faced by Jussieu, precisely because they represent common problems for naturalists who attempted to project the results of their inquiries across great geographic expanses. What was characteristic in the case of Atlantic science was that its many theaters were populated by a broad range of peoples: some descended from Africans, others from Amerindians, and still others from mixtures of these two categories with colonizing Europeans. Each group had its own particular form of natural knowledge, and each contributed to the production of scientific knowledge in an Atlantic context. However, their precise roles were often subsumed in print under the guise of the exotic or expendable supernumerary. And many of these figures did not manage to make their own circuits function effectively: individuals such as Juan de Bordenave, a Spanish priest and close friend of Jussieu's who sought his own advancement by asking Joseph's brother, Bernard, to speak on his behalf to the Spanish ambassador in Paris; or Joaquin de Lamo y Zúñiga, a Spanish American Creole from Cuzco, who railed in a letter to Jussieu against the fact that mapmakers in Europe had disfigured the American continent "with a thousand errors" and nonetheless garnered "public applause" for their work.[47]

It is salutary, then, to recall the stories of individuals whose itineraries challenge the larger currents of commerce and circulation on which our understanding of Atlantic science has traditionally been based. The individual case history in the context of Atlantic knowledge allows us not only to trace the broad and agglomerated flows of peoples, specimens, and knowledge but to specify and interrogate the more idiosyncratic interactions between naturalists and the practical knowledge makers on the ground. Jussieu's activities, it is true, were implicated in the network of imperial and natural historical institutions, and the knowledge he produced did occasionally serve as the raw material out of which imperial knowledge was produced and performed. However, his story does not fit comfortably within a traditional framework of center and periphery because the knowledge extracted from his natural investigations did not always serve to support a particular colonial (or imperial) enterprise. And yet, even with an incomplete historical record and inadequate material traces with which to reconstruct the pathways of circulating knowledge, Jussieu's experience may nonetheless serve as a model of how thwarted Atlantic knowledge was produced and later diverted. Confronting physical challenges and allowing knowledge to be displaced within an imperfect system of communication and exchange was a central feature of natural historical inquiry in the early modern Atlantic context. Rather than seeing Jussieu as an intrepid but ultimately failed *voyageur*, we can now recognize that he was implicated in concentric circles of networks and social layers, some of which inhibited and some of which catalyzed his ability to effect and communicate knowledge both within the Americas and without.

As he reflected back on his career in the comfort of his Lima *cabinet*, Jussieu came to realize that he had failed not only his brothers and those who had placed so much confidence in his abilities but himself as well. In an emotional letter to his brother Bernard, he explained that his greatest sadness during the years he spent in South America was when he discovered that all of the shipments he had sent to Europe during previous five years had been lost:

> I swear to you that this news left me unnerved, and that I was so mortified that I fell ill. All my work! All of my efforts, all lost! How could I dare to show myself in Paris! What was the use of ruining my health, experiencing so much fatigue! Penetrating to the most distant and unhealthy places! Exploring them extensively with the greatest discomfort, if all of my shipments were lost! And above all, if the seeds of the plants from the voyage to Sainte Croix de la Montagne of the Andes were lost! I will feel a mortal chagrin for the rest of my life.[48]

What was ultimately responsible for the tragic state into which Joseph fell at the end of his days was the thwarted knowledge he had produced—

knowledge that was diverted and displaced by circumstances beyond his control. "If my work has been fruitless [*infructueux*]," he continued in his letter, "it is a tragedy that I took all precautions to avoid." It took Jussieu another ten years before he finally managed to traverse the Atlantic and return to France. And perhaps as we recall the trepidation he felt at making this final transatlantic crossing at all, we may better understand why it took him an additional ten years to muster the courage. Broken, despondent, and incapable of further reflection, he finished his days in the company of those he had originally hoped would see him as an intrepid and successful agent of natural historical inquiry.

Despite his own failings and the tragic state into which he ultimately descended, Joseph characterized his botanical observations in South America as a sphere of independence, free from the irritants and constraints that accompanied his participation as a member of the Academy's expedition to measure the shape of the earth. Joseph waxed enthusiastic at having lived a scientific life "without obligations to anyone," and it was precisely this "independence in which I have lived here that has made my stay tolerable."[49] This perceived freedom to botanize without constraint may have been Jussieu's greatest joy, but it also may have served as the symbol of his demise. Independent forces that sent the results of his investigations on a journey of their own, headlong into the jostling waves and uncontrolled currents of eighteenth-century Atlantic circulation.

Notes

1 "Primum est vivere deinde philosophari," Joseph de Jussieu, April 12, 1748, Muséum National d'Histoire Naturelle, Paris (hereafter MNHN), MS 179:31. This tension between a metropolitan center that "philosophizes" and a periphery that collects is admirably discussed in Ralph Bauer, *The Cultural Geography of Colonial American Literatures: Empire, Travel, Modernity* (Cambridge: Cambridge University Press, 2003).

2 At mid-century, the verb "*philosopher*" in French had a broad range of connotations, ranging from "reasoning about moral and physical concepts" to "reasoning in a philosophical manner." One can presume that Jussieu's "*philosopher*" is more closely akin to the former definition than the latter, given the direct relationship to the natural sciences inherent in this first definition. See "Philosopher," *Dictionnaire de l'Académie Française*, 4th edn, 1762.

3 Joseph de Jussieu, April 12, 1748, MNHN, MS 179:31.

4 Londa Schiebinger, *Plants and Empire: Colonial Bioprospecting in the Atlantic World* (Cambridge, Mass.: Harvard University Press, 2004), 226.

5 Joseph was the youngest member of a prominent generation of French botanists, and he was chosen by his brothers Antoine and Bernard to accompany the Franco-Hispanic expedition to colonial Quito in the capacity of company botanist. For

more on the origins and impact of the Quito expedition in the Americas and back in Europe, see Neil Safier, *Measuring the New World: Enlightenment Science and South America* (Chicago: University of Chicago Press, forthcoming); Antonio Lafuente and Antonio Delgado, *Los Caballeros del Punto Fijo: Ciencia, política y aventura en la expedición geodésica hispanofrancesa al virreinato del Perú en el siglo XVIII* (Madrid: Serbal/CSIC, 1987).

6 For the early history of the Jardin des Plantes, see F. Bourdier, "Origines et transformations du Cabinet du Jardin royal des Plantes," *Sciences et l'enseignement des sciences: revue française des sciences et des techniques* 18 (1962), 35–50; Henry Guerlac, "Guy de la Brosse," *Dictionary of Scientific Biography*, 7, 536–41; Yves Laissus, "Le Jardin du Roi," in *Enseignement et diffusion des sciences en France au XVIIIe siècle*, ed. René Taton (Paris: Hermann, 1964).

7 Safier, *Measuring the New World*.

8 The list from Panama, "Plantas circa urbes Panama observavi spont nascentus," listed approximately 50 plants and trees, among which the Tamarind, Guava, Papaya, Manioc, and Cajou would have been the more recognizable American species to botanists who had greater familiarity with European varietals.

9 An inventory drawn up by Don Bernardo Gutierres Vocanegra, Governor of Portobelo, described the content of the trunks hauled by "six blacks [*negros*]" from Portobelo to Panama. This cargo included at least twenty-one trunks of books, including, presumably, at least some of the above-mentioned titles. "Inventario y reconocimiento del equipaje de los Académicos Franceses," June 1, 1736 in José Rumazo, *Documentos para la Historia de la Audiencia de Quito* (Madrid: Afrodisio Aguado, 1948–50), 6:18.

10 Joseph de Jussieu, April 12, 1748, MNHN, MS 179:31.

11 Joseph de Jussieu, MNHN, MS 179:31. "Nous ramassons chemin faisant des memoires instructifs, le tout faira un corps d'ouvrages curieux et bien remplis."

12 Jussieu, not one as we will see who was particularly attuned to financial matters, found the cost for this transport quite expensive: "le transport des marchandises bagages est tres cher il en coute 6 piastres par chaque mule de cavalierie. les mules de domestiques et par chaque mule de bagage." The service of the "nègres" was cheaper: "5 piastres chaqu'un et nous ne fallut une vingtaine." Joseph de Jussieu, MNHN, MS 1626:56.

13 Antoine-Laurent de Jussieu, "Notice sur M. Joseph de Jussieu, contenant un extrait de sa correspondence...," MNHN, MS 179.

14 It is interesting to note Jussieu's description of the itinerary by which these pills would ideally travel, which gives a sense of the circuits that physical objects as small as a pillbox might make from one side of the Atlantic to the other: "… the packet should be addressed to Mr. Maurcit [of Petite Goave], who will ensure it is sent to Mr. Godin in Jamaica. From Jamaica the packet will be addressed to the English director of the assistant who will ensure that it is in turn sent to us in Quito," Joseph de Jussieu, February 17, 1736, MNHN, MS 179:20.

15 Joseph de Jussieu, October, 1735, MNHN, MS 179:11.

16 See the discussion of the "bicho de culo," or "mal do bicho" in Júnia Furtado's chapter in this volume.

17 In a shipment sent to the Frères Jussieu from La Paz in 1749, Joseph includes the "Opuntia peruviana remota rotunda validissimis ad longis spinis…" which

is used as a purgative to give relief for the "bicho de culo." This specimen Jussieu picked up on the road from Quito to La Paz.
18 Joseph de Jussieu, February 26, 1741, MNHN, MS 179:26.
19 Joseph de Jussieu, MNHN, MS 179:30.
20 Charles-Marie de la Condamine, "Sur l'arbre du Quinquina," *Mémoires de l'Académie Royale des Sciences* (1737), 226–43.
21 Joseph de Jussieu, August 31, 1739, MNHN, MS 179:24.
22 Joseph de Jussieu, March 16, 1745, MNHN, MS 179:29.
23 Joseph de Jussieu, March 16, 1745, MNHN, MS 179:29.
24 Joseph de Jussieu, September 25, 1747, MNHN, MS 179:30.
25 Joseph de Jussieu, July 23, 1762, MNHN, MS 179:73.
26 Joseph de Jussieu, March 16, 1745, MNHN, MS 179:29.
27 Joseph de Jussieu, April 25, 1743, MNHN, MS 179:27.
28 Joseph de Jussieu, September 25, 1747, MNHN, MS 179:30.
29 In 1533, Gaspar de Espinosa, a conquistador from Panama, had written that Atahuallpa himself claimed that cinnamon grew wild in his country. Prior to that, the sole reference to a cinnamon-like species in the New World was from Alvarez Chanca, the Sevillian doctor who was present on Columbus's second voyage and who wrote that "[the American cinnamon] is not as fine as that seen back home." As cited in Antonello Gerbi, *Nature in the New World: From Christopher Columbus to Gonzalo Fernández de Oviedo* (Pittsburgh: University of Pittsburgh Press, 1985), p. 24.
30 Joseph de Jussieu, 31 August, 1739, MNHN, MS 179:24.
31 "Canelos," in Antonio de Alcedo, *Diccionario Geográfico de las Indias Occidentales o América*, I: 213.
32 Joseph de Jussieu, April 25, 1743, MNHN, MS 179:27. For the most recent discussion of the links between global botanical resources as objects of interest for state-sponsored scientific activities, see Emma Spary, "'Peaches Which the Patriarchs Lacked': Natural History, Natural Resources, and the Natural Economy in France," *History of Political Economy* 35 (2003): 14–41. Spary emphasizes the tight nexus between global resources, medical botany, and the definition of the nation in early modern France. Because he did not actually publish any of his studies, Joseph de Jussieu's activities do not figure in her study. Likewise, in Spary's *Utopia's Garden* (Chicago: University of Chicago Press, 2000), Joseph de Jussieu is noted for having been "stranded" in South America and for having forgotten his native tongue. On a similar topic, see Staffan Müller-Wille, "Nature as a Marketplace: The Political Economy of Linnaean Botany," *History of Political Economy* 35 (2003), 154–72. This subject has been creatively presented and assessed in the collected volume *Merchants and Marvels: Commerce, Science, and Art in Early Modern Europe*, ed. Pamela H. Smith and Paula Findlen (New York: Routledge, 2001). For a pre-Buffonian perspective on the stakes of exotic biota at the Parisian Jardin du Roi, see Neil Safier, "'…To Collect and Abridge…Without Changing Anything Essential:' Rewriting Incan History at the Parisian Jardin du Roi," *Book History* 7 (2004), 63–96.
33 Joseph de Jussieu, 31 August, 1739, MNHN, MS 179:24.

34 Joseph de Jussieu, August 31, 1739, MNHN, MS 179:24. Joseph would wait thirty years before that happy moment, and the delay in achieving this goal—of being reconnected with his family, a desire he expressly and regularly repeats in his letters—is one of the most perplexing and enigmatic aspects of his correspondence. A year and a half after the Cuenca letter cited above, he writes to his brothers that "I wish for nothing more than the consolation of being among you once again."
35 Joseph de Jussieu, April 12, 1748, MNHN, MS 179:31.
36 "Borbonia peruviana laurifolio cupula fructus ampla expansa aromatica. cortica ligni cinnamomeo," Joseph de Jussieu, April 12, 1748, MNHN, MS 179:31.
37 See Mary Louise Pratt, *Imperial Eyes: Travel Writing and Transculturation* (London: Routledge, 1992). In a similar vein, Johannes Fabian posits the idea that "danger tales" may "reflect generic demands rather than individual experiences." See Fabian, *Out of Our Minds: Reason and Madness in the Exploration of Central Africa* (Berkeley: University of California Press, 2000), 87.
38 Joseph de Jussieu, September 25, 1747, MNHN, MS 179:30.
39 Joseph de Jussieu, April 12, 1748, MNHN, MS 179:31.
40 Joseph de Jussieu, MNHN, MS 179:36.
41 Mary Terrall, "Heroic Narratives of Quest and Discovery," *Configurations* 6 (1998), 226. Joseph's representation of his journey to Canelos, despite being circulated in manuscript form to a limited audience, is nevertheless consistent in many ways with what Terrall has described as a "narrative of [a] heroic quest for truth."
42 See Bernardo Gomes de Brito, *Historia Tragico-Marítima*, recently republished as *The Tragic History of the Sea*, ed. and trans. C. R. Boxer, intro by Josiah Blackmore (Minneapolis: University of Minnesota Press, 2001).
43 Joseph de Jussieu, April 12, 1748, MNHN, MS 179:31.
44 Joseph de Jussieu, February 18, 1736, MNHN, MS 179:21.
45 Joseph de Jussieu, April 12, 1748, MNHN, MS 179:31.
46 Yves Laissus exemplifies this attempt to unearth Joseph's positive achievements, all the while bemoaning the neglect into which the youngest Jussieu has fallen: "Pas un biographe n'a été tenté par cette personnalité forte et curieuse, aucun écrivain n'a conté les aventures de ce voyageur intrépide et seuls les manuscrits, restés inédits, que garde la Bibliothèque du Muséum national d'histoire naturelle, semblent conserver son souvenir." Laissus, "Note sur les manuscrits de Joseph de Jussieu (1704-1779) conservés a la Bibliothèque Centrale du Museum National d'Histoire Naturelle," *89ème Congrès des Sociétés Savantes* (Lyon, 1964), 9-16.
47 Joseph de Jussieu, October 22, 1770, MNHN, MS 179:91; March 13, 1751, MNHN, MS 179:47.
48 Joseph de Jussieu, April 4-10, 1761, MNHN, MS 179:64.
49 Joseph de Jussieu, April 4-10, 1761, MNHN, MS 179:64.

CHAPTER 9

Atlantic Competitions

Botany in the Eighteenth-Century Spanish Empire

DANIELA BLEICHMAR

Because the examination and methodical investigation of the natural productions of my American dominions are advisable for my service and for the good of my vassals, not only to promote the progress of the physical sciences, but also to banish doubts and adulterations in matters of medicines, dyes, and other important arts; and to increase commerce; and in order that herbaria and collections of natural products be formed, describing and delineating the plants that are to be found in those fertile dominions of mine; [and] to enrich my natural history cabinet and court botanical garden…

Royal Order authorizing the Royal Botanical Expedition
to Chile and Peru, 1777[1]

With these words, Charles III commanded Spanish botanists Hipólito Ruiz and José Pavón to embark on a botanical expedition to Chile and Peru. The King had occasion to repeat these exact words multiple times over the following years and to make many other statements to that effect. Between 1760 and 1808, Spain sponsored fifty-seven expeditions to its colonies, eight of them with botany as a sole or central aim.[2] In addition to the Chile and Peru expedition, the crown funded investigations of the Orinoco, the New Kingdom of Granada (corresponding to present-day Colombia, Venezuela, and Ecuador), the Philippines, New Spain (parts of present-day Mexico), Cuba, and Ecuador and a circumnavigation expedition modeled on Cook's first voyage (Table 9.1).[3] These expeditions formed part of a widespread and concerted effort to renew the Spanish empire economically and politically through a thorough transformation of its colonial policies.[4] The so-called Bourbon reforms attempted to revitalize an impoverished Spain by strengthening its industry and reshaping its relationship with the viceroyalties.[5] The Spanish crown sought to repeat its earlier success with mineral riches by exploiting profitable natural commodities. Rather than purchasing spices from European competitors, Spain hoped to locate plants such as cinnamon, tea, or nutmeg in its own colonial territories and to improve its trade in the valuable antimalarial cinchona. A better-known and efficiently administered empire, it was hoped, would furnish rich revenues by providing new natural products that Spain could sell within Europe or use to compete with trade monopolies maintained by other countries.

This chapter discusses metropolitan and colonial efforts to locate, exploit, and promote valuable spices throughout the Spanish empire. It highlights the fiercely competitive nature of Atlantic science, which pitted both European and American players against one another as they scrambled to establish, maintain, or break trade monopolies in natural commodities. In this climate of international economic and political competition, botanical expertise represented a highly valuable form of practical knowledge that created opportunities for naturalists to sell their services to interested patrons.[6] Naturalists operated within dense institutional and administrative networks—especially in the Spanish case, in which the exploration and exploitation of nature had a long and well-established tradition within the structures of colonial governance.[7] For eighteenth-century naturalists and administrators, botany was both big business and big science.[8]

In a first section of this chapter, I examine the transatlantic pursuit of profitable natural commodities, focusing on cinnamon and pepper as representative case studies. Economic botany brought into collaboration traveling European naturalists, colonial *criollo* naturalists, mestizo and indigenous collaborators, and administrators and institutions across the empire (in the Atlantic and in the Pacific).[9] Through physical examination and chemical analysis, naturalists in the colonies attempted to locate regional varieties of cinnamon, pepper, tea, nutmeg, cloves, and other profitable plants. They also explored possibilities of improving the exploitation of the cinchona tree, highly valued for the antimalarial properties of its bark and Spain's only monopoly on a natural commodity at the time.

Table 9.1 Botanical expeditions conducted in the Spanish Empire during the second half of the eighteenth century

Expedition	Dates	Areas	Naturalists	Artists
Limits expedition to Orinoco	1754–1756	Orinoco (Venezuela)	Pehr Löfling	Juan de Dios Castel
Botanical expedition to Chile and Peru	1777–1788	Peru and Chile	Hipólito Ruiz, José Pavón, Joseph Dombey	José Brunete Isidro Gálvez
Botanical expedition to New Kingdom of Granada	1783–1810	New Granada (Colombia, Venezuela, Ecuador)	José Celestino Mutis + multiple associates	Salvador Rizo, Francisco Matis, >40 others
Expedition to the Philippines	1786–1794	The Philippines	Juan de Cuéllar	Anonymous local artist(s)
Circumnavigation (Alejandro Malaspina)	1789–1794	South, Central, North America, Australia, the Philippines	Tadeus Haenke, Luis Née, Antonio Pineda	José Guío, José del Pozo, Francisco Pulgar, José Cardero, Tomás de Suría, Fernando Brambila, Francisco Lindo, Juan Ravenet, José Gutiérrez
Botanical expedition to New Spain	1786–1803	Mexico and Guatemala	José Mociño, Martín de Sessé, Vicente Cervantes	Atanasio Echeverría, Vicente de la Cerda
Expedition to Cuba	1796–1802	Cuba	Baltasar Manuel Boldó, José Estévez	José Guío (prev. Malaspina expedition)
Expedition to Ecuador	1799–1808	Ecuador	Juan Tafalla	Francisco Pulgar Francisco Xavier Cortés

In a second section of the chapter, I situate the Spanish expeditions within the larger system within which they operated. In addition to sponsoring scientific voyages, the Spanish Crown attempted to identify and exploit profitable *naturalia* by deploying the widespread colonial bureaucratic network already in place, turning it into a system of collectors and informants. New peninsular institutions, such as Madrid's Royal Botanical Garden and Natural History Cabinet, coordinated these efforts. Their directors wrote instructions asking local administrators to write reports, commission images, and compile collections. These documents specified the types of plants and animals to be collected and described and provided directions for their safe transportation across land and sea. Although this type of request was common to other European governments and to scientific academies, the Spanish Crown was unique in making it part of its official imperial policy. Royal orders passed in 1789 required colonial administrators at every level to produce reports describing any potentially useful natural productions that existed in their regions and to send back to Spain both information and samples. Once in Madrid, these materials were assessed by naturalists, physicians, and pharmacists belonging to an extensive network that included not only the Botanical Garden and Natural History Cabinet but the Royal Pharmacy and numerous Royal and Naval hospitals.

In a final section of the chapter, I address notions of center, periphery, and circulation. Departing from Bruno Latour's model, I argue that to concentrate exclusively on the movement of information and specimens from colonies to metropolis would imply missing half the story. Trajectories were not only imperial but colonial; initiatives originated not only in Madrid but in places like Bogotá, Lima, and Mexico City.[10] Colonial governors and administrators sponsored investigations of local nature with an enthusiasm that suggests they had their own reasons to be interested in natural history, beyond their duty to ensure the prompt fulfillment of orders from Madrid. Throughout the colonies, local officials actively encouraged the exploration of nature, hoping to identify products that would boost the regional economy by securing profitable trade with the metropolis. As Spanish and Spanish American naturalists became increasingly ensconced in local projects, the relationships they forged in the New World competed with those they maintained across the ocean. Distance from Madrid—in space, time, and experience—granted naturalists considerable autonomy, and favored the development of independent attitudes. Addressing multiple audiences, they were as involved with imperial concerns as with local colonial ones. In close contact not only with Madrid but with Lima, Mexico City, or Manila, naturalists working in the colonies became nodes of a truly global network in which center and periphery were far from clear or stable categories.

Rediscovering the Empire: The Search for Natural Commodities

The work of eighteenth-century Spanish botanical expeditions—and indeed much if not most natural history research throughout the empire—was fueled by competitive worries and aspirations. Peninsular officials, such as the economist and minister Pedro Rodríguez de Campomanes (1723–1802); the Minister General of the Indies, José de Gálvez (1720–87); and his close ally and director of Madrid's Royal Botanical Garden (founded 1755), Casimiro Gómez Ortega (1741–1818), feared that potentially valuable commodity withered away in the colonies while consumption of natural imports in the peninsula and the colonies enriched other European nations at the expense of Spanish pockets.[11] Naturalists throughout the empire, whether Spaniards or colonials, shared this concern. Finding themselves in the midst of abundant and underexploited natural resources, they penned strong critiques of imperial policies and colonial attitudes and offered the rediscovery of American nature as a potential solution to the problem.

The pursuit of American natural commodities was constantly framed in terms of a response to Dutch, British, and French activities. In 1780, the *Real Sociedad Económica de Madrid* calculated that Spain purchased 600,000 pounds of cinnamon from the Dutch every year, with the same amount consumed yearly in the Americas.[12] This expenditure seemed particularly unnecessary given the long-recorded presence of cinnamon in the New World. Conquistadors Sebastián de Benalcázar and Gonzalo Pizarro identified it in Peru in the 1530s; the plant was also found some years later in New Spain, where the colonial government conducted some trade in the mid-1570s.[13] All the major chronicles and natural histories of the New World from the sixteenth and seventeenth centuries described American cinnamon.[14] Nevertheless, for about 150 years, American cinnamon remained largely ignored as a commercial product because the Spanish Crown saw little need for it. Early samples proved to be of noticeably lower quality than Ceylonese cinnamon, a trade dominated in Europe by the Portuguese in the sixteenth century. Spanish access to Asia from the 1530s through the Philippine islands and the unification of Spain and Portugal between 1580 and 1640 ensured Iberian dominance of the eastern trade for many years. With a heavy flow of spices coming from the East and a torrent of precious metals pouring in from the West, the Spanish Crown had little reason to pursue American cinnamon until well into the seventeenth century. It was not until the emergence of the Netherlands as a global player through the East India Company's Asian spice trade that Spain started showing interest in a way to compete with this trade.[15]

In the eighteenth century, Spain changed its earlier neglectful attitude toward American cinnamon, showing enormous interest in its investigation and exploitation. The Jesuit missionary José Gumilla's *El Orinoco ilustrado y defendido* (1741) described rich forests of cinnamon trees in South America and suggested that American cinnamon and Philippine spices gave Spain the perfect weapons with which to seriously maim Dutch trade.[16] Jorge Juan and Antonio de Ulloa, writing after their American travels with the La Condamine expedition (1735–45), critiqued both peninsular and American attitudes toward American cinnamon. They described metropolis and colony as entangled in a destructive loop of negligence: if only cinnamon cultivation were promoted in the colonies, then the product would be more widely available in Spain and would surely come to replace Eastern cinnamon; this success, in turn, would encourage even greater cultivation. No other nation, they claimed, showed such a cavalier attitude toward natural commodities. They cited France as an example of more intelligent policies and as a model of imperial science:

> The French [nation], [being] very passionate about coffee, and seeing that it lost considerable sums in bringing it from Asia, decided to take its plants to the islands of Martinique and Saint Domingue, and in a few years the plantations there have grown so much that with the coffee production from these two islands there is a crop perfectly sufficient for consumption there [in the islands] and for the growing one in France. And when they decided to prohibit completely the entry and sale of Eastern coffee, they did not object to the great difference that exists between one and the other, since they were unable to make coffee from the islands as good as the other. If this nation [France] had a tree as esteemed as cinnamon in the countries that it owns, it would create a high level of trade, and would invest heavily in cultivating and propagating this species in order to increase its utility. So, why should we [Spaniards] show ourselves so careless in profiting from the riches provided lavishly by the extensive forests of Peru?[17]

The biggest problem that American cinnamon posed for the Spanish empire was that, more than three hundred years after it had first been discovered, it remained unclear whether the product was in fact the same plant as Eastern cinnamon from Ceylon and China or a different species altogether.[18] Advocates of American cinnamon maintained that whatever differences could be observed between the varieties were due to the deteriorated state in which the American product arrived in Europe, owing to poor cultivation and shipping conditions. Establishing this fact proved difficult, however, because the trees from different regions were not available for comparison in Europe, where the product arrived as a dried branch or its bark. To prove the quality of American cinnamon, Spanish naturalists and administrators set about obtaining samples of regional varieties. These were assessed both in the colonies and in Madrid, where the Royal Botanical Garden attempted to establish the taxonomical identity of the varieties by comparing dried herbarium specimens and images, and the Royal Pharmacy performed chemical analyses of infusions, distillations, and oil preparations from bark samples.

The expeditions to Chile and Peru and to New Granada, both of which operated in the Quito region, offered excellent opportunities for establishing the taxonomic identity of American cinnamon and exploiting its production. Thus, as Ruiz and Pavón set out for Chile and Peru in 1777, Casimiro Gómez Ortega prepared a report to the Crown discussing the likelihood of breaking the Dutch monopoly on cinnamon and clove with American substitutes from the Quito region, through which the naturalists would be traveling.[19] In it, he argued that replacing Eastern cinnamon with an American variety would benefit both the viceroyalties, which would sell the product to Spain, and the metropolis, which would profit from marketing this commodity domestically and internationally.

However, the story of American cinnamon turned out to be a bitter one for the Spanish empire. Joseph Dombey, the French naturalist who traveled with Ruiz and José Pavón between 1777 and 1784, sent Gómez Ortega samples of cinnamon bark from Quito and Santa Fe de Bogotá. Dombey also promised to ask Antoine-Laurent de Jussieu, professor of botany at the Paris *Jardin du Roi*, to provide Gómez Ortega with samples of Ceylon cinnamon too so that he could compare the two varieties.[20] Dombey later concluded that the two types of cinnamon were in fact different species; José Celestino Mutis, director of the Botanical Expedition to New Granada, concurred. When Mutis's botanical contributor, José Eusebio Ramos, reported finding cinnamon trees in the Bee Mountains in October of 1783, Mutis immediately dispatched a collaborator to the region to investigate these claims.[21] Within a couple of months, Mutis received samples of the tree's leaves, flowers, and bark for analysis.[22] In a report to Viceroy Antonio Caballero y Góngora, Mutis announced that the plant in question was not cinnamon but rather a new genus that he himself had identified in 1772.[23]

Eventually, American cinnamon was classified as *Laurus cinnamomum,* a species and genus different from Ceylonese cinnamon *(Cinnamomum ceylanicum)* and Chinese cinnamon *(Cinnamomum cassia)*.[24]

Pepper posed a different type of problem for Spanish aspirations to compete in the European spice market. Though nobody claimed to have found Eastern pepper in the Americas, the Spanish empire had high hopes for the economic potential of an American plant that could act as a substitute: malagueta or Tabasco pepper (allspice). In this case, the problem was not to establish botanical identity but rather to increase the rate of production and to encourage European consumption. Again, Casimiro Gómez Ortega was optimistic. In a 1777 report to José de Gálvez, Gómez Ortega outlined a full-blown strategy for benefiting from the plant. His plan entailed publishing a memoir announcing the discovery of the plant and advertising its wonderful properties so as to create expectation and demand for the product in Spain, performing public demonstrations at the Madrid botanical garden, and diffusing the commodity through Europe— selling it at a lower price than the Eastern pepper marketed by the Dutch to encourage consumption.[25]

Three years later, Gómez Ortega published the memoir he had proposed, entitled *Historia natural de la Malagueta o Pimienta de Tavasco* (Madrid, 1780). The thirty-four-page memoir opened not with natural history but with a discussion of the expeditions' connection to economic reforms and notions of utility (a term combining ideals of usefulness, profit, and service). Unless natural products were exploited by fostering their extraction, diffusing information, and encouraging consumption through free trade, Gómez Ortega warned, even the most benign and fecund territories would became sterile deserts, useless both to their colonists and to the metropolis. The Royal Order establishing free trade on malagueta (April 23, 1774) would demonstrate this fact by boosting agriculture, industry, and the arts.

Gómez Ortega also provided a patriotic frame for his description of the malagueta plant, linking its contemporary investigation to an illustrious genealogy of Spanish exploration in times that had been decidedly more glorious for Spain. The plant, he noted, had first been described in the 1570s by Francisco Hernández. Just as Spain was honored by having discovered almost all of the countries where *malagueta* grew, he explained, it was also the Spaniard Hernández who captured the glory of being the first to describe this plant. Gómez Ortega's reference to Hernández was by no means casual, as in 1780 he had published an edited collection of Hernández' works and the memoir on *malagueta*, putting in evidence his ambitious pursuit of a unified vision of natural history as an applied and patriotic undertaking.[26]

In the first two chapters of the memoir, Gómez Ortega described the tree with great care, always couched in a patriotic economic and political context. The third chapter of the memoir, entitled "Means of exploiting *malagueta*," addressed the collection of the spice from the tree and its preparation for trade. It also introduced a note of urgency by referring to British cultivation of a pepper tree in Jamaica. Gómez Ortega sought to alarm and entice his public in equal measure: he pronounced it "very worthy of admiration" that the British had reaped great profit from the growth and traffic of Jamaican pepper while the Spanish had failed to do the same with *malagueta*.[27] In the light of this admonition, his description of British efforts made the exploitation of this spice sound extremely simple. Slaves simply needed to shake the tree to make the fruit fall and then dry it in the sun for ten to twelve days. With that, the pepper would be ready for export. Gómez Ortega ended the chapter by indicating the price per pound of Jamaica pepper, in this way underscoring the profit awaiting both future planters in the Spanish colonies and merchants in the continent.

The memoir addressed the comparative merits of *malagueta* and Eastern pepper as condiments—a crucial factor in its potential success as a trade good. Gómez Ortega insisted that American pepper was "very superior to the ordinary eastern pepper that we buy from foreigners, and that food seasoned with *malagueta* tastes like all-spice."[28] The chapter ended with the assertion that, as authorities agreed that American pepper had identical dietetic and medicinal virtues to those of pepper, cinnamon, and cloves, "it is of the greatest importance to the interests of the King and the nation that its consumption and trade be fostered. The total exemption from duties, as granted by His Majesty, is one of the most effective means to accomplish such an important end. Another means may be, one hopes, to generally spread knowledge about it and news of its nature, and to that [end] this little work is directed."[29] Gómez Ortega pronounced it shameful that *malagueta* could not be found in Spanish pharmacies, given its presence in foreign ones. Its similarity to pepper, cinnamon, and cloves implied that it could replace these Eastern spices: any recipe, culinary or medical, calling for one of these species could be prepared by using three times the specified amount of *malagueta*—a measure that would surely contribute to increased economic profits.

In the final chapter of the memoir, Gómez Ortega attempted to recruit readers as collaborators of Madrid's Royal Botanical Garden, requesting that they send *malagueta* samples and providing detailed recommendations and specifications as to how they should do so. The memoir ended with an appeal to the viceroys and local governors of Mexico, Guatemala, Puerto Rico, and Cumana to write to the Ministry of Indies with details about the conditions under which *malagueta* trees grew in the territories they governed, noting that in this way they would serve the country and prove their love of king and homeland. For more information on what exactly was expected of them, Gómez Ortega referred his readers to a set of guidelines for contributors that he had published the previous year, *Instrucción sobre el modo más seguro y económico de transportar plantas vivas por mar y tierra a los países más distantes* (Madrid, 1779).[30]

For Gómez Ortega, natural history was a politically and economically useful science motivated by patriotic and competitive considerations. Thus, his various publications—the *Instructions,* the memoir on *malagueta,* and the edition of Francisco Hernández' works—were part of a larger project that also mobilized traveling expeditions, networks of correspondents and contributors, peninsular and colonial institutions, and administrative structures. Spanish and *criollo* naturalists formed part of this network. The expeditions were not isolated projects; rather, they formed part of an intricate system—what James McClellan and François Regourd have dubbed a "colonial machine."[31]

Natural Commodities in Circulation

Beyond the natural history expeditions, the Spanish Crown attempted to identify and exploit profitable *naturalia* by deploying the widespread colonial bureaucratic network already in place, turning it into a system of collectors and informants. This had been done much earlier with the *Relaciones geográficas* questionnaires, highlighting the continuity of earlier colonial practices within the context of Enlightenment. Now, however, printed instructions replaced questionnaires, and objects, rather than written reports, became the desired contribution. On the founding of the Royal Natural History Cabinet in 1776, its director, Pedro Franco Dávila, authored a set of instructions that were widely distributed throughout the peninsula and the colonies. Addressed to administrative authorities at all levels of government, these directions described in considerable detail over twenty-four printed pages the appropriate manner in which to collect, cleanse, preserve, pack, and properly archive minerals, animals, and plants so that they could be exhibited in the cabinet.[32]

Three years later, Gómez Ortega's own *Instrucción* (1779) similarly sought to complement the operation of the expeditions by enlisting the eyes and hands of administrators and amateur naturalists throughout the colonies. Culled largely from French instructions by Duhamel de Monceau, Gómez Ortega's pamphlet devoted great attention to the minute details of transplanting, packaging, and transporting plants across long distances and of preparing herbaria.[33] Though this practical information was rather standard of this type of instruction, Gómez Ortega penned an original section specifying in twelve headings the most desirable plants expected from the colonies, providing their Latin and vulgar names, known location, and properties. Cinchona topped Gómez Ortega's list, followed by equivalents for Ceylonese cinnamon, Malabar pepper, Ambonese cloves, and nutmeg from the Banda Islands (all of which, Gómez Ortega assured the reader, existed in the Spanish Americas), and then by *malagueta*. The remaining nine items focused on various other choice natural commodities, among them breadfruit ("its multiplication in America would be of the greatest use because it is very nutritious for workers and blacks," he explained, following the British model), mangosteen, sarsaparilla, calaguala, and *cauchú* (rubber).[34] The list thus clearly reflected Spanish priorities in natural history: promoting the Spanish monopoly on cinchona; exploring the existence in the American colonies of natural commodities held in monopoly by European competitors; identifying potential replacements for the latter; and searching for those natural commodities pursued by other European nations.

Gómez Ortega's publication belonged to an existing genre of instructions and requests (printed and manuscript) through which European naturalists and collectors attempted to obtain desirable specimens in suitable condition from contributors living in other regions—especially those stationed in outposts of empire. Text, images, and objects circulated within networks of correspondence, obligation, and patronage that followed an unspoken but enforced etiquette based on civility and reciprocity. Gardens, cabinets, academies, and private individuals throughout and beyond Europe participated in these networks. André Thouin, head gardener of the Paris Jardin du Roi between 1764 and 1793, established a widespread network of contributors that surpassed 400 members at its peak in 1786.[35] Like Gómez Ortega, Thouin provided traveling expeditions with instructions for collecting and transporting vegetable cargoes.[36] The furious pace of Joseph Banks's correspondence is legendary: he is reputed to have written more than fifty letters a day.[37] He also distributed instructions to many of his numerous contributors, who supplied him with countless botanical specimens from around the world.[38] Correspondents writing from British colonies often discussed the likelihood of finding local products to compete with the trade of other European nations, exactly as in the Spanish case.

The Spanish Crown, however, was unique in making this system part of its official imperial policy. Each governor, viceroy, or *intendente* of Puerto Rico, Santo Domingo, Havana, Louisiana, Yucatan, New Spain, Santa Fe, Peru, and Caracas received six copies of the *Instrucción* with orders to distribute them to those individuals better suited to carry them out in the territory under his authority. Then, in July 1789, royal orders required colonial administrators at every level to produce reports on potentially useful natural productions in their areas and send back to Spain both information and samples to be assessed by royal botanists and pharmacists. Colonial administrators quickly put these royal orders into effect, and responses flowed from points throughout the Spanish empire. To name but one example: only four months after the orders were passed in 1789, the Viceroy of New Granada, José de Ezpeleta, transmitted to Mutis the king's wishes to receive notice of all the useful trees growing in his dominions, asking that the naturalist furnish wood samples and indicate their names and all known uses, qualities, and virtues—whether medical, for the dye industry, or for other purposes.

In addition to tapping into the imperial network, over the years Madrid's Royal Botanical Garden dedicated much energy to nurturing relationships with correspondents: Antonio Palau, professor of botany at the garden from 1773 to 1793, was particularly active in this regard.[39] A tally of new contributors between the years 1783 and 1794 totaled eighty-six men.[40] These collaborators furnished the garden with live plants, seeds, and written information. In 1785, Juan del Castillo wrote from Puerto Rico with a list of local plants that might interest the Madrid garden, among them wild nutmeg, guaiacum, cedar and other woods, cotton, cacao, indigo, ginger, and many edible and medicinal fruits.[41] Also from Puerto Rico, Victoriano de Aldea Urries sent five crates of live plants in 1789 and again in 1790, whereas Ramón Hernáiz shipped seeds.[42] Mariano Espinosa tended a botanical garden in Havana, where he suggested planting cinnamon. He also dispatched crates of live plants every year between 1793 and 1796; the shipments included pineapple, tobacco, mamey, guava, peppers, and other tropical fruits.[43]

The results of Palau and Gómez Ortega's efforts at establishing a network of contributors are clearly reflected in the yearly indexes of seeds planted at the garden, which record both a growth in the total number of plants and in the proportion of them that came from outside Europe. In 1772, only 4 of 650 species originated in the Americas, amounting to a mere 0.6 percent of the total.[44] By 1788, 262 of the 1,250 species growing in the garden were American. This represents 21 percent of the total plants and 43 percent of those newly planted over the previous year.[45] In 1781 alone, seventy-seven Peruvian seeds received from Ruiz and Pavón were planted at the garden.[46] The botanists sent twelve more seed shipments between 1782 and 1787.[47] This success, however, needs to be significantly qualified. Though Gómez Ortega did receive many plants from contributors and managed to grow them, the types of American plants cultivated in the garden held little prospects for profit. Few of them were new, many of them were decorative, and those with commercial applications were not likely to be exploited on the scale necessary to fulfill that potential. Their value lay in their foreignness, and the botanical garden used them to strengthen its prestige and its collections by distributing samples among its growing network of European correspondents (which included participants in Britain, France, Sweden, Switzerland, Italy, Denmark, Holland, and Portugal), creating obligations that could be fulfilled only through the exchange of other exotic seeds.[48]

Madrid's botanical garden and natural history cabinet also received objects and information from the natural history expeditions.[49] For example, in April 1779, Ruiz and Pavón sent back to Spain a first shipment consisting of four crates of live plants; six with dried herbarium specimens, seeds, bulbs, roots, and natural curiosities; and one with almost 200 drawings of plants and about sixty floral anatomies. The next major collection was only sent five years after the first, in May 1784. It included six stoves with live trees and shrubs, all of which were lost in a storm off the Chilean coast. The remaining materials—fifty-five crates with natural history samples collected over a period of five years and about 1,000 drawings, 800 of them new and 200 copies replacing images sent in 1780 on a ship that had been seized by the English—did not fare much better, as the ship was wrecked and everything was lost. A third shipment, sent one year later (May 1785), did make it to Madrid, though none of the twenty-nine species of live plants survived the journey. Ruiz and Pavón had time for one last and large shipment, sent in January 1787, with seventy-three boxes of dried plants, almost 600 drawings, and eighteen containers of live plants. Finally, when Ruiz sailed into Cadiz in September 1788, he had with him fifteen more crates of dried plants, almost 600 more drawings (perhaps copies of those sent the previous year, in case of loss), twenty-four crates of live plants, and three volumes of botanical observations totaling 2,000 descriptions.[50] After Ruiz and Pavón's return to Spain, Juan José Tafalla, the pharmacist who carried on with their work in South America, continued sending dried plants, roots, seeds, and drawings to Madrid for many years.[51]

Colonial Projects: Peripheries as Centers

The centripetal trajectory of specimens and information from all points of the peninsula and the colonies to Madrid conformed to Gómez Ortega's understanding of the botanical garden as the nucleus of a unified global natural history network that deployed travelers, administrators, and other institutions, such as the cabinet and the pharmacy. "In this way," he explained to potential contributors in his *Instrucción*, "a Botanical Garden becomes the center of correspondences of its class, of useful experiments on Botany and Agriculture, and of the propagation of plants worthy of being multiplied."[52] If only it were possible to arrange for each ship arriving to Spain from the Canary Islands, Havana, Cartagena, and Buenos Aires to take a crate with natural productions from the region to Madrid, the Royal Botanical Garden could reproduce the plants it received and redistribute them throughout the world. "In a few years," he hoped, "we would become the owners of the greater part of the vegetable riches of Spanish America."[53] Gómez Ortega's vision was directly modeled on the French *Jardin du Roi*. All the coffee trees growing in Martinique and nearby islands, he noted, could be traced back to Paris, where they had originally been planted before being shipped to the Caribbean.[54] Joseph Banks, likewise, looked to the French model and transformed Kew Gardens based on it.[55]

However, when Gómez Ortega enthused in his *Instrucción*, "we would become the owners," his "we" (the key word in that passage) is significantly different from Thouin's or Banks's. Gómez Ortega's "we" includes both Madrid *and* the viceroyalties: it is the imperial we. Addressed to contributors throughout the empire, it promised that economic botany would benefit both the Spanish Americas and Madrid.

Thus, the centripetal trajectory described by Gómez Ortega is pointedly different from the model that Bruno Latour proposes for late-eighteenth-century scientific travel as based on his examination of Jean François de Galaup de La Pérouse's experiences in Sakhalin, an island off the coast of Siberia, in 1787.[56] According to Latour, the production and circulation of knowledge in European expeditions operated cyclically and iteratively. If Parisian cartographers did not know whether Sakhalin was a peninsula or an island, the only way for them to learn this was to dispatch someone there. Once *in situ,* the traveler collected data by conducting observations or measurements and also by talking to local inhabitants. However, data were useless "out there": the success of the enterprise depended on mobilizing it back to Paris. Only there could raw information be processed into knowledge, in the shape of an answer to the original question. That is, the map showing that Sakhalin is actually an island could be produced only in the center, not in the periphery. Not that the story ends there. A new chapter then begins, as new journeys back to the periphery can be undertaken using the map.

Latour's model helpfully points out the cyclical nature of knowledge production and the importance of networks, standardization, accumulation, and connectedness. However, on the basis of problematically rigid notions of center and periphery, it suggests that knowledge production takes place exclusively in the metropolis (as the only possible "center of calculation") and ignores negotiations between metropolis and colony and between different colonial territories.⁵⁷ Latour's acumen in describing the trajectories of information and objects as viewed from Europe by naturalists, such as Thouin or Banks, is one of his model's successes, but it comes at a high price. It is quite conceivable that La Pérouse in Sakhalin or Banks in Australia may have felt themselves to be in a distant periphery, whose value existed only in relation to, and was provided by, the European center. Matters, however, looked rather different to naturalists in the Spanish Empire—who, unlike La Pérouse and Banks, lived permanently or for long periods in the regions they surveyed. For them, the circulation did not resemble the flight of a boomerang, always returning to the center, but rather a more reciprocal paddle game. Every letter or shipment from one side provoked a reply from the other. Seeds, to give one example, traveled in both directions. The Madrid botanical garden not only received specimens from the Americas but sent its own shipments to colonial institutions, such as the Mexico City botanical garden.⁵⁸

Although Europe always remained the ultimate frame of reference as the source of funding, prestige, and significance, Spanish and *criollo* naturalists were as involved in local agendas as they were in metropolitan ones, if not more. Based in colonial capitals such as Mexico City, Lima, and Bogotá, which had institutions of higher education, printing presses, important private libraries, and active intellectual and artistic communities, these men found themselves working very much in the thick of things, not at the edge. For such men as Mutis, Ruiz, Pavón, Sessé, and Mociño, trajectories were multiple, overlapping, and highly sensitive to local interests, emphases, and interpretations.⁵⁹

The fact that Madrid dispatched a large number of travelers, instructions, and orders to its territories does not imply that there was no initiative in the colonies. Mutis, as I have already mentioned, suggested the New Granada expedition. This was also the case with the New Spain expedition and the botanical garden it established in Mexico City, both of them proposed by Sessé.⁶⁰ The botanical garden, Sessé proudly informed Gómez Ortega in a letter requesting European seeds for it, would be equal in size to the one in Madrid, if not larger.⁶¹ The opening ceremony in May 1788 was celebrated with an ambitious firework display in the best tradition of Mexican festivals and celebrations.⁶² The public performance, described a week later in the *Gaceta de México,* began with three papaya trees on stage, two of them female trees decorated with their fruit and flowers and between them a male tree without fruit. Before amazed spectators, sparks blew from the male to the female trees, representing the transfer of pollen to fertilize flowers. Lighting effects at the foot of the male tree suggested a growing garden, illuminating the scene with bright changing colors. As the trees disappeared, an inscription appeared in fireworks, reading *"amor urit plantas"* (love inflames plants), to celebrate the Linnaean system.[63] As in Madrid, the garden employed a botanical lecturer to provide lessons and demonstrations. More than fifty people turned up to watch expedition member Vicente Cervantes show and describe plants, mentioning their medicinal virtues, economic value, and Greek, Latin, and Mexican names.[64] The term concluded—again, as in the peninsula—with a public performance of "botanical exercises" in which top students identified and described plants presented to them before an audience that included local notables.[65] Thus, this colonial garden represented an exact equivalent of the Madrid institution and not a subordinate acclimatization center whose entire purpose consisted in furnishing the metropolis with desirable commodities.

The avidity with which local governors encouraged the investigation of their territories and their level of involvement with these projects suggest that administrators had their own reasons to be interested in natural history, beyond their duty to satisfy orders from Madrid. Investigations of New Granadan nature, for instance, had begun long before the Crown passed its orders to local administrators. Mutis's independent work had started in the early 1760s, and local officials and governors had prepared a series of reports to inform the viceroy about natural commodities. In 1783, Viceroy Caballero y Góngora commissioned Francisco Armero to explore the reputed existence of American cinnamon trees in the Bee Mountains.[66] He also received a report from a Lucas Herzo y Mendigaña on the potentially useful trees in the province of Novita.[67] José García de Léon Pizarro, president of the Quito *Audiencia* between 1778 and 1784, received a similar report describing the medicinal plants and useful vegetation of the region.[68] Throughout the colonies, administrators actively encouraged the exploration of local nature, hoping to identify products that would boost the regional economy by securing profitable trade with the metropolis. The impetus for natural exploration originated *in situ* much more often than in the metropolis. Though Madrid issued rather vague directives to explore everything and identify anything of potential value, administrators and naturalists in the colonies had greater awareness of local circumstances, opportunities, and dangers and had considerably more at stake personally, professionally, and economically. Given the scope of activity with local origins and motivations, Madrid could at times lag years behind colonial projects—twenty years, in the case of Mutis.

The emergence of local projects tended to depend largely on the presence and initiative of interested administrators and naturalists. Local administrators responded to Gómez Ortegas's request not only by circulating his instructions but by issuing their own. Though the former described methods for transporting specimens from the colony to the metropolis, colonial directives focused on the development of local plantations and the improvement of the regional economy. To colonial administrators, Madrid represented above all a market for the natural commodities they intended to produce locally. Colonial reactions by no means reflect an attitude of passive diffusion but rather an active engagement with an imperial project that promised to benefit local interests. If Madrid was, for all involved, clearly the center, the viceroyalties were not the periphery: *they* were the empire.

This is not to imply that Madrid did not matter: it remained the center of validation and funding for the entire empire. However, it did not monopolize power, nor did it always set the most pressing agenda. Naturalists reported both to metropolitan authorities in Madrid and to local ones in the Americas—a situation that at times proved difficult to navigate, especially when salaries came from local coffers, but Madrid offered the recognition they desired for their botanical work or when local interests expected their investment to pay off handsomely and quickly while metropolitan institutions and authorities repeatedly rejected American samples. Naturalists invariably found themselves caught between competing demands from two very different audiences.[69] Often, positioning could depend on geographical location: with Madrid at least a couple of months away by maritime post, colonial arguments often proved more urgently pressing and also tended to be more in accord with naturalists' own situations and opinions.

Furthermore, distance from Madrid—in space, time, and experiences—granted naturalists considerable autonomy and favored the development of independent attitudes. Naturalists usually used their letters to inform metropolitan authorities of decisions already taken, rather than to ask for permission to take them. Distance also resulted in botanists establishing strong relationships and obligations locally as well as across the ocean, becoming in this way nodes of a truly colonial network. Mutis, for instance, developed over the years a widespread circle of contributors in the Spanish Americas—even before the expedition received official authorization from Madrid in 1783.[70] His many students traveled botanizing in South America, and his collaborators provided descriptions and natural history specimens from different regions.[71] As his reputation as an expert on American natural products grew, more and more correspondents submitted specimens for his evaluation. Most often, they asked him to pronounce authoritatively on the identity of a questionable natural product. To give just one example among many, Mutis answered a query from Juan José de Villaluenga, president of the Quito *Audiencia*, about whether a product sold in both Quito and Bogotá as *goma de Guayacán* was actually that substance. Mutis concluded it was not, adding that very often substances sold in the Americas as cinchona, balsam, ipecacuanha, and other *materia medica* were not those products at all.[72] To encourage and reward Villaluenga's efforts, Mutis sent him some of the indigo seeds he had obtained from Caracas.[73] In return, Mutis received alfalfa seeds, which led to a new obligation he fulfilled by sending instructions for using indigo in the dyeing industry.[74] All the other expeditions also benefited from local contributions and were engaged in comparable networks of collaboration, obligation and, often, competition.[75]

As naturalists became more and more deeply ensconced in local projects, the relevance of Madrid to their daily work diminished. Moving subtly and gradually away from Madrid, their priorities and allegiances shifted from the distant metropolis to much more immediate and present concerns. The local character of natural investigations and the economic promise involved often placed naturalists and territories in direct competition with one another: given the Crown's interest in natural commodities, presenting a region as a privileged repository of natural riches offered it advantageous status vis-à-vis other colonies. While concern about Dutch pepper, British tea, or French coffee might be shared in Madrid, Mexico City, Lima, and Bogotá, the economic opportunities presented to each of the colonies encouraged rivalries.

The natural commodity that generated the most competition without a doubt was cinchona.[76] Miguel de Santiesteban had first identified cinchona in Popayan, New Granada, in 1752; until then, the tree was thought to grow exclusively in the Loja region of the Quito *Audiencia*. The following year, Santiesteban presented to the Spanish Crown a proposal for a cinchona *estanco*, that is, a state monopoly on sale of the product. Santiesteban continued his efforts to promote New Granada cinchona over the years, enrolling Mutis in the cause. As the Loja hills became increasingly depleted, New Granada and Peru, where cinchona had also been identified, vied for a share of this profitable trade. In the 1780s, Viceroy Antonio Caballero y Góngora and Mutis repeatedly sent New Granadan cinchona to Spain and attempted to persuade Madrid of its good quality. Caballero y Góngora suggested that cinchona from Loja should be reserved for sale in Peru, the Philippines, and Asia, whereas New Granadan cinchona should supply the European market. Mutis also promoted a cinchona *estanco*, which he claimed would prevent market speculation.[77] This project pitted New Granada, which favored the *estanco* that it would supply with local cinchona, against Peru, which opposed the idea in favor of unrestricted free trade. As for the Quito *Audiencia,* it had no intention of letting either South American competitor interfere with its profitable monopoly. Loja growers resolved the threat posed by cinchona from Huanuco (Peru) by taking control over production in that region and used their connections with commercial houses in Spain to edge New Granadan cinchona out of the market.

The cinchona wars were also waged on the taxonomical front, in both Madrid and the Americas. Beyond competition among regional growers, there existed legitimate doubts about the medical effects and botanical identity of Loja, New Granadan, and Peruvian cinchonas. Throughout the 1770s and 1780s, Madrid's Royal Pharmacy performed multiple analyses to assess whether the different varieties were all truly cinchona, while physicians experimented with patients in the Royal Hospitals to establish whether they had equivalent curative properties.[78] In 1785, the Pharmacy pronounced Loja and New Granadan cinchona a single species with identical effects. The opinion, however, was contested by the Marquis of Valdecazana, supervisor of royal pharmacies and the king's chamberlain, and several medical reports. Further tests resulted in a compromise, with New Granadan cinchona pronounced inferior in quality to the Loja variety but nevertheless suitable for medical use. However, amid contradictory information that made it impossible to resolve the issue, in February 1789 the Crown suspended shipment of New Granadan cinchona, and moved away completely from the product in September 1790.[79] Years later, Alexander von Humboldt described the arbitrariness of this decision by protesting that "physicians, like the Popes, drew lines of demarcation on the map" to establish that cinchona growing north of a certain point ceased to be effective.[80]

At stake in the discussions about the relative merits of regional cinchona varieties were not only local economies but botanical reputations, in the Americas and Europe. Disagreements about cinchona became entangled with old rivalries between Mutis and Gómez Ortega and with the battle for control of Spanish botany being waged in Madrid between Gómez Ortega and Antonio José Cavanilles. The Peruvian contingent—Ruiz, Pavón, and López Ruiz—sided with the former, while Mutis and his former student Francisco Antonio Zea, based in Madrid at the time, backed the latter. In 1792, Ruiz authored the first Spanish publication on cinchona, *Quinología, o tratado del árbol de la quina o cascarilla, con su descripción y la de otras especies de quinos nuevamente descubiertas en el Perú* (Madrid, 1792).[81] The treatise publicized the expedition's discoveries, organizing its findings into seven types of Peruvian cinchona. In 1799, he revised this classification into nine species in the second volume of the *Flora Peruviana et Chilensis*.[82] The following year, Francisco Antonio Zea, Mutis's Madrid-based former student from New Granada, challenged their findings. In an article published in Cavanilles's *Anales de historia natural,* Zea promoted similarities between Peruvian and New Granadan cinchona by eliminating some of the distinctions outlined by Ruiz and Pavón, agreeing with Mutis's decision of limiting the cinchona types to four.[83] Zea also extolled the high quality of New Granadan cinchona, based on reports from foreign botanists, and reasserted Mutis's priority as a discoverer of the tree in that region.[84] This provoked responses from Mutis's rival in that dispute, Sebastián López Ruiz, and from Ruiz and Pavón.[85] Under the guise of introducing the discovery of four new species of cinchona by their collaborator, Juan Tafalla, Ruiz and Pavón used their *Suplemento a la quinología* (Madrid, 1801) to maintain the superiority of Peruvian over New Granadan cinchona and to attack Mutis and Zea as botanists, defending López Ruiz. Ruiz and Pavón derided Mutis's collection as unreliable, citing his long absence from the field and his dependence on students and peons, and also criticized his use of color as a criterion for distinguishing cinchona varieties.[86] The following year, Alexander von Humboldt, traveling in South America and thus able to examine cinchona varieties first-hand, got involved in the discussion. Humboldt sided with Mutis, his host in Bogotá and a correspondent for some time, writing to López Ruiz that both men had discovered cinchona independently, but Mutis in 1772 and López Ruiz three years later.[87] In the end, the two rivals found it more useful to side with one another: in 1803, probably through Humboldt's intervention, López Ruiz suggested to Mutis organizing the publication in a journal of a letter from Humboldt extolling the virtues of New Granadan cinchona, to persuade European readers to purchase the product discredited by Ruiz and Pavón.[88] Meanwhile, back in Spain, Cavanilles ousted Gómez Ortega as director of the botanical garden in 1801, in what Mutis described as "a great revolution."[89] In his inaugural speech, Cavanilles reviewed the history of Spanish botany, naming Mutis as its most noteworthy exemplar.[90] On his death three years later, the New Granada–born Zea took over as director.[91]

The case of cinchona demonstrates not only the interplay among botany, medicine, trade, and politics that characterized the investigation of natural commodities in the Spanish colonies but the way in which colonial agendas vied with imperial ones. If Madrid existed as the center of an extensive network of crisscrossing trajectories, it clearly was not the only participant with much at stake on the success of the expeditions. Mexico City, Lima, Bogotá, and other colonial centers had as much, if not more, to gain. The tension between imperial and colonial interests gradually became untenable. Local traditions of understanding and using specimens and local economic interests in exploiting natural commodities went hand in hand with a proprietary attitude toward landscape and territory. Enlightened *criollos* agreed with European ideals of science as useful to both economic and moral improvement. For them, however, this improvement was connected to an increasing sense of ownership. The *Mercurio Peruano,* a journal published in Lima between 1790 and 1795, voiced this belief in a statement whose eager hopefulness matches the exalted name of the society that published it, the *Sociedad Académica de Amantes del País.* "Scientific expeditions," it declared, "should erase the sad memories of bloody expeditions. They lead far-away towns to culture, order, the arts, and countless goods."[92] If Madrid considered the expeditions a way of reaching back to glorious times of conquest, power, and profit, Spanish Americans found in them the promise of reconfiguring and moving away from that past.[93]

Notes

1 *Real Cédula*, Aranjuez, April 8, 1777. Archivo del Museo de Ciencias Naturales, Madrid (hereafter AMCN), item 13 in María de los Ángeles Calatayud Arinero, *Catálogo de las expediciones y viajes científicos españoles a América y Filipinas (siglos XVIII y XIX)* (Madrid: CSIC, 1984). The New Spain expedition received almost identically phrased instructions, *Real Orden,* El Pardo, March 20, 1787. Archivo del Real Jardín Botánico, Madrid (hereafter ARJBM), V, 1, 1, 17. All translations are mine.

2 Antonio Lafuente and Nuria Valverde, "Linnaean Botany and Spanish Imperial Biopolitics," in Londa Schiebinger and Claudia Swan (eds), *Colonial Botany: Science, Commerce, and Politics* (Philadelphia: University of Pennsylvania Press, 2004).

3 The literature on the expeditions includes *La expedición Malaspina, 1789–1794,* 9 vols (Madrid: Lunwerg Editores, 1987–1996); Daniela Bleichmar, *The Visual Culture of Natural History: Botanical Illustrations and Expeditions in the Eighteenth-Century Spanish Atlantic* (Ph.D. dissertation, Princeton University, 2005) and "Painting as Exploration: Visualizing Nature in Eighteenth-Century Colonial Science," *Colonial Latin American Review,* 15 (2006), 81–104; Andrew David, *The Voyage of Alejandro Malaspina to the Pacific 1789–1794* (London: The Hakluyt Society, 2000); Alejandro Díez Torre et al. (eds), *La ciencia española en ultramar* (Madrid: Doce Calles, 1991); Iris H. W. Engstrand, *Spanish Scientists in the New World: The Eighteenth-century Expeditions* (Seattle: University of Washington Press, 1981); Marcelo Frías Núñez, *Tras El Dorado vegetal: Jose Celestino Mutis y la Real Expedición Botanica del Nuevo Reino de Granada (1783–1808)* (Sevilla: Diputación Provincial de Sevilla, 1994); Antonio González Bueno (ed.), *La Expedición botánica al Virreinato del Perú (1777–1788)* (Madrid: Lunwerg, 1988); A. Federico Gredilla, *Biografía de José Celestino Mutis* [1911] (Bogotá: Plaza & Janés, 1982); Antonio Lafuente and José Sala Catalá (eds), *Ciencia colonial en América* (Madrid: Alianza, 1992); Mauricio Nieto Olarte, *Remedios para el imperio: historia natural y la apropiación del Nuevo Mundo*

(Bogotá: Instituto Colombiano de Antropología e Historia, 2000); B. Sánchez, Miguel Ángel Puig-Samper, and J. de la Sota (eds), *La Real Expedición Botánica a Nueva España 1787–1803* (Madrid: V Centenario/Real Jardín Botánico, 1987); María Pilar de San Pío Aladrén (ed.), *El águila y el nopal: La expedición de Sessé y Mociño a Nueva España (1787–1803)* (Madrid: Lunwerg Editores, 2000); Juan Pimentel, *La física de la monarquía: Ciencia y política en el pensamiento colonial de Alejandro Malaspina (1754–1810)* (Madrid: Doce Calles, 1998); Marie Louise Pratt, *Imperial Eyes: Travel Writing and Transculturation* (London and New York: Routledge, 1992); Arthur Robert Steele, *Flowers for the King: The Expedition of Ruiz and Pavon and the Flora of Peru* (Durham, NC: Duke University Press, 1964); and Benjamín Villegas (ed.), *Mutis y la Real Expedición Botánica del Nuevo Reyno de Granada*, 2 vols (Bogotá: Villegas Editores; Barcelona: Lunwerg Editores, 1992).

4   Paula S. De Vos, "Research, Development, and Empire: State Support of Science in the Later Spanish Empire," *Colonial Latin American Review*, 15 (2006), 55–79.

5   Jorge Cañizares-Esguerra, "Eighteenth-Century Spanish Political Economy: Epistemology and Decline," *Eighteenth-Century Thought* 1 (2003), 295–314.

6   See Schiebinger and Swan, *Colonial Botany*.

7   See Antonio Barrera-Osorio, *Experiencing Nature: The Spanish American Empire and the Early Scientific Revolution* (Austin: University of Texas Press, 2006), and the chapters by Barrera-Osorio and Sandman in this volume.

8   Londa Schiebinger, *Plants and Empire: Colonial Bioprospecting in the Atlantic World* (Cambridge, Mass. and London: Harvard University Press, 2004*)*.

9   Indeed, given the global dimension of the Spanish empire and the interconnectedness through trade and governance of its Atlantic and Pacific territories (as represented, for instance, by the Philippines, Peru, or the port of Acapulco), the story of the Spanish Atlantic is always closely connected to that of the Pacific, and embedded in a global dimension.

10 On the co-existence of imperial and colonial scientific agendas, see Jorge Cañizares-Esguerra, *Nature, Empire, and Nation: Explorations of the History of Science in the Iberian World* (Stanford: Stanford University Press, 2006), as well as the essays by Miruna Achim, Orlando Bentancor, Paula De Vos, Matthew James Crawford, Irina Podgorny, Mauricio Nieto Olarte, and Carlos López Beltrán in the special issue of the *Journal of Spanish Cultural Studies*, vol. 8, no. 2 (July 2007).

11 See for instance Pedro Rodríguez de Campomanes, *Reflexiones sobre el comercio español a Indias* [1762], ed. and with an introduction by Vicente Llombart Roca (Madrid: Instituto de Estudios Fiscales, 1988).

12 Cited in Steele, *Flowers for the King*, p. 90.

13 Correspondence of Martín Enríquez, Viceroy of New Spain, 1574–76, Archivo General de Indias, Seville (hereafter AGI): Mexico, 19, N. 151 and 175–7.

14 Marcelo Frías Núñez and Andrés Galera (eds), *Pedro Fernández de Cevallos: la ruta de la canela americana* (Madrid: Historia 16, 1992), 7–52.

15 Jonathan Israel, *Conflicts of Empires: Spain, the Low Countries and the Struggle for World Supremacy, 1585–1713* (London: Hambledon Press, 1997), esp. 305–

18 and 349–60; Benjamin Schmidt, *Innocence Abroad: The Dutch Imagination and the New World, 1570–1670* (New York: Cambridge University Press, 2001), 123–84.
16 José Gumilla, *El Orinoco ilustrado y defendido* [1741] (Caracas: Academia Nacional de la Historia, 1963), 248–9.
17 Luis J. Ramos Gómez, *Las noticias secretas de América, de Jorge Juan y Antonio de Ulloa (1734–1745)* (Madrid: CSIC, 1991), vol. 1, 586.
18 Establishing botanical identity was a difficult and common problem in early modern botany: see for instance Emma Spary's discussion of Dutch debates about nutmeg, "Of Nutmegs and Botanists: The Colonial Cultivation of Botanical Identity," in Schiebinger and Swan, *Colonial Botany*, 187–203.
19 Casimiro Gómez Ortega, "Informe al Duque de Losada sobre la canela y el clavo de Quito," January 15, 1777, cited in Francisco Javier Puerto Sarmiento, *Ciencia de cámara: Casimiro Gómez Ortega (1741–1818), el científico cortesano* (Madrid: CSIC, 1992), 153–5.
20 Joseph Dombey to José de Gálvez, Lima, March 16, 1780. AMCN, item 37 in Calatayud Arinero *Catálogo de las expediciones.*
21 José Celestino Mutis to J. E. Ramos, Santa Fe, October 15, 1783. ARJBM, III, 1, 2, 58; reproduced in *Archivo Epistolar del Sabio naturalista Don José C. Mutis*, ed. Guillermo Hernández de Alba, 4 vols (Bogotá: Instituto Colombiano de Cultura Hispánica, 2nd edn, 1983), vol. 2, 270.
22 Antonio Caballero y Góngora to Mutis, December 1783?, *Archivo Epistolar*, vol. 3, 41.
23 Draft of letter from José Celestino Mutis to Juan Casamayor, Caballero y Góngora's secretary, Santa Fe de Bogota, December 23, 1783. ARJBM, III, 1, 2, 22; *Archivo Epistolar*, vol. 1, 144.
24 Frías Núñez and Galera, *Pedro Fernández de Cevallos*, pp. 28–51. This Atlantic failure led to a Pacific push: in the 1780s, as it became clear that American cinnamon and eastern cinnamon were in fact different species, Spain explored the possibility of propagating Ceylonese and Chinese varieties in the Philippines.
25 Casimiro Gómez Ortega to José de Gálvez, February 23, 1777. AGI, Indiferente General, 1544, cited in Puerto Sarmiento, *Ciencia de cámara*, 154–6.
26 Casimiro Gómez Ortega (ed.), *Francisci Hernandi...Opera* (Madrid, 1780).
27 Gómez Ortega, *Historia natural de la Malagueta*, 17.
28 Ibid., 22.
29 Ibid., 25.
30 There is a modern edition, with an introduction by Francisco Javier Puerto Sarmiento (Madrid: Biblioteca de Clásicos de la Farmacia Española, 1992).
31 James E. McClellan and François Regourd, "The Colonial Machine: French Science and Colonization in the Ancien Régime," in Roy Macleod (ed.), *Nature and Empire: Science and the Colonial Enterprise*, *Osiris*, 2nd series, vol. 15 (2000), 31–50.
32 [Pedro Franco Dávila], *Instrucción para que los Virreyes, Gobernadores, Corregidores, Alcaldes mayores e Intendentes de Provincias en todos los Dominios de S.M. puedan hacer escoger, preparar y enviar a Madrid todas las producciones curiosas de la Naturaleza que se encontraren en las Tierras y Pueblos de sus*

distritos, a fin de que se coloquen en el Real Gabinete de Historia Natural que S.M. ha establecido en esta Corte para beneficio a instrucción pública (Madrid, 1776).

33 Duhamel de Monceau, *Avis pour transport par mer des arbres, des plantes vivaces, des semences, des animaux, et de differents autres morceaux d'Histoire Naturelle* (Paris, 1752).

34 Gómez Ortega, *Instrucción*, 37–45; quote p. 41. In contrast to this level of specificity, most instructions remained rather vague; for instance, James Petiver's *Brief Directions for the Easiest Making and Preserving Collections of all Natural Curiosities* (London, 1709?) merely requested that contributors "gather whatever you meet with, but if very common or well known, the fewer of that Sort."

35 Emma C. Spary, *Utopia's Garden: French Natural History from Old Regime to Revolution* (Chicago: University of Chicago Press, 2000), 49–98.

36 André Thouin, "Mémoire pour diriger le jardinier dans les travaux de son voyage autour du monde," in Louis Antonie Milet-Mureau, *Voyage de La Pérouse autour du monde* (Paris, 1797), vol. 1.

37 Spary, *Utopia's Garden*, 50–1. See Neil Chambers (ed.), *The Letters of Sir Joseph Banks: A Selection, 1768–1820* (London: Imperial College Press, 2000).

38 John Gascoigne, *Science in the Service of Empire: Joseph Banks, the British State and the Uses of Science in the Age of Revolution* (Cambridge: Cambridge University Press, 1998); David Mackay, "Agents of Empire: The Banksian Collectors and Evaluation of New Lands," in David Philip Miller and Peter Hans Reill (eds), *Visions of Empire: Voyages, Botany and Representations of Nature* (Cambridge: Cambridge University Press, 1996), 38–47 and David Philip Miller, "Joseph Banks, Empire, and 'Centers of Calculation' in late Hanoverian London," in ibid., 21–37.

39 Antonio Palau, correspondence 1783–86, ARJBM, I, 21, 1–12.

40 "Copia del listado de los títulos despachados a los correspondientes del Real Jardín Botánico desde el año de 1783," Madrid, 1794?–1801?. ARJBM, I, 13, 6, 15; and "Razón de los títulos despachados a los correspondientes del jardín desde el año de 1783," Madrid, 1794?–1801?. ARJBM, I, 13, 6, 16.

41 Puerto Rico, June 8, 1785, Juan del Castillo to Gómez-Ortega. ARJBM, I, 20, 2, 1.

42 Victoriano de Aldea Urries to Gómez Ortega, Puerto Rico, April 20, 1790. ARJBM, I, 20, 1, 2. Ramón Hernáiz to Gómez Ortega, Puerto Rico, May 12, 1790. ARJBM, I, 20, 2, 30.

43 ARJBM I, 5, 10, 1; I, 5, 9, 5, items 1–8; I, 5, 9, 7, items 7–10; I, 7, 5, 1; I, 7, 5, 5; I, 8, 6, 3; I, 9, 4, 1; and "Borrador del informe de Mariano Lagasca sobre la actividad de Mariano Espinosa en La Habana," Madrid, 1817?–1820?. ARJBM, I, 30, 4, 3.

44 Puerto Sarmiento, *Ciencia de cámara*, p. 333.

45 "Catálogo de las plantas cultivadas en el Real Jardín Botánico procedentes de América con expresión de sus lugares nativos y nombres vulgares," Madrid, June 9, 1788. ARJBM I, 4, 9, 8.

46 Madrid, June 3, 1781. ARJBM, I, 3, 2, 1.

47 Hipólito Ruiz, "Lista de las semillas remitidas por Hipólito Ruiz y José Pavón para sembrar en el Jardín Botánico de Madrid, desde el año 1782 a 1787," Peru, 1787. ARJBM, IV, 7, 3, 8; there is another list for 1788. ARJBM IV, 7, 3, 11.

Several *catálogos de siembras* record the planting of these seeds, ARJBM IV, 7, 3, 13.
48 Yearly lists of seeds sent to foreign correspondents 1786-95, ARJBM I, 4, 4, 4; I, 4, 6, 1; I, 4, 9, 1-7; I, 5, 1, 3-5; I, 6, 4, 1-3; I, 7, 4, 1-3; and I, 8, 4, 1. List of foreign correspondents 1785-1800, ARJBM, I, 11, 1, 8. Exchanges with European correspondents, ARJBM, I, 3, 4, 1; I, 3, 5, 2; I, 3, 6, 14-17; I, 3, 6, 20; I, 4, 4, 3; I, 5, 3, 5; I, 6, 4, 5; I, 8, 6, 6; I, 20, 1, 4-11; I, 20, 1, 13-14; I, 20, 1, 17-25; I, 20, 2, 2; I, 20, 2, 4-8; I, 20, 2, 10; I, 20, 2, 18-24; I, 20, 3, 3-14; ARJBM, I, 20, 5, 1-15; I, 20, 5, 29; I, 20, 6, 8-13; I, 20, 7, 2-7; I, 21, 11, 7-8; I, 21, 11, 13-14; I, 21, 11, 16-17; I, 21, 12, 8-9; I, 22, 4, 14.
49 See Paula S. De Vos, "'Curiosities of Nature and of Art': The 'Curious' Side of Natural History Collecting in the Spanish Empire," in Daniela Bleichmar, Paula De Vos, Kristin Huffine, and Kevin Sheehan (eds), *Science in the Spanish and Portuguese Empires (1580-1800)* (Stanford: Stanford University Press, forthcoming).
50 Ruiz and Pavón's shipments are described in great detail in Hipólito Ruiz, *Relación histórica del viage, que hizo a los reinos del Perú y Chile el botánico D. Hipólito Ruiz en el año 1777 hasta el de 1788, en cuya época regresó a Madrid*, ed. Jaime Jaramillo Arango, 2 vols (Madrid: Real Academica de Ciencias Exactas Físicas y Naturales, 2nd edn, 1952), 430-76, and Steele, *Flowers for the King*, 137-55. The live plants are itemized in Hipólito Ruiz, "Lista y Razón de las plantas vivas remitidas en varias ocasiones a España entre los años 1779-1787," Peru, 1787. ARJBM, IV, 7, 3, 9.
51 Lists of shipments from Juan Tafalla in Peru, 1780?-1802. ARJBM, IV, 12, 1, 6, and "Listas de las Semillas que tiene remitidas Dn. Juan Tafalla en diferentes remesas y en diferentes años y otras curiosidades que tiene enviadas en varios cajones desde el año de 1790 hasta el de 1799." ARJBM, IV, 13, 5, 1.
52 Gómez Ortega, *Instrucción*, pp. 9-10.
53 Ibid., p. 22.
54 Ibid., pp. 7-8. For André Thouin's description of this centripetal network, see Spary, *Utopia's Garden*, 90-1.
55 Richard Drayton, *Nature's Government: Science, Imperial Britain, and the "Improvement" of the World* (New Haven: Yale University Press, 2000), ch. 2.
56 Bruno Latour, *Science in Action: How to Follow Scientists and Engineers through Society* (Cambridge, Mass.: Harvard University Press, 1987), 215-57.
57 Latour's notion of "immutable mobiles" is also both helpful and problematic. It does not address the transactions through which information is produced *in situ*, which often involve the erasure of local knowledges (as described in various essays in Schiebinger and Swan, *Colonial Botany*); nor does it address the question of interpretation: a map might be a map, but meanings are created according to who views it, where, when, and to what purposes, destabilizing any presumed "immutability."
58 Martín de Sessé to Casimiro Gómez Ortega, Mexico, April 26, 1786, ARJBM, V, 1, 1, 7, and Mexico, March 27, 1788. ARJBM, V, 1, 1, 18. Gómez Ortega to Pedro Acuña, Madrid, November 5, 1792. AMCN, item 509 in Calatayud Arinero (1984). Sessé to Viceroy Revillagigedo, Mexico, March 15, 1793. ARJBM, V, 1, 4, 4.

59 Along similar lines, Susan Scott Parrish argues for mutuality rather than hierarchy in her discussion of natural history exchanges between North America and Britain at the time: see *American Curiosity: Cultures of Natural History in the British Atlantic World* (Chapel Hill: University of North Carolina Press, 2006).

60 Martín de Sessé to Casimiro Gómez Ortega, Habana, January 30, 1785. ARJBM, V, 1, 1, 1.

61 Martín de Sessé to Casimiro Gómez Ortega, Mexico, March 27, 1788. ARJBM, V, 1, 1, 18, ff. 1r–1v.

62 *Gaceta de México,* May 6, 1788, pp. 75–7.

63 The Linnaean system provoked fierce debates in Mexico; see Roberto Moreno, *Linneo en México: las controversias sobre el sistema binario sexual, 1788-1798* (Mexico: UNAM, Instituto de Investigaciones Históricas, 1989).

64 Martín de Sessé to Antonio Porlier, Mexico, May 27, 1788. ARJBM, V, 1, 1, 22. Sessé to Gómez Ortega, Mexico, June 27, 1788. ARJBM, V, 1, 1, 23.

65 Announcement of the public botanical exercises to be performed on December 11, 1788 by José Vicente de la Peña, Francisco Giles y Arellano, and José Timoteo Arsinas. ARJBM, 5, 1, 1, 27.

66 ARJBM, III, 2, 5, 19.

67 ARJBM, III, 2, 5, 24. Investigations of New Granadan nature were conducted as early as the 1730s: ARJBM, III, 2, 5, 1.

68 Apolinar Díez de la Fuente to José García de Léon Pizarro, Archidona, January 18, 1784. ARJBM, III, 2, 5, 20.

69 An experience not unique to the Spanish world; see D. Graham Burnett, *Masters of All They Surveyed: Exploration, Geography, and a British El Dorado* (Chicago: University of Chicago Press, 2000).

70 Policarpo Fernández, for instance, sent Mutis a list and observations of 117 plants from the Hacienda de la Vega de San Juan in March 1772. ARJBM, III, 4, 11, 4.

71 Antonio Gago, to give but one among many examples, traveled through the provinces of Darien, Portobello, Panama, and Veragua as commissioned by the expedition, Santa Fe, January 8, 1795. ARJBM, III, 2, 6, 84; *Archivo Epistolar*, vol. 2, 111–12.

72 Mutis to Juan José de Villaluenga, Mariquita, July 10, 1786. ARJBM, III, 2, 2, 196 and 197; *Archivo Epistolar*, vol. 1, 312–15.

73 José Celestino Mutis to Juan José de Villaluenga, Mariquita, December 26, 1786. ARJBM, III, 1, 2, 75; *Archivo Epistolar*, vol. 1, 359–61.

74 José Celestino Mutis to Juan José de Villaluenga, Mariquita, July 11, 1787; October 26, 1787; and December 11, 1787. ARJBM, III, 1, 2, 77 to 79; *Archivo Epistolar,* vol. 1, 392–4, 396–8, and 403.

75 The New Spain expedition counted on the collaboration of Ignacio de León y Pérez," Cacique principal de San Juan Acazingo y profesor de farmacia por el Real Tribunal del Protomedicato." Ignacio de León y Pérez to Martín de Sessé, Valle de Santa Rosa, Mexico, November 27, 1792; March 3, 1793; March 18, 1793; and April 30, 1793. ARJBM, V, 1, 3, 17; V, 1, 4, 2; V, 1, 4, 10; and V, 1, 4, 15.

76 On competition between Quito, New Granada, and Peru about cinchona, see Aymler Bourke Lambert, *An Illustration of the Genus* Cinchona (London, 1821);

Gonzalo Hernández de Alba, *Quinas amargas, el sabio Mutis y la discusión naturalista del siglo XVIII* (Bogotá: Academia de Historia de Bogotá, 1991); Nieto Olarte, *Remedios para el imperio,* pp. 184–232; and Steele, *Flowers for the King,* 187–211. The Spanish exploitation of cinchona is the subject of a dissertation in progress by Matthew James Crawford, History Department, University of California San Diego; see his essay, "'Para desterrar las dudas y adulteraciones': scientific expertise and the attempts to make a better bark for the royal monopoly of Quina (1751–1790)," *Journal of Spanish Cultural Studies* 8 (2007), 193–212.

77 José Celestino Mutis, "Real proyecto del estanco de la quina," in *Flora de la Real Expedición Botánica del Nuevo Reino de Granada (1783–1816),* 50 vols (Madrid: Ediciones de Cultura Hispánica, 1954–2001), vol. 44.

78 "Papeles referentes a quinas, 1771–1786," Archivo de Farmacia, Archivo General de Palacio, Madrid (hereinafter AGP), C-2-16 and C-3-16.

79 "Informe de Quina," August 1789 and "Quina. Noticias particulares, muy circunstanciadas que deberán tenerse presentes siempre que se tratare de este ramo," 1789, both in AGP, Reinados, Carlos IV, Cámara, *legajo* 10, *caja* 1 (*legajo* 4650).

80 Alexander von Humboldt, "Cinchona forests of South America," in Lambert, *An Illustration of the Genus* Cinchona, 32–3.

81 An earlier French report by La Condamine, "Sur l'arbre du quinquina," was translated by Sebastián José López Ruiz in 1778 as "Estudio sobre la quina," but remained unpublished until recently (facsimile reproduction, Barcelona: Editorial Alta Fulla, 1986). The manuscript is held in ARJBM, III, 4, 11, 10.

82 Hipólito Ruiz and José Pavón, *Flora Peruviana et Chilensis,* 3 vols (Madrid, 1798–1802), vol. 2 (1799).

83 Mutis discussed the plant in a 1790 memoir, published two years later as *Instrucción formada por un facultativo existente por muchos años en el Perú, relativa de las especies y virtudes de la quina* (Cadiz, 1792), note the geographical misattribution by the printer; "El arcano de la quina: discurso que contiene la parte médica de las cuatro especies de quinas oficiales," *Papel periódico de la ciudad de Santa Fe de Bogotá,* 1793–94, reprinted in Madrid, 1828, and available in an edition by Manuel Hernández de Gregorio (Burgos: Fundación de Ciencias de la Salud, 1994); and an article published in the *Diario de Madrid,* November 11, 1880, no. 315, reproduced in *Flora de la Real Expedición Botánica,* vol. 44, 42–3.

84 Francisco Antonio Zea, *Anales de historia natural,* vol. 2 (1800), 196–235.

85 Sebastián José López Ruiz, *Defensa y demonstración del verdadero descubridor de las quinas del Reyno de Santa Fé, con varias noticias útiles de este específico, en contestación a la memoria de don Francisco Antonio Zea, su autor el mismo descubridor* (Madrid, 1802).

86 Hipólito Ruiz and José Pavón, *Suplemento a la quinología* (Madrid, 1801), 35–6 and 111.

87 Alexander von Humboldt to Sebastián López Ruiz, Quito, February 4, 1802. ARJBM, III, 1, 5, 44.

88 José Celestino Mutis to Sebastián López Ruiz, Santa Fe, March 22, 1803. ARJBM, III, 1, 2, 45; *Archivo Epistolar,* vol. 2, 197–9 (where it is mistakenly dated May 22).

89 José Celestino Mutis to Alexander von Humboldt, Bogotá, May 21, 1802. ARJBM, III, 1, 2, 38; *Archivo Epistolar,* vol. 2, 175–6. Gómez Ortega used the third volume of the *Flora Peruviana et Chilensis* (Madrid, 1802) to attack Cavanilles, Mutis, and Zea, as described in a letter from Cavanilles to Mutis, Madrid, January 22, 1803. ARJBM, III, 1, 1, 79; *Archivo Epistolar,* vol. 3, 214–15.

90 Antonio José Cavanilles to Mutis, Madrid, March 2, 1802. ARJBM, III, 1, 1, 77; *Archivo Epistolar,* vol. 3, 211–12.

91 *Oficio* from Pedro Cevallos, Palacio de Aranjuez, May 11, 1804, naming Francisco Antonio Zea director of the Madrid Royal Botanical Garden on Cavanilles's death. ARJBM, I, 22, 1, 2.

92 *Mercurio Peruano,* IX, 25, reproduced in Jean-Pierre Clément, *El Mercurio Peruano, 1790-1795,* 2 vols (Madrid: Iberoamericana and Frankfurt am Main: Vervuert, 1997), 118.

93 The connections among natural history, nationalism, and the Latin American wars of independence are explored in Thomas Glick, "Science and Independence in Latin America (with special reference to New Granada)," *Hispanic American Historical Review* 71 (1991), 307–34.

SECTION IV

Contested Powers

CHAPTER 10

The Electric Machine in the American Garden

JAMES DELBOURGO

A few days previous to my arrival in Egypt, two Germans, who travelled about with an electrical machine, had made some experiments at Alexandria, Rossetta, and Cairo. They had imagined, that this would be a method of picking up a great deal of money; but, except for a small number of Europeans residing in these three towns, with a few Greeks and Syrians, they had not many spectators. They were even advised not to attempt to display the effects of their machine, and excite the astonishment of the people, who would infallibly raise against them an outcry of sorcery, which might be attended with very unpleasant consequences to the electrifiers, and perhaps to the other Europeans.

—Charles Sonnini, *Travels in Upper and Lower Egypt*
(London, 1800)

Electricity in Eden

An electrical machine was a moral instrument. Made of a glass globe or plate fixed in a wooden frame four or five feet tall, generators required the laborious mechanic-like rotation of the glass to produce friction and collect charge for experiment or display (Figure 10.1).

These were no mere material engines, however. What they offered was, precisely, access to the immaterial: sparks of "electric fire" made tangible the forces that mediated between spirit and matter, animating the Creation. More than the sum of their glass, wood, and metal parts, electrical generators were moral machines for cultivating reason, sensibility, and virtue through physical experiences of divine power.[1]

As moral machinery, generators found a home in the American garden long before writers such as Nathaniel Hawthorne described shrieking locomotives shattering the edenic peace of a pastoral civilization. Under European eyes and the eyes of its Christian settlers, America had always been a garden, where Fallen man would labor to redeem the sin of tasting forbidden knowledge. Particularly in Puritan New England, spiritual election required the cultivation of land and the soul; agriculture and botany were sacred tasks. Leo Marx's classic *The Machine in the Garden* (1964) made machinery synonymous with large-scale industrial production and the eclipse of Jefferson's virtuous agrarian republic. With an Emersonian dread of industrial machinery as the destroyer of human virtue at its

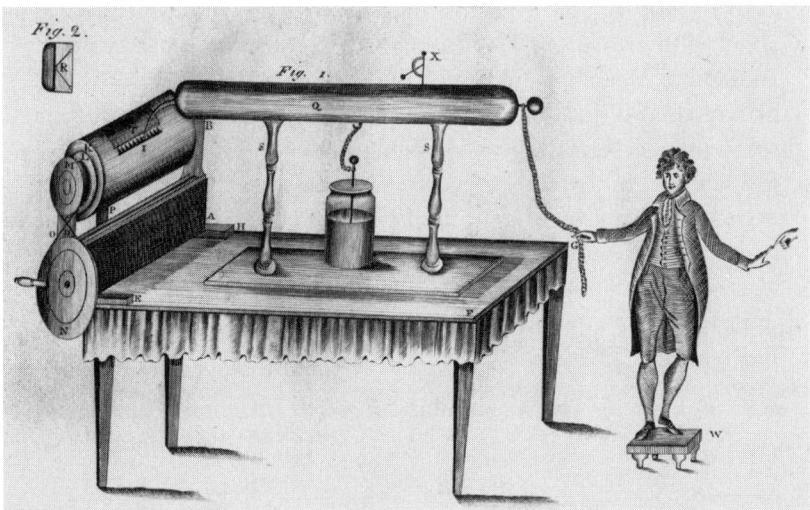

Figure 10.1 The Electric Machine. This is a classic electrostatic generator with Leyden jar attached and an experimental volunteer on an insulating stool, in one of the first illustrations of a complete electrical apparatus published in North America. Frontispiece of *An Epitome of Electricity and Galvanism, By Two Gentlemen of Philadelphia* (Philadelphia: Jane Aitken, 1809), courtesy of the Bakken Library and Museum, Minneapolis.

argumentative center, Marx's study cast technology and the pastoral ideal in largely oppositional terms. However, his analysis passed over an entire era in which machinery was seen to serve the cultivation of spirit and virtue. The two orreries built in the early 1770s by the Pennsylvania astronomer-artisan David Rittenhouse are exquisite examples of virtuous machinery. These mechanical planetaria demonstrated the motions of the planets of the orderly Newtonian solar system. In so doing, they taught valued truths about the rational and benevolent character of the Creator. As Provost of the College of Philadelphia and experimental philosopher William Smith put it in a demonstration using Rittenhouse's machine, the orrery was "a System of mimic worlds, by which all that Newton taught, & which so few are able fully to comprehend, is made to speak to the Eyes in a Language truly intelligible." Experimental machines made nature speak.[2]

Philosophical apparatus also played an important role in projects to exert metropolitan control over distant peripheries. Experimental machines were a means of exporting laboratory technique to discipline experience in the world "outside." They translated potentially chaotic natural phenomena into a recognizable series of signs—numerical measures or standardized languages, verbal or visual—for aggregation into a coherent and useful system. "Metrology," in Bruno Latour's words, "is the name of this gigantic enterprise to make of the outside world a world inside which facts and machines can survive." This metrological order, it has been persuasively argued, was tightly linked with Europeans' attempt to impose political order and, indeed, moral order over their dominions. Imperial networks can usefully be seen as metrological ones, and empire can be said to break down where networks end and exotic experience resists conversion to standardized signs. Early modern Europeans saw the ability of philosophical apparatus to discipline experience in non-European climates and geographies as vouchsafing their natural theology. They would demonstrate both the universal regularity of the natural world and the moral reliability of a divine order that yielded stable truths.[3]

However, as with other philosophical apparatus, the truths that electrical machines yielded proved variable. The epigraph to this chapter by the French naturalist Charles Sonnini nicely stages a common metropolitan view of resistance to enlightenment in the periphery, in which European demonstrators are threatened as unwelcome sorcerers by a hostile superstitious province. The machines of enlightenment could inspire mistrust, provoke resistance, and result in "very unpleasant consequences" for Europeans. When provincials themselves turned experimental demonstrators, such as in colonial British America, the politics of enlightenment could be equally volatile. This was in no small part because electrical machines themselves were ambiguous in their effects: was the experience of bodily electrification they afforded one of discipline or indiscipline? Did such experience constitute participation in an orderly political cosmology uniting provincial Americans with Britons or did the electric fire's shocks and sparks offer a dispensation into a new order of being? This chapter examines electrical machines as engines both of unity and disunity around the Atlantic world, as imagined agents of imperial integration, racial differentiation, and political separation. It ends with an unlikely machine that had lurked in the American garden all along: the electric eel. In attempting to handle the eel like an electric machine, experimenters engaged with the central tension at work in the relation between electricity and empire: could exotic natural phenomena be reduced to predictable effects through experimental discipline?[4]

The Province of Enlightenment

In the mid-eighteenth century, Benjamin Franklin produced an entirely original language for conceptualizing electricity and conducting electrical experiments, yet the "Philadelphia experiments" (as they became known) produced no redrawing of the map that placed European cities at the center of scientific production and American ones at the margins. Why was this so? Atypically, patronage rather than commerce brought the first electrostatic generator to the colonies in 1747, when the Pennsylvania Proprietor Thomas Penn sent the dexterous printer Franklin a machine to continue experiments already begun with glass tubes at the Library Company of Philadelphia. Franklin and his experimental circle followed European electricians' lead in two crucial ways. Conceptually, they subscribed to a view of electricity, established by Newton in his optical writings, as an active principle inhabiting all gross matter, partly material and partly immaterial, and an essential means of divine animation in the cosmos. Bodily experiences of electricity playfully discharged in makeshift laboratories were thus already disciplined ones, in that they referred back to divinely ordained relations between matter and spirit. Following codes of bodily experiment was, precisely, the second important means by which the Philadelphians inserted themselves within networks of licit electrical practice. By reading their own bodies as instruments for detecting the presence and quantity of electric charge, they could make reasoned guesses about the behavior of the electric fluid in an auto-experimental idiom intelligible to the transatlantic community of electricians.[5]

Dismissing the notion of different kinds of electricity and rejecting a strictly mechanical belief in the existence of microscopic electrical particles (a theory supported by the abbé Nollet, the leading French electrician), Franklin reduced all manifestations of the electric fluid to variations in quantity of charge. All bodies possessed variable charges that, if brought into sufficient proximity, sought an equilibrium—hence, the production of shocks and sparks that realized electricity's "desire" for equalization between "positive" and "negative" bodies. Having successfully used this quantitative scheme to account for the play of different charges in the Leyden jar (an early capacitor), Franklin ventured the virtuosic analogical gambit that it could also account for the action of thunderclouds. If lightning were in reality only electricity on a vast scale, a large metal wire should be able to conduct it out of harm's way. "Let the reader judge of the exquisite pleasure he must have felt," exalted Franklin's friend and fellow electrician Joseph Priestley, when "he perceived a very evident electric spark" in his hand from the string of a kite he claimed to have flown during a storm in 1752. Passing lightning through a body already disciplined by experiment enabled the recognition of atmospheric electricity. The lightning rod thus became plausible in an instant of sublime metrological experience that made the sky an extension of Franklin's laboratory and lightning into a recognizable laboratory phenomenon—static electricity.[6]

Astoundingly, the Philadelphian's economic language of plus and minus quantities became the standard lexicon of electricity, validated both by the effectiveness of the lightning rod (despite the fact that conductors often, in fact, failed as protective devices) and his system's utility in predicting electrical effects in and between charged bodies. This did not mean that European experimenters automatically embraced Franklin's results or did not possess local political motives for appropriating them. Indeed, the French, rather than the British, became among the first to perform Franklin's sentry-box test on lightning, led by the comte de Buffon, who was motivated in part by his personal rivalry with Nollet. Although such experimenters as Franz Aepinus later successfully challenged elements of the system, Franklinism became the dominant language of electricity from British North America to the kingdom of New Spain and from Paris and London to Bologna and St. Petersburg. With the utility of electrical experiment so vividly dramatized, the ranks of itinerant demonstrators touring these extended geographies soon swelled. This was not the controlled diffusion of knowledge from an institutional center to its peripheries but a voluntaristic and commercial appropriation of philosophy, technology, therapy, and display, driven in no small part by eager consumers on the periphery.[7]

Artisanal replication and commercial importation spread generators around British America as engines of novel physical experience. Sheer spectacle and the market for diversion took the lead. Tellingly, before Franklin's associate Ebenezer Kinnersley began demonstrating the Philadelphia system, the horologist William Claggett was already showing shocks and sparks in Boston taverns with a machine evidently of his own devising. Replication was not dependent on membership in established experimental networks but could happen, as it were, in the "wilds of America." Traveling through New Granada at century's end, the Prussian natural philosopher Alexander von Humboldt met a man named Carlos del Pozo "in the middle of the *llanos*," who "enjoyed astonishing uneducated people" with a machine "as complete as any found in Europe" built solely from his reading of electrical handbooks. Independent artisanal construction meant that electrical machines were not simply in the hands of elites but among the middling sort, too. Lists of merchants' wares neatly capture the ongoing traffic in apparatus as transatlantic commodities that washed ashore in British America: "imported from London … and to be sold at the very lowest rates … tea, coffee and chocolate, cotton-wool, best French indigo, electrical globes, crow quills, hautboy reeds … with other articles too many to be here enumerated." Machines followed itinerant demonstrators on tours that crossed national lines, often well beyond the confines of the Atlantic. The Viennese Sigismund Niderburg, a medical electrician who claimed to have worked with Galvani in Bologna in the 1790s, set up an electrotherapy clinic in New York before moving his practice to Havana, Cuba at the turn of the century. The Transylvanian Samuel Domjen toured through Maryland, Virginia, the Carolinas, and Jamaica and planned to return home, he told his mentor Franklin in a letter, via Havana, Mexico, Manila, China, India, Persia, and Turkey—all to be paid for through spectacular demonstrations.[8]

Agriculturalists took the relation between electricity and gardening literally. In his *Phytologia* (1800), Erasmus Darwin reported confidently on the electrification of plants as a means to stimulate vegetable growth. In French Saint Domingue, Baudry des Lozières used an electric machine in experiments with cotton. However, in British America, the garden most cultivated by electricity was the garden of the soul. Where the likes of Claggett and Domjen mostly sold corporeal thrill, Franklin's associate Kinnersley was a more disciplined kind of showman. Touring from New England to the West Indies between 1749 and 1774, the former Baptist minister delighted audiences by drawing sparks from their hands and lips, shocking circuits of hand-holding volunteers, discharging guns, and killing small animals. However, the electrification he served up also offered both discipline and refinement for self-consciously improving British Americans: the rational refinement of Franklinist philosophy, which explained active powers and how the lightning rod could tame them; conversational lessons in polite sociability for ladies, gentlemen, and their children; and "more Noble, more Grand, and Exalted Ideas of the Author of Nature" through a pious understanding of the divine natural order. The shocks and sparks they played with were not trivial parlor tricks; they were physical participations in a social, political and cosmological order anchored in piety and driven by provincial desire for genteel self-fashioning. Enlightenment counted as provincial improvement. The experience of electrification would later be linked to challenges to the religious and political order. The skeptic Valentine Rathbun, for example, used electrification as a means to discipline enthusiasm, suggesting that Shakerism resembled "nothing nearer its feeling than the operations of an electerising *machine*." And, as we shall see, patriot writers in the American Revolution conjured the notion of republican electrical convulsions in reaction to British tyranny. However, Kinnersley's demonstrations of Franklin's mastery of the electric skies were carefully un-revolutionary. He insisted that lightning rods were evidence of the benevolence of the deity who remained sovereign over his creation and, prideful talk of the achievements of "ingenious Americans" notwithstanding, proffered no message of political or cultural independence from the mother country.[9]

Since the seventeenth century, Francis Bacon's call for the empirical refoundation of natural knowledge had projected a division of intellectual labor that seemed to map neatly onto the Atlantic world. Colonials would provide information and specimens for organization and interpretation in the metropolis's great clearing house—ultimately the Royal Society. Theories of bodily difference naturalized this epistemological hierarchy. Metropolitans argued that American constellations and climates made the New World's Creole inhabitants humorally degenerate and intellectually inferior. Surely Franklinist electricity and the lightning rod stood this hierarchy on its head? Yet, Franklin's achievement was not heralded as a reversal of this division of labor but a confirmation of it. One important contextual factor was the lack of any independent national consciousness in British America during the 1750s. To the contrary, the Seven Years War produced unprecedented American pride in the British empire's constitutional freedoms and global military supremacy over the French, not an anti-imperial animus. As late as 1772, the Venetian Jesuit Giambattista Toderini praised Franklin's "grand esprit, propre de la Nation Angloise," an identity Franklin would still then have embraced.[10]

Equally important was the positive mediating work between center and periphery performed by discourses of cosmopolitanism. In emphasizing Creole resistance, recent accounts of transatlantic scientific relations risk underestimating American desires to participate as provincial cosmopolitans in European knowledge networks with the prestige and connections they offered. Those who attended Kinnersley's shows in colonial America wanted *in* to European enlightenment, not independence from it. London savants, for their part, took pains to claim British credit for the electrical genius of Philadelphia. Although "not a Fellow of this Society nor an Inhabitant of this Island," President Macclesfield declared in awarding the Royal Society's Copley Medal to Franklin in 1753, he was "a Subject of the Crown of Great Britain," and his invention ought to be "universally diffused" to the "honour of this Society and of the British Nation." Franklin, meanwhile, was happy to paint himself as a modest experimental witness of nature and no proud (metropolitan) theorist, pointedly distinguishing between Nollet's "Theory of Electricity," a "System" the Parisian was busily "defending," from his own humbly empirical "Observations and Experiments." On his own terms, the sage of Philadelphia posed no threat to imperial hierarchies but carefully observed them.[11]

The American Revolution later invested the lightning rod with a seemingly self-evident revolutionary status. In reality, conductors were a conservative technology most often used to protect private property and the material interests of empire. Although Franklin included instructions for mounting conductors in *Poor Richard's Almanack*, most of the people he gave advice to on such matters were wealthy friends and aristocrats wishing to safeguard their mansions against the elements. In Saint Domingue, jewel of France's colonial crown, Parisian itinerants earned handsome fees for mounting rods on private houses, military installations, and government buildings, protecting the plantation complex that worked slaves to death to produce sugar. Lightning rods also facilitated oceanic trade and exploration as technological enlightenment in action on board ship. In the early 1760s, the leading English electrician, William Watson, F.R.S., wrote to Commodore Anson, First Lord of the Admiralty, calling for metal chains to be fixed to the masts of all British ships, a suggestion Franklin had made to Peter Collinson, F.R.S., as early as 1751, and in which their rivals the French soon also became interested. One of several articles in the Royal Society's *Philosophical Transactions* that discussed damage from lightning to ships and military installations from the West Indies to the East Indies, Watson's argument vividly undermines traditional diffusionist accounts of knowledge moving from center to periphery. Citing Franklin and the use of conductors at Philadelphia, Watson urged taming St Elmo's fire by mounting conductors on all British ships and even on domestic military sites, such as the royal powder magazine at Purfleet on the River Thames (scene of a famous quarrel over alleged Franklinist duplicity when rods later failed to prevent damage from lightning in the 1770s). Enlightenment about lightning thus circled back to the center from the inventive periphery, only to travel out again with the maritime agents of imperialism. James Cook, for example, chauvinistically praised the "electric chain" that saved the *Endeavour* in the Batavia Road in 1770 while an unprotected Dutchman perished in the same storm. Johann Forster likewise noted how at Tahiti "fire was seen running down the chain into the water without doing any damage" (Figure 10.2).[12]

The networks that produced, sold, and circulated generators and conductors were thus informal and commercial and crossed national lines among Europeans and American Creoles. Lines of exclusion were firmly drawn, however, to place non-Europeans outside enlightenment's networks. Priestley put the issue forcibly: natural philosophy's capacity to produce "great inventions" was one of "the capital advantages of men above brutes, and of civilization above barbarity." At St John's, Antigua, Kinnersley dramatized the opposition between slavish ignorance and settler inventiveness by blowing up a model "negroe" with the same electricity whites could easily tame with their lightning rods. Back in Philadelphia, the *Pennsylvania Gazette* reported the "Admiration" of the Seneca chief, Kayashuta, at "seeing Thunder and Lightning produced by human Art" in Kinnersley's well-rehearsed displays. Londoners told stories of a "*negro* servant" called Mungo, recently brought to England, who fled an electrical performance by the impresario Benjamin Rackstrow, thinking the electric fire shooting from his fingers the work of the devil. In Jamaica, the British saw their electric machines as devices for expelling the devils they saw at work in resistance to slavery. In a trial between experimental philosophy and the shamanistic practice of "obeah" among rebel slaves, Jamaica's masters turned generators into instruments of torture to punish the leaders of Tacky's Rebellion in 1760, the most serious slave uprising in the eighteenth-century British Caribbean. "On the other Obeah-men, various Experiments were made with Electrical Machines and Magic Lanthorns, which produced very little Effect," ran a report submitted to the House of Commons, "except on one who, after receiving many severe Shocks, acknowledged his Master's Obeah exceeded his own." Though they often portrayed non-Europeans as fetishists who wrongly ascribed spiritual agency to inanimate objects, British-American colonizers invested their machines with the capacity to correct the erroneous supernatural beliefs of others and bring them to obedience through right understanding of natural powers—and of *who* wielded them.[13]

"Providence," opined William Smith in 1772, "called Great Britain (a nation enjoying liberty, religion and science, in their purest and most

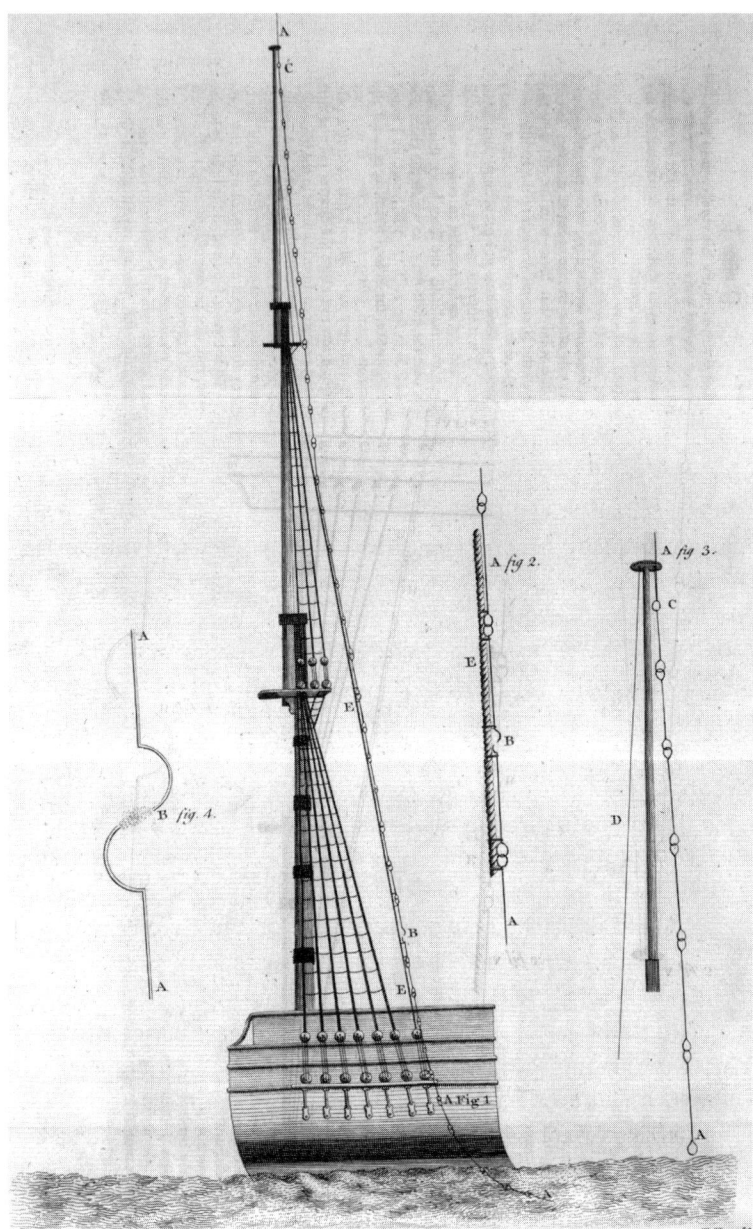

Figure 10.2 Enlightenment on board ship: since 1751, Franklin had suggested attaching chains to ships to act as protective conductors. J. L. Winn, "A Letter to Dr. Benjamin Franklin, F.R.S., Giving an Account of the Appearance of Lightning on a Conductor Fixed from the Summit of the Mainmast of a Ship," *Philosophical Transactions of the Royal Society* 60 (1770), Rare Books and Special Collections Division, McGill University Libraries, Montreal.

improved state) to the possession of that part of America ...[which seems] ordained to empire." Although he spoke on the eve of revolution, Smith's theme was empire—the British empire of liberty and knowledge. He identified philosophical machinery as key motors of the British empire's happy westward course, pointing proudly to the college's "compleat Apparatus for Experimental Philosophy; some parts of which, particularly the Electrical and Astronomical apparatus, are not to be equaled, perhaps, in any other part of the world." "Liberty," he insisted, "will not deign to dwell, but where her fair companion Knowledge flourishes by her side." The electric machine was an idol of enlightenment, marking the "untutored continent" of America as civilized territory, and conferring moral authority on its operators. Empire followed science.[14]

Revolutionary Powers

Electric machines traveled the world in the eighteenth century: North America, the French and British West Indies, Mexico, Venezuela, Africa, India, and China. European traders and colonizers imbued them with the capacity to awe, if not enlighten, other peoples. Such machines did not always function predictably, however. Sometimes, as in the Caribbean and Mexico, excess humidity simply made it impossible to generate sparks. The punitive shocks the British meted out in Jamaica did nothing to deter slaves from believing in Obeah or rebelling against oppression. When the McCartney Embassy carted an array of philosophical instruments to Peking in 1792 to broker better trade relations with the Chinese, naturally assuming the superiority of British technical ingenuity, its members were dismayed to find their hosts already possessed similar astronomical and electrical apparatus, brought by Jesuit missions years before. Machines conferred authoritative access to natural powers on their operators, but claims to moral authority over nature provoked political suspicion. When conductors failed to protect the cathedral at Puebla in 1791, the Mexican polymath and publisher of the *Gaceta de Literatura*, José Antonio de Alzate y Ramírez, blamed the shoddy workmanship of foreign (Spanish) demonstrators who claimed expertise merely by showing off their electric machines. Conversely, Europeans encountered such expertise in the hands of indigenous and Creole populations, and in the form of philosophical prowess, not just mechanical displays. The Scottish physician William Hunter recounted a British military officer's comment on experiments carried out by a Mahratta officer he met in India in 1792: "[H]e even proposed sensible queries, on the nature of the electric fluid ... which showed that he did not look upon the experiments with an eye of mere childish curiosity, which is amused with novelty, but a desire to investigate the cause of this phenomena."[15]

Generators were literally machines of physical communion. A standard display at demonstrations was human circuit-shocking: experimental volunteers joining hands and taking a shock from a generator in the same instant. In the course of the American Revolution, however, patriots suddenly figured generators as engines of separation. "Dr. Franklin," an unnamed revolutionary predicted in the *Virginia Gazette* in 1777, "intends shortly to produce an electrical machine, of such wonderful force, that instead of giving a slight stroke to the elbows of fifty or an hundred thousand men, who are joined hand in hand, it will give a violent shock even to nature her self, so as to disunite kingdoms, join islands to continents, and render men of the same nation strangers and enemies to each other." Electricity could unite the Atlantic world, and it could sunder it. In the ferment of revolution, commentators dramatically intensified the political significance of electrical machines. Suddenly, the political warning issued by Priestley—a liberal Dissenter who linked the truths of experiment with the progress of science and religion—approached momentous fulfillment: "[T]he English hierarchy (if there be anything unsound in its constitution) has equal reason to tremble even at an air-pump, or an electrical machine."[16]

This was no throwaway. Experimental philosophy was increasingly perceived as a liberal bourgeois challenge to the British establishment, one now fatally personified by the traitor Franklin. In British portraits of the 1760s, he had appeared as a bewigged gentleman in a book-lined study, the essence of gentility and learning, who conquered the storm outside through force of disembodied intellect. By contrast, post-revolutionary iconography turned the intellectualist scorn for things and machines on its head, celebrating the corporeal virtuosity of the republican electrician. Jean-Honoré Fragonard's *Au Génie de Franklin* (1779), and later Benjamin West's *Benjamin Franklin Drawing Electricity from the Sky* (c. 1805) were heroic visual representations that launched Franklin into the sky, each underscoring his metrological success in translating lightning bolts into electrical phenomena manipulable by laboratory practice. Such images were paradoxical sublimations of a rising American genius in technological invention. They celebrated the lightning rod as *the* machine of revolution, yet rendered it invisible, preferring to naturalize Franklin's command of heaven's thunder as a power that flowed, unmediated, through his very person. Republican electricity was not a mechanically generated power but a spontaneous sacred fire that streamed through the American body politic, announcing a new moral order that would smash the British mercantilist machine.[17]

The new agents of politics were men of the hand—not monarchs or aristocrats but bourgeois men of social action who made machines go: artisans, farmers, and merchants. Franklin, the son of a tallow chandler, was after all a printer by trade. Invoking machines materialized the agency of rising American genius. "The Doctor, from his uncommon abilities, perhaps not a little whetted by the treatment he met with here," a British sympathizer quipped just after Lexington and Concord, "will no doubt, contribute in some small measure to the winding up the feelings of his oppressed countrymen to a proper electrical charge, for a grand explosion throughout the continent of North America." The embittered colonials, chafing at the burden of new imperial taxation, constituted a giant machine primed for discharging. Just as American provincial cosmopolitanism was yielding to a romantic republican nationalism, British cosmopolitanism was dissolving into cries for the suppression of treason. On this latter view, Franklin was a lowly mechanic who had gotten ideas above his station; a man of things, indeed, as opposed to ideas. A venomous attack on his "pretensions to the title of natural philosopher" in 1777 denied his experiments were "philosophy" at all because, unlike Newton's work on gravity, they offered no method to mathematically compute electrical forces. The sight of such "illiterate men" attended by "wagon[s] loaded with things," stepping "out of their road to turn Philosophers," was a pitiful spectacle that now resulted in a disastrous civil war.[18]

Franklin was little more than a degenerate Creole machine. He was the "the true incendiary" of American affairs, as Solicitor General Wedderburn insisted in denouncing Franklin in 1774 for publishing letters concerning the suppression of Massachusetts, using a "secret cabal" to "blow up the province into a flame." The American, he raged, was "the first mover and prime conductor of this whole contrivance against his Majesty." The "coolness and apathy of the wily American" during the trial itself demonstrated his sheer "bloodthirstiness." The loyalist Peter Oliver confirmed that Franklin had indeed given "such a Shock to Government, and brought on such Convulsions, as the English Constitution will not be cured of in one Century, if ever." The rebellion was a mechanically engineered delusion. The colonists, Oliver lamented, "were like the Mobility of all Countries, perfect machines, wound up by any Hand who might first take the Winch." Thus, "the Wheel of Enthusiasm was set on going, and its constant Rotation set the Peoples Brains on Whirling; and by a certain centrifugal Force, all the Understanding which the People had was whirled away." The prophetic London print *Political Electricity* (1770) confirmed this explosive link between conspiracy, mechanism, and revolution. In one extraordinary panel in a tableau of imperial disintegration that was connected by an "electrical chain," the corrupt minister, Lord Bute, appeared in the form of an electrostatic generator being cranked by Britain's enemies: "his Body ye Electrical Machine shaking hands with ye Principal Nobles in France" (Figure 10.3).[19]

British critics' view of American resistance as the mechanical puppetry of popular passions found its nemesis, however, in the patriot discourse of electrical politics, a language of sentimental revolution founded in quasi-mystical American experience. This language took up the figure of the electrified body and refashioned it into an emblem of revolutionary transgression. In 1776, John Adams wrote of the creation of republican governments as a spreading "electric fire"; American newspaper writers

Figure 10.3 Political Automatism: Lord Bute depicted as an electric machine, near a Franklinesque figure flying a kite between England and France, while a ship bears the electric chain of conspiracy that threatens to undo the empire. "Political Electricity" (London broadside, 1770), courtesy American Antiquarian Society, Worcester, MA.

urged political leaders to "mix with the mass of the people, and get again electrified with a portion of that stern and republican virtue"; and they spoke of military leaders who "kindled that flame of military enthusiasm, which, like electricity, immediately seized the inhabitants of this city." Thomas Jefferson would later recall Virginia's virtuous resistance to Britain as "a shock of electricity, arousing every man and placing him erect and solidly on his centre." Where counter-revolutionaries invoked machines to materialize the dark arts of enthusiastic manipulation, patriots celebrated unmediated electrical experience as a metaphor for certain knowledge of natural and divine will. In this evangelical idiom of electric republican awakening, machines did not drive revolutions; divine powers did. In the end, provincial enlightenment in electricity did not sustain cosmopolitan unity between Britain and America via mechanical discipline. Instead, its technologies and practices generated irreducible experience that figured as an aperture into a new dispensation of moral and political being.[20]

* * *

New World travelers, meanwhile, found that an electrical machine had lurked in the American garden all along: the electric eel. Numb-fish, such as the ray or torpedo, had been known to Europeans since ancient times as wonderful creatures who stunned fishermen at a distance, cured a plethora of ailments, and even stayed ships. By 1800, the Pavian natural philosopher, Alessandro Volta, had constructed the first battery that generated a constant electric "current." Remarkably, this construction of zinc and copper discs soaked in brine was modeled on precise anatomical studies of the internal organs of electric fish, an analogy Volta explicitly invoked in his work. This path to the battery, ostensibly a metropolitan episode in the history of modern physics was, however, first laid in the natural histories of colonial travelers in the Atlantic world. South American voyagers from Jean de Léry to Aphra Behn and Charles Marie de La Condamine had all remarked on numb-fish with stunning powers. The first to recognize such powers as electrical, however, appears to have been the French naturalist Michel Adanson, whose encounter with a catfish in Senegal in 1751 reminded him of the shock from the Leyden jar. Adanson did not press his analogy, but the Leyden connection was precisely what made the American eel's electric powers experimentally recognizable in the Dutch plantation settlements of Surinam, Berbice, Essequibo, and Demerara in Guiana. In 1754, the Essequibo governor, Laurens van 's Gravesande, nephew of the leading experimental philosopher, Willem van 's Gravesande, confirmed in a letter to the Leyden electrician Jean Allamand that the *sidder-vis* encountered in

Guiana produced a shock identical to one he, too, had experienced from the Leyden jar.[21]

In a vivid example of the cross-peripheral production of knowledge, a journeyman from New England named Edward Bancroft became the first to argue for the eel's electricity in English, a claim made possible by his residence as a plantation physician in Dutch Guiana during the 1760s. As with the Dutch experiments, the key for Bancroft was to treat the eel as much as possible like an artificial electric apparatus: if the same effects could be produced as from a generator or capacitor, the eel's electric powers would be demonstrated. This was how Bancroft proceeded. When the eel is "touched with an iron rod, held in the hand of a person, whose other hand is joined to that of another," he recounted in his *Essay on the Natural History of Guiana* (1769), it "communicates a violent shock to ten or a dozen persons thus joining hands, in a manner exactly similar to that of an electric machine." With assistance from native fishermen and slaves, he repeated other standard trials, such as inducing the fish to shock volunteers at a distance through water without direct physical contact. Reasoning backward from effect to cause, electricity and the eel shared an "affinity," he concluded, "not only in the sensations which they communicate, but in the medium through which they are conveyed." Thanks to Franklin, with whom he became friends in the midst of the revolutionary crisis, his work had direct influence in London. John Walsh acknowledged in 1773 that Bancroft's account, to which Franklin had drawn his attention, had informed his own demonstration of the electricity of the torpedo fish, a crucial precursor to Volta's replication of ichthy-electricity in the battery.[22]

Experimental handling of the eel as an electric machine was a precious instance of tropical sensory discipline in Bancroft's Guiana. "All the several senses, and their organs [were] either disordered or violently affected, without being able to determine to which of the many subjects of my examination, I ought to attribute these uncommon effects," he wrote of his sampling of the local flora. Guiana "humble[d] the pride and arrogance of man, by convincing him, that all things are not made obedient to his will, nor created for his use." Turning its rivers into laboratories and using human bodies as instruments to translate the eel into a recognizable electric phenomenon was a rare triumph. It was also a moment wherein peripheral Creole experience explicitly challenged remote metropolitan conjecture. Intensifying Franklin's challenge to Nollet, Bancroft claimed his modest experience overturned the view of the Parisian naturalist René-Antoine de Réaumur (Nollet's patron) that the eel's shock resulted from direct physical contact. "You may, perhaps, think it an act of presumption in me, to dispute the authority of a man, whose literary merit is so universally acknowledged," Bancroft contended, but deference to "great names" was a less sure defense against "the charms of novelty" than personal sensory experience. "Whilst I have sense and faculties of my own, I am resolved to use them with that freedom for which they were given." For Bancroft the Creole and seeming critic of improvement, peripheral epistemology legitimately checked what those at the center could claim to know.[23]

Was the eel, however, truly like a machine? In the late 1790s, Alexander von Humboldt and Aimé Bonpland became the first foreign travelers since La Condamine to journey with official sanction through the South American interior, effecting a remarkable intensification of the project to convert exotic natural electricity into stable mechanical signs. Though superficially similar to Bancroft's, Humboldt's itinerary signified something quite different. Bancroft was a Creole who was seeking his fortune across the periphery, accidentally encountered the eel, and opportunistically published his results to metropolitan attention in London. By contrast, despite originally intending to join the Napoleonic expedition in Egypt, Humboldt made his journey from Paris to South America with profound philosophical self-consciousness. Deeply influenced by German *Naturphilosophie* and by Creole perspectives on American nature, he was a new kind of philosophical traveler, one who regarded *in situ* experience of exotic non-European worlds as fundamental to grasping the unity of nature as a whole. Precisely because the eel was a living thing, whose force drained from its body in the course of laborious attempts to ship it to Europe, Humboldt had to travel to experience its true powers. "I burned with desire to correct my ideas through experience," he confessed: "to make experiments, in the open air, on the banks of these waters where the *Gymnoti* abounded." Turning centuries of natural history upside down, he ventured to America not as a colonial journeyman aiming to reap economic fruits from a botanical garden but, armed with a multitude of precision instruments, as a natural philosopher who dreamed of converting the extravagant biodiversity of the New World to a set of numerical measurements for theorizing *all* of nature anew. In this instrumentalist vision of a new "global physics," measuring temperature, humidity, atmospheric electricity, and other local conditions would make it possible to explain variations of local plant and animal life and ultimately human cultural development.[24]

This was a dramatic intensification of belief in the existence of a moral order in which reliable natural knowledge could be made using philosophical instruments, and it was brought to bear on the electric eels that Humboldt secured for experiment in Venezuela (New Granada), with the crucial assistance of the Guaiqueri Indians at Cumaná on the coast and through indigenous fishermen further South, in and around Calabozo. Humboldt measured these fish, and used gold-leaf electrometers to quantify their charge and eudiometers to analyze the chemical composition of their internal gases. He galvanized dead eels to see whether their electrical motion mechanically exceeded their lifespan (they did not), and put an iron wire to his tongue while touching the eel with the other end, testing for the customary acidic taste of galvanic activity (this was absent). Following the Dutch and Bancroft, he worked tirelessly to read his eels as electric machines but, like his predecessors, he noticed certain disanalogies between mechanical and ichthyological electricity. His fish defied his instruments, for one thing, their electricity registering only inconsistently in his electrometers. More strikingly, he used his own body as the finest tuned instrument of all, preprogrammed by experience with torpedo fish back in Europe, as a basis for comparison with the sensations produced by the American eel. Hours of experimenting with the "Temblador" left his joints stinging; the Tamanac Indians, he noted, called the creature "arimna"— "something that deprives you of movement." He nevertheless declared the shock from the eel "different from that caused by the conductor of an electric machine, the Leyden jar, and even the Voltaic pile." More than this, the disanalogies between mechanical and organic electricity resulted from a singular source: the creature's will. "The convulsion depends solely on the will of the animal," he concluded, continuing, "they are not Leyden jars or perpetually charged galvanic phials, which one discharges merely on establishing communication between the different poles." This was why fully charged eels did not always shock when handled, whereas apparently weakened fish could surprise the unwary and knock them unconscious. Such behavior constituted an *"anomalie très-frappante"*: immechanical action dependent on conscious will and the force of emotions such as anger and fear.[25]

Prized from its habitat, Humboldt's eel thus lay suspended between experimental automatism and willful self-animation. Positioned as mechanical, but defying mechanical manipulation, it became an uncanny living machine. In successive twists of characterization, the eel appeared machine-like, then anthropomorphically human. Subsequent comparison with the structure of Volta's battery now made its organs seem like *"appareils electro-moteurs"*; yet the eel could plunge into "nervous debility, into a veritably asthenic state of being; [where] it seemed tired, exhausted, as would a man whose imagination had been over-stimulated." Though travelers sought to turn the eel into an intelligible and useful object by transforming it into a form of experimental machine, they also

The Electric Machine in the American Garden • 273

regarded it as an irreducible subject: a creature of will and agency that generated spectacles testifying to the sublime power of American nature. Humboldt's vivid description of the process known as "*embarbascar con caballos*" epitomizes this perspective (Figure 10.4). This was the method used by indigenous fishermen to capture eels through "poisoning" them with stampeding wild horses, causing them to exhaust their charge, thus rendering them temporarily harmless.

> I will only imperfectly paint the interesting spectacle of the eels' struggle against the horses, these horses who, their manes standing on end, terror and pain in their eyes, wished to flee the storm which has overtaken them; those yellow and livid eels who, resembling great aquatic serpents, swim on the surface of the water, and pursue their

Figure 10.4 Gymnotus in the Garden: in a scene that struck Humboldt with its sublime aesthetic force, indigenous fishermen employ the method known as "embarbascar con caballos" to render electric eels susceptible to capture through the use of stampeding horses. Frontispiece to volume 2 of Robert Schomburgk, *The Natural History of the Fishes of Guiana* (Edinburgh: W. H. Lizars, 1841–43), Rare Books and Special Collections Division, McGill University Libraries, Montreal.

quarry: all these objects offer, without doubt, the most picturesque of scenes. I remember the superb picture which shows a horse entering a cave, stunned by the sight of a lion! The expression of terror is not greater than that which we witnessed in this unequal struggle."[26]

If Humboldt was prototypical of an instrument-based approach to scientific travel that would come to dominate in the nineteenth century, he was yet a traveler at a crossroads where one of those instruments remained individual aesthetic sensibility. Though it sought to reduce American nature to measurement, this sensibility also amplified the sublime power of the New World and valorized its aesthetic force in a romantic manner unthinkable to previous colonial travelers, who tended to conceptualize their New World ventures more purely in terms of economic botany and represent the power of American nature in negative terms as danger, violence, and chaos. It was precisely such valorizations of American nature that Creole British and Spanish Americans began to appropriate and refashion in the decades around the turn of the nineteenth century to construct romantic nationalist ideologies to drive revolutions against their old European masters. There was, in other words, sublime political power in the natural powers that European scientific instruments and practices could not completely contain.[27]

Epilog: Kasum

Those crucial mediators who gave travelers access to electric eels had their own understanding of such powers. Many years after Europeans and Americans plunged their hands into the rivers of South America longing for experimental communion with organic electric machines, the noted Harvard anthropologist and geneticist, William Curtis Farabee, ventured among the Wapisiana, descendants of the ancient Arawak inhabitants of Guiana, and brought back a story about "the Marriage of the Electric Eel." On Farabee's retelling, an Arawak Marinau, or medicine man, sent word that only the strongest would marry his daughter. Many animals came as suitors, including the jaguar, but none compared to Kasum—the electric eel. The Marinau was skeptical, but Kasum was undeterred: "Touch me and see for yourself how powerful I am." The Marinau did so and was thrown unconscious. "You are very powerful indeed," he conceded, regaining his senses. "You are able to do things even I cannot accomplish. I cannot command the thunder, the lightning or the rain." At the old man's behest, Kasum used his powers to divide the rain clouds to the South and the North and thus claimed the Marinau's daughter's hand in marriage. To this day, Farabee reported, the Marinau invokes Kasum to part the clouds of menacing storms and protect his people. "It is interesting to note," Farabee reflected, "that the medicine man associated the shock of the eel with the thunder, lightning and rain."

Europeans often claimed that only they knew how to move beyond the local to think metrologically; only they possessed cognitive regimes for translating discrete natural phenomena into a coherent system of useable signs. Empire followed metrology, with the enlightened invaders dividing the world between universal "science" and local "beliefs." Colonizers insisted obsessively on irrational indigenous fear of the eel's prodigious powers; terror at lightning, meanwhile, was one of the stock examples of native superstition. By contrast, Farabee's tale suggests that the improbable connection between fish and lightning that Europeans and Americans discovered in the midst of their electrical enlightenment had been known all along. Known by magic rather than machinery, but no less successfully manipulated for that: "[T]he eel facing the storm, or the east, turned away the clouds to the right, or south, and to the left, or north. The medicine man today does exactly the same thing in the same locality and the clouds obey him as they did the eel in ancient times and for the same reason." In finding their own path to understanding the relationship between fish and lightning, the achievements of Euro-American experimenters were dramatic in the field of electricity. However, the point of telling these stories of electric revolutionaries, electric eels, and Kasum together has also been to point out the existence and persistence of competing moral orders of electrical knowledge in the American garden, ones founded on irreducible experience rather than mechanical discipline.[28]

Acknowledgments

Thanks to Simon Schaffer, Kristen Keerma, Nicholas Dew, Neil Whitehead, Vincent Brown, Fiona Clark, Fredrik Jonsson, John Pollack, Elizabeth Ihrig, and Richard Virr.

Notes

1. See Willem Hackmann, *Electricity from Glass: The History of the Frictional Electrical Machine, 1600–1850* (Alphen aan den Rijn: Sijthoff and Noordhoff, 1978).
2. Leo Marx, *The Machine in the Garden: Technology and the Pastoral Ideal in America* (New York: Oxford University Press, 1964); see also John F. Kasson, *Civilizing the Machine: Technology and Republican Values in America, 1776–1900* (Harmondsworth: Penguin, 1977); Richard H. Drayton, *Nature's Government: Science, Imperial Britain, and the "Improvement" of the World* (New Haven:

Yale University Press, 2000), 3–25; William Smith, Lecture Notes on Natural Philosophy, 1768–c.1778, Ms. Coll. 599, folder 11: 32–3, William Smith Papers, University of Pennsylvania; on the Princeton orrery, see Howard C. Rice, Jr, *The Rittenhouse Orrery: Princeton's Eighteenth-Century Planetarium, 1767–1954* (Princeton: Princeton University Library, 1954); Bruno Latour, *We Have Never Been Modern*, trans. Catherine Porter (Cambridge, Mass.: Harvard University Press, 1993), esp. 22–4. By law, British America was a technological colony of Great Britain, where finished manufacturing was restricted to provide an artificial market for British goods. This meant that American electricians tended to be small-scale artisans without connections to larger projects in industrial production. On philosophical machinery's link with industrial productivity in eighteenth-century Europe, by contrast, see Simon Schaffer, "Enlightened Automata," in William Clark, Jan Golinski and Simon Schaffer, eds, *The Sciences in Enlightened Europe* (Chicago: University of Chicago Press, 1999), 126–65. On electricity generally in British America, see James Delbourgo, *A Most Amazing Scene of Wonders: Electricity and Enlightenment in Early America* (Cambridge, Mass.: Harvard University Press, 2006).

3 Bruno Latour, *Science in Action: How to Follow Scientists and Engineers through Society* (Cambridge, Mass.: Harvard University Press, 1987), esp. 215–57, quotation at 251; introduction to Marie-Noëlle Bourguet, Christian Licoppe and Heinz Otto Sibum, eds, *Instruments, Travel and Science: Itineraries of Precision from the Seventeenth to the Twentieth Century* (New York and London: Routledge, 2002), 1–19; Simon Schaffer, "Golden Means: Assay Instruments and the Geography of Precision in the Guinea Trade," *Instruments, Travel and Science*, 20–50; Nicholas Dew, "*Vers la ligne*: Circulating Measurements around the French Atlantic," in this volume.

4 Clark, Golinski and Schaffer, "Provinces and Peripheries," *The Sciences in Enlightened Europe*, 307–11; Ned C. Landsman, *From Colonials to Provincials: American Thought and Culture, 1680–1760* (New York: Twayne, 1997), 57–63.

5 Benjamin Franklin, *Experiments and Observations on Electricity, Made at Philadelphia in America*, 5th edn (London: F. Newbery, 1774); Isaac Newton, *Opticks, Or A Treatise of the Reflections, Refractions, Inflections and Colours of Light*, 4th edn (1730; New York: Dover Books, 1952), esp. queries 8 and 31: 341 and 376; Ernan McMullin, *Newton on Matter and Activity* (Notre Dame: University of Notre Dame Press, 1978), 94–101; John L. Heilbron, *Electricity in the Seventeenth and Eighteenth Centuries: A Study of Early Modern Physics*, rev. edn (Mineola: Dover Books, 1999), ch. 8; on Franklin and science, see Joyce E. Chaplin, *The First Scientific American: Benjamin Franklin and the Pursuit of Genius* (New York: Basic Books, 2006).

6 Joseph Priestley, *The History and Present State of Electricity, with Original Experiments*, 2nd edn (London: J. Dodsley, etc., 1769), 172; Heinz Otto Sibum, "The Bookkeeper of Nature: Benjamin Franklin's Electrical Research and the Development of Experimental Natural Philosophy in the Eighteenth Century," in J. A. Leo Lemay, ed., *Reappraising Benjamin Franklin: A Bicentennial Perspective* (Newark, DE: University of Delaware Press, 1993), 221–42; Simon Schaffer, "Self Evidence," in James Chandler, Arnold I. Davidson, and Harry Harootunian,

eds, *Questions of Evidence: Proof, Practice, and Persuasion Across the Disciplines* (Chicago: University of Chicago Press, 1994), 56–91.

7   Cohen, *Benjamin Franklin's Science*, ch. 6; Heilbron, *Electricity*, chs 14–16.

8   *Independent Advertiser* (Boston), February 8, 1748; Alexander von Humboldt, *Personal Narrative of a Journey to the Equinoctial Regions of the New Continent*, trans. Jason Wilson (1814–25; Harmondsworth: Penguin, 1995), 168; advertisement by Cyrus Baldwin in *Boston Gazette*, January 1, 1770; Sigismund Niderburg, *Improved Galvanismus, and Its Medical Application* (New York, 1803), and *Cultivo del Galvanismo y Uso de sus Virtudes por la Medicina* (Havana, 1807); for Domjen, see Benjamin Franklin to John Lining, March 18, 1755, *Experiments and Observations on Electricity*, 328–9; on the instrument trade in Europe, see James E. Bennett, "Shopping for Instruments in Paris and London," in Pamela H. Smith and Paula Findlen, eds, *Merchants and Marvels: Commerce, Science, and Art in Early Modern Europe* (New York and London: Routledge, 2002), 370–95.

9   Erasmus Darwin, *Phytologia: Or the Philosophy of Agriculture and Gardening* (Dublin: P. Byrne, 1800), 282–3; James E. McClellan, III, *Colonialism and Science: Saint Domingue in the Old Regime* (Baltimore: Johns Hopkins University Press, 1992), 345, n. 60; *New-York Gazette, revived in the Weekly Post-Boy*, June 1, 1752; J. A. Leo Lemay, *Ebenezer Kinnersley: Franklin's Friend* (Philadelphia: University of Pennsylvania Press, 1964); Valentine Rathbun, *An Account of the Matter, Form, and Manner of a New and Strange Religion* (Providence: Bennett Wheeler, 1781), 10.

10 Giambattista Toderini to Franklin, August 15, 1772, *The Papers of Benjamin Franklin*, ed. Leonard Labaree, *et al.* (New Haven: Yale University Press, 1959–), 19: 242; Jorge Cañizares-Esguerra, "New World, New Stars: Patriotic Astrology and the Invention of Indian and Creole Bodies in Colonial Spanish America, 1600–1650," *American Historical Review* 104 (1999), 33–68; Ralph Bauer, *The Cultural Geography of Colonial American Literatures: Empire, Travel, Modernity* (Cambridge: Cambridge University Press, 2003); Joyce E. Chaplin, *Subject Matter: Technology, Science, and the Body on the Anglo-American Frontier, 1500–1676* (Cambridge, Mass.: Harvard University Press, 2001); on the British-American empire of liberty, see Landsman, *From Colonials to Provincials*, 149–80; and Anthony Pagden, *Lords of All the World: Ideologies of Empire in Spain, Britain, and France, c.1500–c.1800* (New Haven: Yale University Press, 1995), ch. 7.

11 Lord Macclesfield, Copley award speech, November 30, 1753, *Papers of Benjamin Franklin*, 5, 130–1; Benjamin Franklin, "Autobiography," *Writings*, ed. J. A. Leo Lemay (New York: Library of America, 1987), 1454; Antonello Gerbi, *The Dispute of the New World: The History of a Polemic, 1750–1900*, trans. Jeremy Moyle (Pittsburgh: University of Pittsburgh Press, 1973), and Jorge Cañizares-Esguerra, *How to Write the History of the New World: Histories, Epistemologies and Identities in the Eighteenth Century Atlantic World* (Stanford: Stanford University Press, 2001); Lorraine Daston, "The Ideal and the Reality of the Republic of Letters in the Enlightenment," *Science in Context* 4 (1991), 367–86; Margaret C. Jacob, *Strangers Nowhere in the World: The Rise of Cosmopolitanism in Early Modern Europe* (Philadelphia: University of Pennsylvania Press, 2006); Jessica

Riskin, *Science in the Age of Sensibility: The Sentimental Empiricists of the French Enlightenment* (Chicago: University of Chicago Press, 2002), 91.

12 *Poor Richard's Almanack*, 1753, *Papers of Benjamin Franklin*, 4: 308; on installing conductors for aristocrats, see, for example, the Duc de Villequier to Franklin, September 23, 1779, American Philosophical Society Manuscript Collection, and as referenced in *Papers of Benjamin Franklin*, 30: 425; McClellan, *Colonialism and Science*, 172–3; Franklin to Peter Collinson, June 29, 1751, *Experiments and Observations on Electricity*, 90; William Watson, "Some Suggestions Concerning the Preventing the Mischiefs, Which Happen to Ships and their Masts by Lightning," *Philosophical Transactions of the Royal Society* 52 (1761–62), 629–35; Louis Simon to Benjamin Franklin, February 14, 1777, *Papers of Benjamin Franklin*, 23: 335; J. L. Winn, "A Letter to Dr. Benjamin Franklin, F.R.S., Giving an Account of the Appearance of Lightning on a Conductor Fixed from the Summit of the Mainmast of a Ship, down to the Water," *Philosophical Transactions of the Royal Society* 60 (1770), 188–91; James Cook, *The Journals of Captain James Cook on His Voyages of Discovery: The Voyage of the Endeavour, 1768–1771*, ed. J. C. Beaglehole (Cambridge: Cambridge University Press, 1955), 433; Johann Reinhold Forster, *Observations Made During a Voyage Round the World*, ed. Nicholas Thomas, Harriet Guest and Michael Dettelbach (Honolulu: University of Hawai'i Press, 1996), 88. The classic statement of the diffusionist model is George Basalla, "The Spread of Western Science," *Science* 156 (May 1967), 611–22.

13 Priestley, *History and Present State of Electricity*, xvii; Kinnersley advertisement, St. John's (Antigua) April 25, 1753, reproduced in Douglas C. McMurtrie, *Early Printing on the Island of Antigua* (Evanston: privately printed, 1943); Kayashuta in *Pennsylvania Gazette*, November 11, 1772; "Mungo" in *Breslaw's Last Legacy; or, The Magical Companion* (London: T. Moore, 1784), 91–2; *House of Commons Sessional Papers of the Eighteenth Century*, 140 vols, ed. Sheila Lambert (Wilmington: Scholarly Resources, 1975), 2: 219; on early English views of fetishism in America and Africa, see Chaplin, *Subject Matter*, ch. 8, and Schaffer, "Golden Means," 31–4.

14 William Smith, speech in Charleston, South Carolina, printed in *Pennsylvania Gazette*, March 12, 1772; Henry F. May, *The Enlightenment in America* (Oxford: Oxford University Press, 1986), 80–6.

15 McClellan, *Colonialism and Science*, 174; Simon Schaffer, "Machine Philosophy: Demonstration Devices in Georgian Mechanics," *Osiris* 9 (1994), 157–82; Fiona Clark, "Nothing Ventured, Nothing Gained: Lightning and Enlightenment in the *Gazeta de Literatura de México* (1788–1795)" (unpublished paper); Bruno de Vecchi Appendini and Carmen Espinosa de los Monteros de de Vecchi, "La Difusión de los Adelantos de la Electricidad en la Nueva España," *Quipu* 13 (September–December 2000), 359–77; Simon Schaffer, "Instruments as Cargo in the China Trade," *History of Science* 44 (2006), 217–46; Hunter quoted in J. M. Steadman, "The Asiatick Society of Bengal," *Eighteenth-Century Studies* 10 (1974), 480–1. On the difficulty of replicating experimental machinery, see Steven Shapin and Simon Schaffer, *Leviathan and the Air-Pump: Hobbes, Boyle, and the Experimental Life* (Princeton: Princeton University Press, 1985), ch. 6;

on eighteenth-century European views of Chinese and Indian science and technology, Michael Adas, *Machines as the Measure of Man: Science, Technology, and Ideologies of Western Dominance* (Ithaca: Cornell University Press, 1989), 79–108, 122–7.

16   *Virginia Gazette*, December 12, 1777; Joseph Priestley, *Experiments and Observations on Different Kinds of Air*, 2nd edn (London: J. Johnson, 1775), xiv.

17   Verner Crane, "The Club of Honest Whigs: Friends of Science and Liberty," *William and Mary Quarterly* 23 (April 1966), 210–33; Wayne Craven, "The American and British Portraits of Benjamin Franklin," *Reappraising Benjamin Franklin*, 247–71; Mary D. Sheriff, "'Au Génie de Franklin': An Allegory by J.-H. Fragonard," *Proceedings of the American Philosophical Society* 127 (June 1983), 180–93; Alfred O. Aldridge, *Franklin and his French Contemporaries* (New York: New York University Press, 1957); Trent A. Mitchell, "The Politics of Experiment in the Eighteenth Century: The Pursuit of Audience and the Manipulation of Consensus in the Debate over Lightning Rods," *Eighteenth-Century Studies* 31 (1998), 307–31.

18   *Pennsylvania Packet and General Advertiser*, July 4, 1775; anonymous, *A Letter to Benjamin Franklin, LL.D., Fellow of the Royal Society, in which his Pretensions to the Title of Natural Philosopher are Considered* (London: J. Bew, 1777), reprinted in *Benjamin Franklin's Experiments: A New Edition of Franklin's Experiments and Observations on Electricity*, ed. I. Bernard Cohen (Cambridge, Mass.: Harvard University Press, 1941), 422–35.

19   Alexander Wedderburn, speech of January 29, 1774, *Papers of Benjamin Franklin*, 21, 47–56; *Peter Oliver's Origin and Progress of the American Rebellion: A Tory View*, ed. Douglass Adair and John A. Schutz, (Stanford: Stanford University Press, 1961), 79, 65, 145–6; James Delbourgo, "Political Electricity: The Occult Mechanism of Revolution," *Common-Place: The Interactive Journal of Early American Life* 5:1 (October 2004), http://www.common-place.org.

20   John Adams to James Warren, April 22, 1776, *Letters of Delegates to Congress*, 26 vols, ed. Paul H. Smith (Washington, DC: Library of Congress, 1976–), 3: 569; *Pennsylvania Gazette*, December 15, 1778 and February 13, 1782; Thomas Jefferson, "Autobiography," *Writings*, ed. Merrell D. Peterson (New York: Library of America, 1984), 9.

21   Michel Adanson, *A Voyage to Senegal, the Isle of Goree, and the River Gambia* (London: J. Nourse, 1759), 244–5; Laurens van 's Gravesande to Jean Allamand, quoted in Philip C. Ritterbush, *Overtures to Biology: The Speculations of Eighteenth-Century Naturalists* (New Haven: Yale University Press, 1964), 36; Giuliano Pancaldi, *Volta: Science and Culture in the Age of Enlightenment* (Princeton: Princeton University Press, 2003), ch. 6; Heilbron, *Electricity*, xxv; Simon Schaffer, "Fish and Ships: Models in the Age of Reason," in Nick Hopwood and Soraya de Chadarevian, *Models: The Third Dimenson in Science* (Stanford: Stanford University Press, 2004), 71–105.

22   Edward Bancroft, *An Essay on the Natural History of Guiana, in South America* (London: T. Becket and P. A. de Hondt, 1769), 196, 199; John Walsh, "Of the Electric Property of the Torpedo," *Philosophical Transactions of the Royal Society* 63 (1773), 461–80.

23 Bancroft, *Natural History of Guiana*, 58, 203, 195. On Bancroft's multiple careers, see James Delbourgo, "The Natural History of a Go-Between: Edward Bancroft's Atlantic Revolutions" (forthcoming); on peripheral epistemology, see Susan Scott Parrish, *American Curiosity: Cultures of Natural History in the Colonial British Atlantic World* (Chapel Hill: University of North Carolina Press, 2006).

24 Alexander von Humboldt and Aimé Bonpland, *Recueil d'Observations de Zoologie et d'Anatomie Comparée* (Paris: F. Schoell and G. Dufour, 1811), vol. 1, "Observations sur l'Anguille Electrique" (all translations mine), 52, 55; Michael Dettelbach, "Global Physics and Aesthetic Empire: Humboldt's Physical Portrait of the Tropics," in David Philip Miller and Peter Hanns Reill, eds, *Visions of Empire: Voyages, Botany, and Representations of Nature* (Cambridge: Cambridge University Press, 1996), 258–92; Marie-Noëlle Bourguet, "Landscape with Numbers: Natural History, Travel and Instruments in the Late Eighteenth and early Nineteenth Centuries," *Instruments, Travel, and Science*, 96–125.

25 Humboldt and Bonpland, "Observations sur l'Anguille Electrique," 59–61, 81–2, 64, 73, 81, 52, 66, 71, 76, 77–8; Humboldt, *Personal Narrative*, 171.

26 Humboldt and Bonpland, "Observations sur l'Anguille Electrique," 87, 72, 55–6; Bill Brown, "Thing Theory," *Critical Inquiry* 28 (Autumn 2001), 1–16.

27 On Spanish-American uses of Humboldt, see Mary Louise Pratt, *Imperial Eyes: Travel-Writing and Transculturation* (New York and London: Routledge, 1992), 111–97; conversely, on Humboldt's debt to Creole Spanish-American valorizations of American nature, see Jorge Cañizares-Esguerra, "How Derivative was Humboldt? Microcosmic Nature Narratives in Early Modern Spanish America and the (Other) Origins of Humboldt's Ecological Sensibilities," in Londa Schiebinger and Claudia Swan, eds, *Colonial Botany: Science, Commerce and Politics in the Early Modern World* (Philadelphia: University of Pennsylvania Press, 2004), 148–65; on the valorization of the landscape in revolutionary North America, see Joyce E. Chaplin, "Nature and Nation: Natural History in Context," in Sue Ann Prince, ed., *Stuffing Birds, Pressing Plants, Shaping Knowledge: Natural History in North America, 1730–1860* (Philadelphia: American Philosophical Society, 2003), 75–95.

28 William Curtis Farabee, *The Central Arawaks* (1918; Oosterhout: Anthropological Publications, 1967), 119–22; Latour, *Science in Action*, ch. 5 and 216.

CHAPTER 11

Diasporic African Sources of Enlightenment Knowledge

SUSAN SCOTT PARRISH

The American colonies are especially worth studying for the history of the Enlightenment because they were places of intense epistemological struggle and negotiation. British America, we now appreciate, was not necessarily exceptional as an inventor of representational government or republicanism; rather, it was exceptional as a meeting place or battleground for once distant peoples, microbes, plants, and animals that produced a strange new world for all. These complex interwoven movements of knowledges and biota made America, not a naked continent awaiting European cloth—as many promoters of colonization represented it—but a place for the fabrication of facts that traveled eastward to avid consumers. America did not passively receive modern civilization (represented by capital-financed global trade networks, far-flung centrally administered European settler colonies, publics linked by print culture, and "universal" scientific systems). Various people in the Americas participated not only in the creation of material prosperity in Europe through their labors with American natural resources but in the creation of an empirically based and hence locally divergent and complex type of nature-knowledge. Because the various branches of natural history were based on sense perception built on extensive experience in an environment, local people, whether indigenous or of European or African descent, all bound together in various chains of face-to-face and written testimony, were crucial testifiers and collectors. The specimens that tacked eastward across the Atlantic—hummingbirds, American Ginseng, giant bones, potatoes, tobacco, Brazilwood, hammocks, and skirts wrought from porcupine quills—were a major material source, from 1492 onward, for the development of botany, pharmacology, zoology, paleontology, geology, and ethnology, among other sciences. Because Europe came in many ways to depend on the matter the Americas provided not only to drive its economies forward but simultaneously to expand its knowledge of the complexity and variety of nature, European elites needed, despite their propagandistic vision of metropolitan centrality, to accept and to credit the hybrid knowledge that emerged from the Americas. Colonial subjects in America were not mere collectors for the knowledge makers of the metropole. European correspondents depended on locals for their kinds of expertise: identifying a novel specimen, understanding its properties or behavior, reporting on or depicting the specimen in its live and natural context, or seeing the interdependence of plants and animals.[1]

When one does not take the European print record of the Enlightenment as autonomous and *sui generis* but instead traces the sources of that final public product westward across the Atlantic (or west and then east back to Africa), one finds the origins of that public record in the epistolary and face-to-face networks of knowledge located in the American colonies. Colonials and European travelers with access to metropolitan scientific institutions, such as the Royal Society of London, reported on the knowledge they garnered in the colonies from white non-elites (farmers, farriers, midwives), Indians, and Africans. Because natural history was predicated on locally embedded observation and experience (as opposed to technologically sophisticated laboratory or astronomical sciences), empiricism often gave authority where political empire took it away. For empiricism was made up of what Francis Bacon called the difficult "woods and inclosures of particulars." Royal Society promoter Thomas Sprat directed naturalists to "heap up" nature's particulars into "a mixt Mass of Experiments, without digesting them into any perfect model" and to render "bare unfinished Histories." Because "Histories" were being formulated in an ever-expanding field of specimens, systems of facts had to continually submit themselves to matter's surprises. This fragmentary, theoretically inchoate, specimen-centered quality of empiricism made local expertise and local access in non-European places crucial to the Enlightenment's laborious reckoning of worldwide matter.[2]

Throughout the British colonies of America, but especially in the southern and Caribbean regions, after coastal and tropical Indian populations had been decimated, Africans became a crucial source for colonial knowledge of the natural world. This fact is often obscured in the promotional iconography and narratives associated with the New Science in which Africans are represented as worshipful and inferior votaries of the new, Eurocentric order of things (Figure 11.1).

Despite this promotional image, Europeans who operated the epistolary networks from their institutional perch in London or Uppsala understood, and admitted in letters, their reliance on colonial correspondents and, beyond that, on the Africans who did the collecting for them. Indeed, there were some facts that, unless they originated with Africans, had no credibility. Though real, this colonial reliance on African testifiers was both ideologically and rhetorically complex. Because of the political conditions of empire and plantation culture, European and colonial acknowledgment of African expertise often came hand in hand with a demonization of diasporic African culture and spirituality.

Africans brought with them to the Americas beliefs in the magical potencies of the natural world and a respect for those adepts who showed knowledge and control of natural processes. In the southern colonies of North America and in the Caribbean, plantation and maroon slaves had more intimate contact with both their masters' agrarian property and forest hinterlands than did their masters. As a result, white colonials believed that Africans possessed zones of knowledge that they themselves did not. Colonial attitudes toward Africans' nature expertise were, as with Indian sources, ambivalent. Colonials credited Africans with the capacity to perform empirical work more extensively than they themselves could but also cast African knowledge as potentially subversive. Unlike the fear of Indian contamination of nature that peaked during warfare, whites began to view the Africans and Creole blacks around them as a toxin in the environment on the establishment of black majorities in the southern and Caribbean colonies. They hence became more suspicious of African knowledge of that environment. This diffuse colonial anxiety became concentrated in an exaggerated alarm about the possibility of slaves using (or misusing) plants to poison their masters. In a paradoxical response to this fear, institutions and print sources on both sides of the Atlantic put a particular trust in African testifiers on the subject of cures for humanly administered plant poisons.

The concept of conjuration—the art of hurting and healing that drew on a spiritually infused natural world—though apparently outmoded in white colonial society after the witch-hunting excesses of the late seventeenth century, had not in fact disappeared but had instead become racialized. Moreover, the category or space of "the hidden" or "the secret" in nature,

Figure 11.1 Though Eurocentric, this engraving does demonstrate that collectors and their specimens must come from all quarters of the globe for the nativity of New Science to occur. "Industria," Frontispiece to Albert Seba, *Locupletissimi rerum naturalium thesauri accurata descriptio, et iconibus artificiosissimis expresso, per universam physices historiam* (Amsterdam: Jansson-Waesberg et al., 1734–65). Courtesy Special Collections Library, University of Michigan, Ann Arbor.

which had been associated with the aboriginal female body in early modern exploration rhetoric and with native American men during times of war, became likewise associated with the black slave in the eighteenth-century colonies. The colonials' perception of black epistemology was not a mere projection of either their fears of slave insurrection or their own outmoded magical knowledge; it *was* loosely based on New World slave societies. Extensive knowledge and use of plant poisons and antidotes and a belief in the manipulability of a spiritualized nature were common to the plantation and even urban cultures of African slaves in the colonies. Religious historian Theophus Smith has described this African and African American conception of the natural world as simultaneously tonic and toxic as a "pharmacopeic cosmos," or a "pharmacosm."[3]

The printed word has been credited as the key conduit of personhood for diasporic eighteenth-century Africans. The cultures of natural history, however, show that knowledge and control of the natural world, demonstrated in oral and empirical ways, was a more open channel for Africans and Creole blacks. It was the talking woods more than the "Talking Book" that was the "ur-trope" of the Anglo-African experience. If the book represented a threshold to civility and personhood that appeared—mystifyingly—to repel African entrance, by contrast the hidden signs and silences of the woods seemed, according to both colonial and slave testifiers, to become readily apparent and audible to Africans (especially after native populations had been driven from the eastern coastal areas). The knowledge and manipulation of the natural world thus offered some measure of efficacy and continuity in the formation of African identity. Numerous figures—collectors, poisoners, healers, adepts—between Boston and the sugar islands attest to this pattern.[4]

In England, a geographical rendering of various peoples' mental and cultural capacities was initiated in the early modern period in tandem with England's nascent global expansion. English (and other northern) writers began to distinguish between a "torrid" and a "temperate zone," and they alleged that people in hot climates could not create great civilizations. Africans would gradually fall under this tropical stigma. Before this shift, the humoral model Renaissance England had inherited from ancient Greece cast both northern Europeans and Africans as equally deviant from the Mediterranean norm of humoral balance and, hence, cultural achievement. If northerners had "gyantly bodies and yet dwarfish wits," Africans were cold, melancholic, and dispassionate. In the seventeenth century, however, as the English colonized the New World and began to trade in enslaved Africans, they manipulated the authorities on humoral theory to rewrite themselves into temperate perfection and the hotter climates into cultural torpor. David Hume's and later Thomas Jefferson's well-known castigations of the mental capacities of Africans are symptoms of what would become toward the end of the eighteenth century a racial binary of the mind.[5]

Though this cultural geography was being formulated but had not yet solidified into a "fact" on which racial binaries rested, Anglo-Americans constructed African expertise spatially, topically, and temporally rather than in an essential hierarchy of superior and inferior. Particularly in the late seventeenth century, before enslaved Africans were a majority in many southern colonies and during the early phases of reducing the coastal wilderness to cultivated land, African knowledge—of rice and tobacco cultivation, of freshwater marsh and coastal fishing, of pathfinding and river navigation, of cattle herding and vaccination—was consciously imported or at least put to use in the colonies. In general, British elite colonials saw Africans as operating within a magical worldview (that had characterized their own officially rejected but still familiar past) and as understanding the sequestered zones of nature (that colonials desired to know without the risks of physical proximity). Colonials needed access to these potent epistemic and geographic zones and, hence, needed African testifiers. Enslaved Africans, in turn, needed to manipulate every advantage they could for some modicum of agency within plantation society and thus frequently provided testimonies about the natural world. An African figure one finds throughout the records of colonial British America, therefore, is the canny and cunning adept, the slave fully initiated into the secrets of nature.[6]

A perceived African expertise became troubling for the white population —especially after black majorities were established—in the area of mineral and, especially, plant poisons. Because colonials acknowledged the keener African capacity to read the signatures (the toxins and virtues) of plants, Africans parsed and mediated a plant sign system for white settlers and, ultimately, a white metropolitan audience. Though the Christian virtuoso was meant to work through disinterested experiments aided by the latest technology to interpret God's natural text, paradoxically heathen adepts, without any modern tools, were allowed by God to decipher his most hidden glyphs. When greater African expertise became troubling to colonials, colonials responded by casting African knowledge as not merely secretive but also as subversive—as "cunning."

If African knowledge could seem subversive to Anglo-Americans, British science—involving technological "improvements," the collection and display of human "curiosities," physical experiments, and pseudoethnography— often seemed oppressive to enslaved Africans. Gilbert Mathison observed in his *Notices respecting Jamaica* (1808–10) that the slaves "are so accustomed to be the subject of exaction that every innovation, though intended for their benefit, gives rise to a suspicion that it is intended for their oppression." On Jamaican plantations, slaves developed the practice of Quashee, playing into their masters' stereotypes of incompetence to stall the exaction of labor or expertise. Africans with aberrant physical traits, such as white spotting of their skin, were examined, transported, and displayed at the Royal Society and other venues, functioning within British transatlantic culture as both a labor-producing property and a curiosity-inducing collectible. In medical experimentation, slaves were often the objects of trials. In 1760, Alexander Garden wrote to his London correspondent, John Ellis, that he had arrived at a new method of treating smallpox "partly owing to some hints given me by my learned and ingenious friend Doctor Adam Thompson, of New York, and partly by some bold trials on a negro of my own." The *Pennsylvania Gazette* and the *Maryland Gazette* ran this advertisement in 1745: "This is to certify, that I had a Negroe Woman bit by a Rattlesnake, and in all Appearance was dying; but applying one the Chinese Stones, sold by Mr. Torres, to the Wound, she in two Hours had no more Pain, and recovered perfectly. Witness my Hand, Francis Bremar. (Orange Quarter, South Carolina, July 24, 1744)." Though diasporic Africans may have at times been healed through these experimental medical regimes, they must have often seen Anglo-American science as inducing yet another painful form of embodiment.[7]

The necessary episteme of subjects within a slave society, moreover, was far from that disinterested curiosity that Royal Society apologists had claimed for themselves at the end of the seventeenth century. In the mid-nineteenth century, both Frederick Douglass and Harriet Jacobs would attest to the many ways in which knowledge within a slave society was linked to subterfuge for both the slave and the slaveholder, thus making disinterest impossible for both races. A key example of this shift in perspective is how the quote from Daniel—"Many shall run to and fro, and knowledge shall increase" (or, more accurately, its Latinate version, "Multi pertransibunt et augebitur scientia")—that had emblazoned the frontispiece to the founding work of the New Science, Bacon's *Instauratio Magna* (1620), and that was again used as an epigram for Sir Hans Sloane's 1707 *Voyage to the Islands Madera, Barbados, Nieves, S. Christophers, and Jamaica*, was later redeployed by the literate North American enslaved community. A divine visitor in the shape of man reads to Daniel from "the book of truth" of the final time when there will be a great war between the king of the south and the king of the north, who

> shall rush upon him like a whirlwind, with chariots and horsemen, and with many ships. ... And tens of thousands shall fall... . But at that time your people shall be delivered, every one whose name shall

be written in the book. And many of those who sleep in the dust of the earth shall awake, some to everlasting life, and some to shame.... But you, Daniel, shut up the words, and seal the book, until the time of the end. Many shall run to and fro, and knowledge shall increase.

Between the Pillars of Hercules, predicting the course of the westering ships, these words had for Bacon called for the end of scholasticism's slavish obedience to classical Mediterranean authority by equating England's maritime expansion with the global and revolutionary collection of empirical testimonies. The "time of the end" that Daniel augured signified to the early Royal Society apologists an English-led pansophic reform of the world that would bring about Christ's return. Harriet Jacobs, by contrast, understood the apocalyptic knowledge of "the time of the end" as the practical knowledge needed to subvert the southern "demon" of slavery: "I wished also to give information to others, if necessary; for if many were 'running to and fro,' I resolved that 'knowledge should be increased.'" The many running to and fro were not Atlantic vessels bearing distant specimens. Rather, they were fellow slaves and sympathetic whites bearing rebellious and clandestine information. The slave network of informants could afford to be neither disinterested nor public, for it worked continually to elude a demonic authority and effect secret deliverance.[8]

Jacobs seems relevant here (even though she lived during a later period of southern slavery than the eighteenth-century subjects of central concern) because she enunciated—by reinterpreting the New Science's key scriptural quote—the impossibility of open "disinterested" knowledge in a society with an extreme imbalance of power. She demystified the association between "hidden" knowledge and paganism so often made by colonials and showed the social necessity of covert investigation and covert networks of information within a slave society. Just as information about American nature flowed from Native Americans to colonials always in a troubled political context that influenced both what natives were willing to tell and how colonials framed that knowledge, the political subjugation of African and black Creole informants conditioned how slave knowledge was anticipated by whites and communicated by blacks. If Jacobs and Douglass later showed how both the slave's and the slaveholder's knowledge was forced by the unnatural institution of slavery to become "cunning," most colonial commentators associated only the slave with this quality without pointing to circumstantial causes. *Cunning*, that term especially associated with women healers and fortune-tellers accused of witchcraft in seventeenth-century New England, had attached itself to a new and likewise vulnerable subject.[9]

Colonials associated cunning with the possession of a secret knowledge of processes, events, and essences not apprehensible to the ordinary mind and the will to use that knowledge for self-interested ends. That a secretive botanical knowledge became the major manifestation of such cunning to colonial observers had to do in part with elite colonials' distance from and slaves' intimacy with woods, marshes, swamps, waterways, and mountains. It also reflected how colonials anxiously read the African concept of the pharmacosm in light of the presence of slave majorities in the South and Caribbean and the troubled history of the supernatural in seventeenth-century England and New England.

In the pre-1720 period of colonial history, when this slave familiarity with marshes, swamps, rivers, and woods was still regarded as a manageable asset, slaves were called on frequently to collect natural specimens for colonials, travelers, and metropolitan correspondents. James Petiver instructed his apprentice, George Harris, who was traveling up the coasts of the Americas in 1698: "Procure Correspondents for me wherever you come, and take directions how to write them, and procure something from them [with whom] you stay, showing their Slaves how to collect things by taking them along with you when you are abroad." A Dutch correspondent at the Cape of Good Hope, John Starrenburgh, replied to Petiver's suggestions with the ripe dismissal: "As for sending a Hottentot, or Some other body a Collecting, is Impossible, for a hottentot and a hog is the Same, and other people are not to be had for that purpose." Nevertheless, colonial collectors in North America routinely used Africans or, sometimes more ambiguously, "servants" to gather specimens for them. Petiver urged Byrd to use his slaves to find specimens. And he wrote in 1701 to the South Carolina widow, Hannah English Williams, who owned a large plantation outside Charleston and was a frequent correspondent, about collecting "small Flies, Moths, Bees, Wasps, Beetles, Grasshoppers, etc.":

> I do not, Madam, expect that you should give yourself the trouble of getting these things but that you would be pleased to let any servant for an hour or two once or twice in a weeke when fair to goe into the fields and woods to bring home whatever they shall meet with, and as many of each sort as they can gett, having severall friends both in England and abroad to obleige.

Because Petiver was the London virtuoso with the best colonial connections of his generation, having correspondents from the Caribbean to New England, his practice of encouraging and trusting enslaved collectors would have set an influential example.[10]

On his second trip to America in 1722, Mark Catesby, overwhelmed by the labor of collecting specimens, wrote to one of his patrons, the noted English botanist William Sherard, to ask for twenty pounds to buy a "Negro Boy" to assist him. And Alexander Garden's trust in a slave of his own was thorough enough to send him on a specimen-gathering expedition. He wrote to Carolus Linnaeus in 1771:

> To procure these ["Chigoes" or chiggers], and other natural curiosities, I sent a black servant last summer to the island of Providence. During his stay there, he collected and preserved some fishes amongst other things; but meeting with tempestuous weather in his return, and being, for several days together, in dread of immediate shipwreck, he neglected all his specimens, many of which perished. Some were fit only to be thrown away, and others were greatly damaged. What remain, such as they are, I shall, by this opportunity, send for your examination.

John Ellis, writing to Garden of the botanizing labors of another African slave, described some seeds he had received from Pensacola, Florida from a curious tree that had been "collected by accident, by the Chief Justice, Mr. Clifton's, black servant." If slaves were enjoined to perform this work often enough, no doubt many came to understand which specimens would pique the curiosity of their masters and even their masters' distant correspondents. Garden clearly trusted his own slave to discern the collectible from the common in the botanical world even if his London correspondent diagnosed another slave's discovery as an "accident." Other times, it appears that slaves pretended not to understand what specimen hunting was wanted of them. Garden wrote to Ellis in 1755: "I spoke to several of the fishermen, but as yet to no purpose. Most or indeed all of them are Negroes, whom I find it impossible to make understand me rightly what I want; add to this their gross ignorance and obstinacy to the greatest degree; so that though I have hired several of them, I could not procure any thing [fish]." By 1771, Garden had changed his mind about using Africans as collectors; perhaps these fishermen found catching fish for specimens rather than for food to be illogical, or perhaps they resented Garden's interfering in their own practiced methods.[11]

In 1761, Mr. D'Ahlbergh, a Swedish immigrant living in Dutch Surinam, wrote a letter to his countryman Linnaeus, telling him of the discovery of a South American root that was esteemed "for its efficacy in strengthening the stomach, restoring the appetite, &c." The discoverer of the root had been a slave named Kwasi or, more importantly, Graman (Great Man) Quacy. Linnaeus named the tree in the African's honor

Quassia amara. In Surinam, where it became a major pharmacological export, it was called "Quassiehout" or "Kwasi-bita." Kwasi was rewarded for his works by his colonial masters: around the time of his root discovery in 1730, he was given a golden breastplate with the inscription "Quassie, faithful to the whites" by a member of the Surinam Council and became the slave to Governor Mauricius in 1744. Then, John Gabriel Stedman related in his Surinam *Narrative* (1796): "By his insinuating temper and industry, this Negro not only obtained his freedom from a state of slavery time out of mind, but by his wonderful artifice and ingenuity has found the means of acquiring a very competent subsistence." In sum, Stedman took Kwasi to be "one of the most extraordinary black men in Surinam, or perhaps in the world." The governor sent Kwasi to the Hague in 1776 to pay a visit to Willem V, Prince of Orange, where the prince gave him, among other presents, a suit of clothes made of blue and scarlet with gold lace trim. Stedman painted a watercolor of Kwasi in this suit, from which the poet-engraver William Blake made the somewhat satirical engraving included here (Figure 11.2).

In this image, Kwasi would have been about eighty-six. By the end of the African's turbulent life, when he was a free man living in his own planter's house in Paramaribo with his own three slaves, he received letters from such places as the Hague and Uppsala addressed to "The Most Honorable and Most Learned Gentleman, Master Phillipus of Quassie, Professor of Herbology in Suriname."[12]

Enslaved Africans such as Kwasi saw distinct advantages to mastering the art and methods of collecting for white colonials. Kwasi knew that his European masters were foremost dedicated to turning the natural realm in Suriname into commodities that could be traded within a world market; he perceived too their interest in placing natural specimens into a universal system of organization. Kwasi knew how to switch, opportunistically, from the cosmology he brought with him from Africa as a seven-year-old and developed within a polyglot Afro-Surinam plantation culture to the episteme of his masters. Biota that slaves found and gathered made their way to the likes of Petiver, Collinson, Ellis, and Linnaeus. In the Americas, then, slaves were often the origin point in the Enlightenment enterprise of the universal collection and systematization of nature even if that enterprise was otherwise managed by well-connected white naturalists. Though their discoveries were sometimes presented as accidental by colonial go-betweens, it is clear that Africans used their own culture's insistence on knowing one's spiritual-natural surroundings to develop a familiarity with American specimens that would be advantageous in their dealings with Europeans.

The celebrated Graman Quacy.

Figure 11.2 Drawing satirically from the Harlequin tradition of the black-masked sly servant, Blake nevertheless stations Graman Quacy amid the architectural locations of power (the Governor's mansion and Fort Zelandia), which Quacy's work as a potent conjurer in charge of the black patrols helped to stabilize. William Blake, *The Celebrated Graman Quacy*. Engraving from a watercolor by John Gabriel Stedman printed in Stedman, *Narrative, of a Five Years' Expedition, against the Revolted Negroes of Surinam...*, 2 vols (London, 1796), II, facing 348. Courtesy Special Collections Library, University of Michigan, Ann Arbor.

At the same time that colonials used slaves as collectors, they also associated slaves with poison. Although the use of enslaved collectors waned somewhat with the establishment of black majorities, fear of Africans' capacities to use plants and minerals to harmful ends was present throughout the eighteenth century. This fear took hold of the white imagination, first because of the slaves' greater knowledge of many—especially sequestered—parts of the southern and Caribbean landscape, and second, because slaves, after their numbers surpassed that of Euro-Americans, were feared as a toxin in the colonial body politic. Garden, despite his trust in his own "black servant" on the Caribbean voyage in 1771, nevertheless expressed to John Ellis in 1760 his anxiety about the large African population within the colony of South Carolina. The colony has "a double enemy within ourselves to fear, viz. the small pox and the negroes." And, repeating his metaphor of the African as a harmful element within the colonial body, he wrote, We have "about 70,000 negroes in our bowels!" "This is our happy situation!" Third, though the stage was set by the early eighteenth century for the rule-bound, divine, and transparent nature of physico-theology, there was a residual Anglo belief that the invisible parts of nature could be controlled to harmful ends by those close enough to nature to know its secrets. Two of the main types of Renaissance magic—"natural magic," involving the knowledge and manipulation of the occult properties of the natural world, and "ceremonial magic," the practice of seeking help from the spirit world—were associated with African slaves. The third type was "celestial magic," involving the divination of astral influences, which was not typically attributed to diasporic Africans.[13]

Though complex poisoning and healing practices were part of the diasporic African pharmacosm and though individual whites were at times the victims of their slaves' poisoning, colonials presumed incorrectly that they were the main target of their slaves' botanical subterfuge. Intercolonial print culture, moreover, helped to spread and give authority to the misguided alarm. Richard Ludlam, the Society for the Propagation of the Gospel (SPG) missionary for Goose Creek, South Carolina, told in 1724 of slave treacheries "by secret poisonings and bloody insurrection." The *South-Carolina Gazette* published accounts in 1735 and 1738 about slave poisonings in other colonies and in 1749 published an editorial that condemned the "horrid practice of poisoning White People." Reverend William Cotes, another SPG missionary, in a 1751 letter explaining planter resistance to slave baptism, wrote of the "horrid practice of poisoning their Masters or those set over them, having lately prevailed among [the slaves]." "For this practice, 5 or 6 in our Parish have been condemned to die, altho 40 or 50 more were privy to it." Garden himself wrote that African poisoning was "so certain … as to render the use of Medicines entirely ineffectual even when given by the ablest practitioners in the province." A South Carolina statute of 1751 stated:

> That in case any slave shall teach or instruct another slave in the knowledge of any poisonous root, plant, herb, or other poison whatever, he or she, so offending, shall, upon conviction thereof, suffer death as a felon.... And to prevent, as much as may be, all slaves from attaining the knowledge of any mineral or vegetable poison, it shall not be lawful for any physician, apothecary or druggist, at any time hereafter, to employ any slave or slaves in the shops or places where they keep their medicines or drugs.

And many slaves were indeed executed "by burning, gibbeting, [and] hanging." In an early indication of this wider colonial anxiety about nefarious slave knowledge, one South Carolinian told the SPG minister Francis Le Jau in 1707, "The Negroes are generally very bad men, chiefly those that are Scholars." In eighteenth-century Virginia, more than 175 slaves were brought to court on charges of poisoning. In Orange County in 1746, a slave named Eve was "drawn on a hurdle to the place of execution and there burnt at the stake." And a report coming from Alexandria, Virginia, about four slave poisoners whose severed heads were displayed on the courthouse chimneys appeared in the *Pennsylvania Gazette* and the *Georgia Gazette* in 1767 and 1768, fanning an intercolonial fear about African knowledge of poison.[14]

In the Caribbean, where there were more active maroon insurgencies and more formidable slave rebellions, the poisoning of whites was envisioned on a larger scale than in the mainland colonies. In the last quarter of the eighteenth century, a young domestic enslaved girl on Barbados poisoned a number of her mistress's children because she "disliked the employment so." In his *History of Jamaica*, Edward Long related the observations of a Jamaican doctor named Barham, who told how "a practitioner of physick was poisoned" with a savanna flower common to Jamaica, Sloane's *Apocynum erectum*, "by his Negroe-woman, who had so contrived it, that it did not dispatch him quickly." After a 1739 treaty was struck between colonial authorities and the Leeward maroons in Jamaica, a treaty highly unpopular with many of the maroons and with plantation slaves because it demanded the return or death of all future runaways, the planter-commander in charge of the military campaign that had secured the treaty, Colonel John Guthrie, was sent to negotiate a treaty with the Windward maroons. On his way, he was "seized with a most violent griping pains in the bowels" and died soon thereafter; "it was strongly suspected that he was poisoned by one of the many slaves who 'were in the utmost despair' over the settlement." On French Saint Domingue (now Haiti), the practices of voodoo, marronage, and "the attempt by maroon leader Francois Makandal to poison every white person on the island in 1757 confirmed that planters daily walked 'on powder kegs.'" Slaves also used poison in this case to "enforce conspiracy in the slave community." Saint Domingue's scientific society, inaugurated in 1784, offered a prize in 1789 for anyone who could solve the cause of death owing to a number of poisons. Poisoning made whites suspicious of all black involvement in the occult; authorities cracked down on the black Creole and slave practitioners of mesmerism, a practice fashionable among white islanders. White fear of poisoning on Saint Domingue was pervasive, affecting scientific and medical experimentation and legislation regarding even the European-derived occult.[15]

Slaves were associated with poisons and botanical charms not only in the South and the Caribbean but in the metropolitan centers of the North as well. In New York City's slave rebellion of 1712, according to the SPG minister John Sharpe, "a free Negro who pretends sorcery gave them a powder to rub on their clothes which made them so confident," echoing the Coromantee tradition of a priest's anointing warriors before battle. The slaves tied "themselves to secrecy by sucking the blood of each others hands" before setting fire to a house on the night of April 1 and killing eight of the whites who answered the alarm. "Peter the Doctor," a free black laborer, was believed to be the leader of the conspiracy but was never convicted. After an alleged conspiracy was detected in New Jersey in 1734, the charged conspirators were arrested; some "had their ears cut off and the others whipt ... Several of them had poison found about them." Two slaves were executed for attempting to poison a master near Trenton, New Jersey, in 1738 using "Arsenick and an unknown kind of root." In the more consequential New York conspiracy of 1741 involving slaves, free blacks, and whites in which Fort George and several other buildings were burned down, a key figure was again a black doctor. "Doctor Harry," who had been banished from the city a few years before for his physic, was burned on suspicion of poisoning. Poisoning continued to be perceived as an element of rebellion in New York and New Jersey until the Revolution.[16]

In the foiled Charleston slave uprising of 1822 led by free black Denmark Vesey, poison and the supernatural were again represented as playing a part. Though Vesey used the twin messages of late Enlightenment revolution and radical Christianity he had imbibed as a transatlantic sailor to spur his fellow Africans in bondage, another leader, "Gullah" Jack, gave confidence to the slaves through his perceived "mastery over the supernatural environment" and his claims that he could not be killed by white men because of his magical powers. According to captured rebel Harry Haig, Jack was to send Haig and others to pour "a bottle of poison" into as many wells around the city as they could a few days before the insurrection. At the trial, a slave testifying against "Gullah" Jack had to be persuaded that "he need no longer fear Jack's conjurations (as he called them)."[17]

The colonial recording of African and black Creole poisoning was not merely a projection of white anxiety. Poisoning and Obeah (rituals using harming medicines) were practiced with regularity. Yet poisoning, and sorcery in general, was more often an expression of interblack conflict—owing to African ethnic rivalry, African–black Creole rivalry, or conflicting interests within the plantation economy—than a means of insurrection. Obeah in Jamaica used shadow catching, poisons, and charms against a particular person by a practitioner hired by a client. Obeah men were typically older and sometimes deformed African-born blacks who were skilled in poisonous and medicinal herbs. The chief sorcerer, or Obeah, of the Windward rebels in Jamaica, however, was a woman, "Nanny," after whom the rebels' town was named. Bryan Edwards wrote that they are "all of them attached to the gloomy superstitions of Africa (derived from their ancestors) … The Gentoos of India are not, I conceive, more sincere in their faith than the negroes of Guinea in believing the prevalence of Obi (a species of pretended magick), and the supernatural power of their Obeah men." For the Matawais, a Surinam maroon people, older Africans were understood to have a larger knowledge of the *óbias*, a term that for Surinam maroons refers more widely to many types of magical and divinatory powers and lesser deities and the protective bundles of herbs and other matter worn to ward off danger. A modern informant related of the period from 1700 to the 1740s, when this group of slaves was escaping up the Saramaka River: "There were older people with them who knew all kinds of óbia. For example, there was one that prevented defeat, and another that let them go for long periods without food." Kwasi, the Afro-Surinam collector known to Linnaeus, was considered to be "faithful to the whites" not only because of his root discovery but because he was the chief "*lockoman*, or sorcerer" on the plantations, where he divined criminals and emboldened black rangers to fight against maroons by providing them with protective natural charms. Not all diasporic Africans, therefore, had access to and control over the volatile pharmacosm; rather, distinct individuals, often possessing a tie with older, African-derived botanical-religious practices, were the sources for the general population to tap into the healing and harming powers of the natural world. Though elite colonials purported to believe in a more visible, rational Nature, they worked with and through these African adepts to control the plantation environment.[18]

Colonial exaggeration or misinterpretation of the target of Obeah, though, indicates that African poisoning, in particular, seemed the inevitable expression of all slaves' mangled gnosis. It seemed the result of a plant knowledge unredeemed by God and of slaves' dwelling at the source of secrets without scientific and Christian protection in the presence of secrecy's power to deform. Moreover, the "horrid practice of poisoning White People" enacted the fear of the white body politic plagued by a toxic "enemy within ourselves." However, then the next question must be: if white perceptions of African poisoning kept alive and made synonymous with African descent the early modern idea of harming, or "black" magic (which the New Science had theoretically excised from its own episteme), why did colonials believe that God had given slaves the peculiar power to heal?

Folk medicine played an important part in slave reminiscences of the nineteenth century. An unidentified slave commented in the 1840s on herbalism: "I knows t'ings dat de wite folks wid all dar larnin' nebber fin's out, an' nebber sarches fo' nudder." Another slave, Silvia King, discussing her uses of roots and barks, observed: "White folks just go through de woods and don't know nothin.'" What became a central feature of nineteenth-century American-born or Creole slave culture had its beginnings in earlier generations of African-born slaves who brought with them their customs of looking for dynamic sources of both harming and healing within their spiritual-botanical world, or pharmacosm, and who also learned a great deal from natives encountered on plantations, on the frontier, or in wilderness flight. Enslaved Africans, deriving their knowledge from generations of oral tradition and face-to-face tutelage, looked on white nature knowledge and saw it as insufficient.[19]

In the early or frontier period of slavery, African knowledge of medicinal plants in the southern colonies and Caribbean often came from native sources; indeed, Sloane believed that what Africans in the Caribbean knew of simples came from Indians. In early Surinam society, the close contact between Indians and Africans, both on the plantations and in communities of rangers and rebels, would have been significant. The Lángu clan of the Saramaka (their name deriving from Loango at the mouth of the Congo, where its two leaders originated) tells a story of its early history (1690s–1731) in which its leader, Kaási, travels with an Indian, his personal óbiama called Piyái (meaning Shaman). Piyái knew of Kaási's Loango god and shared his own knowledge of a magic that provided invisibility (dúngara óbia). A 1970s Lángu informant, in telling this story, remarked that "Indian óbia is the strongest of all óbias!" Inasmuch as knowledge of the spirit world and physical healing were inseparable, Africans in Surinam would have combined their African healing methods with Indian ones. The black Methodist preacher Boston King wrote in 1798 that his mother, who had been a seamstress and nurse on a South Carolina plantation, had "some knowledge of the virtue of herbs, which she learned from the Indians." Mark Catesby explained that the snakeroot of Virginia, a root used often by native tribes, had passed into slave usage. Because of the European belief that God placed antidotes in the same environs as toxins, colonials would have seen African access to locally and experientially derived knowledge of the botanical world as a legitimate means to discovering antidotes.[20]

In letters back to England and in their publications, colonials and travelers, especially physicians, acknowledged African herbal knowledge, usually in conjunction with that of older white females and in sharp distinction from the generality of practicing Anglo-American male doctors. Sloane, though himself skeptical of the breadth and systematization of African physic, recorded the reliance of Caribbean colonials on black doctors: "One Hercules, a lusty Black Negro Overseer, and Doctor, was not only famous amongst the Blacks in his Master Colonel Fuller's Plantation, but amongst the Whites in the Neighborhood, for curing several Diseases, and particularly Gonorrhaeas ... There are many such Indian and Black Doctors, who pretend, and are supposed to understand, and cure several Distempers."

Captain John Walduck, who lived in Rupert's Fort, Barbados, sent Petiver in 1712 "one book of plants, with as many of their names and virtues as I could learne; their uses I have gott from our Physicians (shall I call them), nurses, old women and Negroes." Edward Long told of an antidote ("nhandiroba kernels") for poisoning that an Anglo-Jamaican doctor had learned from the printed European sources of Piso and Linnaeus (who described it as "the antidote cocoon of Jamaica") but also from slaves "that he employed to gather it, [who] called it sabo" and who "esteem [the kernels] antidotes to poison." Garden wrote in 1753 that, if it were not for what the Charleston doctors "learn from the Negroe Strollers and Old Women, I doubt much if they would know a Common Dock from a Cabbage Stock." Metropolitan sources, moreover, requested that colonials get information from native and African sources: for example, in the "Inquiries Recommended to Colonel [Sir Thomas] Linch Going to Jamaica, London, Decem. 16, 1670," Royal Society secretary Henry Oldenburg asked the governor "whether it be true, that the Indians and Negro's make the leaves of it [palma Christi], applied to the Head, the only remedy for their Head-ache?"[21]

In *Friendly Advice to the Gentlemen-Planters of the East and West Indies* (1684), a tract critical of Caribbean plantation society, the Anglo-Barbadian author Thomas Tryon uses an African speaker, who goes from being "Slave" to "Negro" during his persuasive discourse, to criticize British medicine as falling short of its empirical program. Africans, Tryon argues via his black ventriloquist, are the true Baconians and better healers:

> Mast[er]: Though you pretend to do Cures, yet you never read Galen nor Paracelsus, nor have any Apothecaries to make a Trade of the Materia Medica, nor Chymists to tell you the Medicinal Vertues of Minerals… .
>
> Negr[o]: It is also true, that we have no Lip-learned Doctors, nor are confined to the old musty Rules of Aristotle or Galen, nor acquainted with the new Fancies of your modern Fire-working Chymists, or Vertuosi, nor will we compare our selves to you in those things; but we have so much understanding, as not to content ourselves to see with other mens Eyes, and put out our own, as many of your learned Rabbies do; nor want we amongst us those that God and Nature have endued with Gifts of knowing the Vertues of Herbs, and that can by genuine Skill, administer good Medicines, and perform greater Cures, than your famous Doctors with their hard Words and affected Methods.

British medicine bows to both ancient authority and modern fashion. Anglo doctors have only lip learning without eye-derived "understanding" and God-given "Skill." African healing, according to Tryon's critique of slave society, is instead based on direct observation and actually effects cures. Still more, "God and Nature" have given to Africans the knowledge of plant virtues that remains occulted to all other observers.[22]

Slaves on southern plantations often refused colonial medical attention in favor of African practitioners. Kinship, a greater trust in fellow black practitioners' herbal knowledge, and a shared belief that physical illness was rooted in human ill will and conjuration rather than merely in physiological causes made slaves desire African over Anglo healers. In the nineteenth century, planters by and large calibrated slave health in terms of "soundness"—or the physical and mental fitness to labor, reproduce, and obey—an economically based concept that not only failed to recognize slaves as fully human but did not incorporate slave culture's sense of health and illness as produced from a communally mediated botanical-spiritual matrix. Planters continued paradoxically to employ enslaved healers on plantations throughout the life of slavery, seeking to achieve the masters' concept of soundness through the slaves' realm of the pharmacosm. Planters in the earlier period, however, because the ideological (as opposed to the legal) structures defending the institution of slavery were more porous in the eighteenth century and because elite settler society's own magical worldview had not yet faded into distant memory, would have been better situated not only to give credence to African methods but likewise to fear their potency. Indeed, if one looks in the most popular colonial print medium of all, consumed by both wealthy planters and struggling tradesmen, the almanac, with its omnipresent "man of signs" that displayed how a body's anatomical parts were influenced by the zodiac, one can see how many colonial households still operated within a pagan-Christian pharmacosm.[23]

In late-eighteenth-century Virginia, William Dawson asked his neighbor Robert Carter whether he would send his black coachman, "Brother Tom," because "the black people at this place hath more faith in him as a doctor than any white doctor." In 1786, Carter sent one of his slaves, who was "very desirous of becoming a Patient of Negroe David," to a slave healer owned by William Berry of King George County. The South Carolina planter Henry Ravenel wrote that his slave and root doctor, Old March, was so trusted by the slaves that the white doctor "complained that his prescriptions were thrown out of the window, and March's decoctions taken in their stead." Several Virginia planters in the eighteenth century, including Thomas Jefferson, employed black male and female slaves as healers. In many cases, the treatments of white and black practitioners were used jointly. The Bermudan slave Mary Prince told of her suffering from rheumatism from standing too often in cold water while living in Antigua:

> The person who lived in next yard, (a Mrs. Greene) could not bear to hear my cries and groans. She was kind, and used to send an old slave woman to help me.... When the doctor found I was so ill, he said I must be put into a bath of hot water. The old slave got the bark of some bush that was good for the pains, which she boiled in the hot water, and every night she came and put me into the bath, and did what she could for me.[24]

Of Barbados, Griffith Hughes wrote that slaves "stand much in Awe of such as pass for Obeah Negroes, these being a sort of Physicians and Conjurers, who can, as they believe, not only fascinate them, but cure them when they are bewitched by others." Hughes also wrote of Africans: "The Capacities of their Minds ... are but little inferior, if at all, to those of the Europeans," though slavery, he argued, "brutalizes human Nature." In Jamaica, Edward Long caustically described the group healing ritual known as "myalism":

Not long since, some of these execrable wretches in Jamaica introduced what they called the myal dance, and established a kind of society, into which they invited all they could. The lure hung out was, that every Negroe, initiated into the myal society, would be invulnerable by the white men; and, although they might in appearance be slain, the obeah-man could, at his pleasure, restore the body to life.

They used an infusion of the herb "branched colalue," which threw the party into a "profound sleep" until rubbed with another infusion "as yet unknown to the Whites," at which point "the party, on whom the experiment had been tried, awoke as from a trance." Myalism appears to have been a form of "anti-witchcraft," and the myal men might have been used in plantation hospitals because of their familiarity with herbal medicine. Olaudah Equiano told in his autobiography of the magicians-cum-physicians who practiced in his (alleged) birthplace in the remote interior of the kingdom of Benin. They "were very successful in healing wounds and expelling poisons." "They had likewise some extraordinary method of discovering jealousy, theft, and poisoning" ... "still used by the negroes in the West Indies": bearers take up the corpse of the victim and allow themselves, by some occult force, to be physically compelled to the home of the guilty party. In French Guiana, a captured runaway testified in 1748 of maroon medical treatment and the magical etiology of illness: a number of maroons were "being treated with herbs in their houses, to wit André for yaws, Rémy for pain in his foot, which he attributes to sorcery, Félicité for pains throughout her body, which she also attributes to a spell.... It is Couacou who is the herbalist." African healers, rather than being unconcerned with causation, used a combination of herbal remedies and divination that interpreted both physical signs and the social pressures weighing on a sick person.[25]

Despite diasporic Africans' preferences for practitioners who worked from a pharmacosm familiar to them and the sanctioning of some of these practitioners by slave owners, legislatures sought more generally to control inter-slave medical treatment, signifying the profound concern such African knowledge produced in white authorities. In 1748, the Virginia legislature passed a law stating that no slave should administer medicine to another slave unless both masters should give their approval. The penalty was death. In South Carolina, slave doctors could not administer medicine to fellow slaves without the supervision of a white. Punishment for disobeying this law was fifty stripes. The whites' impulse to regulate inter-slave medical practices registered how reliant both blacks and whites were on African knowledge.[26]

Antidotes discovered by Africans that seemed widely applicable were not restricted, however, but instead rewarded and made public in local and transatlantic publications. Although slaves' testimonies against whites in courts were customarily disallowed, slaves' verifiable testimonies about botanical cures were encouraged. Writing from Boston in 1716 to Dr. John Woodward, the secretary of the Royal Society, Cotton Mather described the procedure of inoculation he had learned from his Coromantee slave. In an earlier tract on Christian conversion, Mather had written with fully racialized disdain of Africans: "Indeed their Stupidity is a Discouragement. It may seem, unto as little purpose, to Teach, as to wash an Aethiopian." In this letter to Woodward, however, he explained that he had first learned of the practice of inoculation from his slave, Onesimus, before he had read about it in the pages of the *Philosophical Transactions* (communicated by a Greek correspondent working in Turkey, Dr. Emanuel Timonius, but propounded in various settings by Lady Mary Wortley Montagu after observations made in Turkey and by Dr. Thomas Sydenham). Mather named Onesimus (meaning "helpful") after the fugitive slave whom Saint Paul had converted while he himself was in prison, whom Paul called "my own heart" and whom Paul helped to free from his master Philemon (Phil. 1:8–14). Mather wrote:

> Many months before I met with any intimations of treating the smallpox with the method of inoculation anywhere in Europe, I had from a servant of my own, an account of its being practised in Africa. Inquiring of my Negro-man Onesimus, who is a pretty intelligent fellow, whether he ever had the smallpox, he answered, both yes and no; and then told me that he had undergone an operation which had given him something of the smallpox, and would forever preserve him from it.... . He described the operation to me, and showed me in his arm the scar which it had left upon him.

In his medical manuscript, *The Angel of Bethesda*, Mather reports on slave testimony from a larger group of Boston Africans, this time letting go of his own baroquely learned style to mimic their patois: "I have since mett with a Considerable Number of these Africans, who all agree in one Story; That in their Countrey grandy-many dy of the Small-Pox: But now they Learn This Way: People take Juice of Small-Pox; and cutty-skin, and putt in a Drop; then by'nd by a little sicky, sicky: then very few little things like Small-Pox ... any more." In attempting to quote their patois, Mather authenticated and made distinctive his source and, in turn, represented his place in the British periphery as a center of exotic knowledge surpassing, in this instance, even the Royal Society. In the pamphlet wars that surrounded the smallpox controversy in Boston, however, Mather's chief opponent, Dr. William Douglass, castigated Mather for relying on "negroish" evidence and, in mockery, some Bostonians renamed their slaves "Cotton Mather." Mather was willing to court controversy in Boston both because of his perpetual self-command to "do good" publicly and also because he understood the high currency this exotic knowledge would have in the metropole. He also hoped the Royal Society's (and learned London's) high valuation of his communication would redound upon his own status as a communicator of scientific news.[27]

On occasion, colonial authorities rewarded enslaved Africans for divulging the secrets of their cures. In 1729, a "very old" African slave named James Papaw was given his freedom by Virginia authorities for revealing his "many wonderfull cures … in the most inveterate venerial Distempers," which he had kept "a most profound Secrett." These root and bark decoctions proved so successful that the colony offered him an annuity of twenty pounds if he would "make a discovery of other secrets for expelling poison." In 1750, the *South-Carolina Gazette* published the story of a South Carolina slave who revealed a cure to a poison and received not only his freedom but £100 a year for life. "Caesar's Cure" of plantain roots and wild horehound was reprinted in local almanacs for more than thirty years and copied into plantation "scrapbooks" into the nineteenth century.[28]

The November 7, 1750, *Maryland Gazette* printed an account by the author "A. B.," who offered to pass on "Physical Secrets" that a freed slave had obtained from "an old, skilful, experienced Guinea Doctor, his Predecessor; in particular, a Cure for the Stone and Gravel, the Bite of a Snake, Dry Gripes, and Fluxes of both Kinds." "A. B." attested to the ex-slave's credibility by stating that he has "liv'd a regular, Christian, blameless Life" and then further defends his source: "Useful Knowledge has been often communicated to the World by the Simple and Ignorant, from whence great Advantages have accrued." As with James Papaw, these "physical secrets" originated in an old African, as opposed to a Creole slave. Here they pass through a Christianized slave apprentice, then through a presumably white author, and finally into colonial print. Not only did the African source need legitimation through these conduits but, conversely, the information had to originate in an old and experienced African for a southern colonial audience to trust it. This audience would have witnessed the regulated but pervasive and efficacious culture of African healing; they would have therefore trusted the knowledge itself but would have needed the human source to be socially vindicated through Christian and more assimilated mediators.[29]

A letter concerning the "specific Antidote against the Indian or Negro Poison" of the West Indies, "esteem'd the most destructive of any," was printed in the *Philosophical Transactions* in 1742 and reprinted in the *South-Carolina Gazette* in 1749. The metropolitan author Dr. Edward Milward explained to the president of the Royal Society, Martin Folkes, that knowledge about this antidote was "purchas'd from a famous Negro Poisoner, at a great Expence" by a Doctor Burgess, a Caribbean colonial, who hoped that it would help those English people traveling in the West Indies "amongst the Spaniards." "The Negroes ... use a Poison of a strange and extraordinary Nature," and he described the range of effects, from horrid evacuations to death; from the poison, "the Negroes turn white." Mobilizing the English myth of the Black Legend and the Protestant association of Catholic countries with unreformed magical tendencies, he opined: "I know that the Spaniards have Knowledge of this very Poison, and am satisfied, that I have seen several Bocaneers die of it, given them by Spanish Women. I am also persuaded that it is the same Poison used in Spain and Italy." He explained that the antidote is the root of the "Sensible Weed ... , or Herba Sensitiva," and gives Sloane's *Voyage* as a potentially corroborating source. He concluded: "The Remedy deserves ... a fair and impartial Trial, as the Author has not indulged in any rhetorical Flourishes, or Theory, but seemingly confin'd himself to Truth, and plain matter of Fact." Though the quoted author is Doctor Burgess, the testifier or source of the unadorned fact is the African poisoner himself. Again, this "Fact" moved from an African to a colonial to an Englishman to the Royal Society president to the *Transactions*. Though these sources had increasing social credibility the closer they moved to metropolitan print, they nevertheless had to originate in a testifier whose access to the occulted zones of nature, to the magical itself, was assumed. These facts had no epistemic authority if they did not originate with an African. Not only did southern and Caribbean colonials form a wary trust in African sources based on lived experience of plantation pharmacosmic culture but this experientially embedded trust was sufficiently strong to persuade British metropolitan scientific institutions to credit the African sources at a great distance. Colonial go-betweens were vital enough links to American curiosities that they had acquired the social credibility, in turn, to win the metropolitan legitimation of their otherwise suspect sources.[30]

Nature's secreted zone, which Baconian scientific and Renaissance exploration rhetoric had constructed and sexualized in the late sixteenth and seventeenth centuries, was racialized as well during the late seventeenth and eighteenth centuries in the New World. Generally, the sequestered parts of the landscape visualized as Indian "haunts" in seventeenth-century wartime writing became associated in the South and the Caribbean with Africans and Creole blacks. Related to this overlay of blackness and natural topoi was the phenomenon that plant poisons and antidotes, understood as encrypted in and by nature, eventually became the exclusive epistemic provenance of black testifiers. The ambivalent Anglo attitude toward this knowledge—wherein slaves could be executed for the use of poisons but manumitted for the disclosure of antidotes—both reflected a self-interested set of rewards and punishments and indicated the Enlightenment's lingering belief that what was invisible or hidden in nature could be potentially manipulated for dangerous or curative ends by adepts closer to nature than themselves. Though colonial authorities sought to control secret African manipulation of the natural world—fearing the subversive application of such knowledge—they nevertheless tried in every way they could to use such African expertise to their own advantage. Colonials used Africans as discoverers of curative specimens and procedures, as collectors of such specimens, as magical manipulators of their fellow slaves' environments, as river navigators and forest pathfinders and, finally, as importers and adaptors of agricultural methods. Though they did not—for the most part—share the magical worldview of the Africans on whom they relied, colonial Europeans recognized the botanical expertise that arose out of such a worldview. Moreover, colonials understood that to transform their disadvantageous distance from the metropolis into an advantage, they needed not only to emphasize their closeness to novel specimens but to exhibit their access to the distinct knowledge of Africans.

Notes

1   On the conceptual dependence of the U.S. documents demanding representational government on those of the Glorious Revolution in England, see Pauline Maier, *American Scripture: Making the Declaration of Independence* (New York: Knopf, 1997). John Bartram first noted animal-plant in the American colonies and communicated it to Peter Collinson, who wrote back to Bartram in 1737: "The balance kept between the Vegitable & Animal productions is really a fine Thought & what I never met with before, but it is more remarkable with you than with us for you have Wild animals & mast in greater plenty than Wee have," Edmund Berkeley and Dorothy Smith Berkeley, eds, *The Correspondence of John Bartram, 1734–1777* (Gainesville: University Press of Florida, 1992), 67.
2   Bacon, quoted in Mordechai Feingold, "Mathematicians and Naturalists: Sir Isaac Newton and the Royal Society," in Jed Z. Buchwald and I. Bernard Cohen, eds, *Isaac Newton's Natural Philosophy* (Cambridge, Mass.: Massachusetts Institute of Technology Press, 2000), 80; Francis Bacon, *The Advancement of Learning*, ed. William Aldis Wright (Oxford, 1876), Book One, 30-2 and *Novum Organum*, trans. Robert Ellis and James Spedding (London, n.d.): aphorism

XIX declares that the "true way" of knowledge "as yet untried" "derives axioms from the senses and particulars, rising by a gradual and unbroken ascent. So that it arrives at last at the most general axioms last of all," 64; Thomas Sprat, *History of the Royal Society of London, for the Improving of Natural Knowledge,* eds Jackson I. Cope and Harold Whitmore Jones (London, 1667; reprinted in St. Louis: Washington University Press, 1959), 115; altering Bacon's metaphor of a gradual ascent toward axioms, Sprat imagined cycles in which a prior synthesis of matter would be overturned by the new questions and searches generated out of that synthesis, p. 31.

3   Theophus H. Smith, *Conjuring Culture: Biblical Formations of Black America* (Oxford: Oxford University Press, 1994), 44 (quotations), 76. Smith argues that, especially in nineteenth- and twentieth-century African American culture, the Bible was "reconfigured as a *pharmacopeic cosmos*... . [or a] 'pharmacosm,' which designates a world capable of hosting myriad performances of healing and harming." Smith's term "pharmacosm" is ideal for describing the magical natural worldview of enslaved and freed Africans in the eighteenth century as well but, in the eighteenth century, black writers and subjects were not yet participating in a communal black religious culture that had reenvisioned the Bible as a pharmacosm; their conversion to Christianity, therefore, represented more of a renunciation of their former worldview.

4   Henry Louis Gates, Jr., has argued that "the production of literature was taken to be the central arena in which persons of African descent could, or could not, establish and redefine their status within the human community"; see Gates, *The Signifying Monkey: A Theory of African-American Literary Criticism* (New York: Oxford University Press, 1988), 129, 131 (quotations).

5   Thomas Walkington, *The Optick Glasse of Humors...* (1631; rpt. Delmar: Scholars' Facsimiles & Reprints, 1981), 29; Aimé Césaire has termed this equation of warm climates with mental inferiority the "geographical curse" of the tropics in *Discourse on Colonialism*, trans. Joan Pinkham (New York: MR, 1972), 34; Mary Floyd-Wilson, *English Ethnicity and Race in Early Modern Drama* (Cambridge: Cambridge University Press, 2003), 2, 4.

6   See Peter H. Wood, *Black Majority: Negroes in Colonial South Carolina from 1670 through the Stono Rebellion* (New York: Norton, 1974), 30, 59–62, 120–2; Lorena S. Walsh, *From Calabar to Carter's Grove: The History of a Virginia Slave Community* (Charlottesville: University of Virginia Press, 1997), 63; Judith A. Carney, *Black Rice: The African Origins of Rice Cultivation in the Americas* (Cambridge, Mass.: Harvard University Press, 2001) (a letter sent to England as early as 1648 stated: "We perceive the ground and Climate is very proper for it [rice cultivation] as our *Negroes* affirme, which in their Country is most of their food" [90]); Joyce E. Chaplin, *An Anxious Pursuit: Agricultural Innovation and Modernity in the Lower South, 1730–1815* (Chapel Hill: University of North Carolina Press, 1993); Richard Price and Sally Price, eds, *Stedman's Surinam: Life in an Eighteenth-Century Slave Society: An Abridged, Modernized Edition of "Narrative of a Five Years Expedition against the Revolted Negroes of Surinam" by John Gabriel Stedman* (Baltimore: Johns Hopkins University Press, 1992), 132, 169–170, 189, 210; W. Jeffrey Bolster, *Black Jacks: African American Seamen*

in the Age of Sail (Cambridge, Mass.: Hrvard University Press, 1997), 45. Paul Gilroy, in *The Black Atlantic: Modernity and Double Consciousness* (Cambridge, Mass.: Harvard University Press, 1993), sees Africans within British America as more limited to types of irrationality and innocence (45, 57). See also Mechal Sobel, *The World They Made Together: Black and White Values in Eighteenth-Century Virginia* (Princeton: Princeton University Press, 1987), 5, 78, 97–9; Yvonne Chireau, "The Uses of the Supernatural: Toward a History of Black Women's Magical Practices," in Susan Juster and Lisa MacFarlane, eds, *A Mighty Baptism: Race, Gender, and the Creation of American Protestantism* (Ithaca: Cornell University Press, 1996).

7 Gilbert Mathison, *Notices Respecting Jamaica, in 1808–1809–1810* (London, 1811), 101, quoted in Orlando Patterson, *The Sociology of Slavery: An Analysis of the Origins, Development, and Structure of Negro Slave Society in Jamaica* (Princeton: Princeton University Press, 1969), 176 (Patterson discusses Quashee on p. 180); William Byrd II, "An Account of a Negro-Boy That Is Dappel'd in Several Places of His Body with White Spots," RS, *Philosophical Transactions*, XIX (1695), 781–2; Alexander Garden to John Ellis, April 1, 1760, *Correspondence of Linnaeus* I, 483; *Pennsylvania Gazette*, October 17, 1745; *Maryland Gazette*, September 13, 1745.

8 Harriet Jacobs, *Incidents in the Life of a Slave Girl; Written by Herself*, ed. L. Maria Child (Boston, 1861), 84, 287; Dan. 10, 11: 40–1, 12: 1–4.

9 On cunning women, see David D. Hall, ed., *Witch-Hunting in Seventeenth-Century New England: A Documentary History, 1638–1692* (Boston: Northeastern University Press, 1991), esp. 170–84; and Keith Thomas, *Religion and the Decline of Magic* (London: Weidenfeld and Nicolson, 1971), 212–52.

10 James Petiver to George Harris, October 18, 1698, Sloane MS 3333, 235–6, British Library, London and John Starrenburgh to Petiver, March 27, 1701, Sloane MS 4063, fol. 74; both in Raymond Phineas Stearns, "James Petiver: Promoter of Natural Science, c.1663–1718," American Antiquarian Society *Proceedings* LXII, pt. 2 (1952), 269, 280, 334; Petiver to Hannah English Williams, May 22, 1701, Sloane MS 3334, fol. 67v–8, British Library, London, in Beatrice Scheer Smith, "Hannah English Williams: America's First Woman Natural History Collector," *South Carolina Historical Magazine* 87 (1986), 84–6.

11 See Alan Feduccia's introduction to Feduccia, ed., *Catesby's Birds of Colonial America* (Chapel Hill: University of North Carolina Press, 1985), 4; Garden to Ellis, 1755, Ellis to Garden, January 14, 1770, Garden to Carolus Linnaeus, June 20, 1771, *Correspondence of Linnaeus* I, 331, 349, 571.

12 J[ohn] G[abriel] Stedman, *Narrative, of a Five Years' Expedition, against the Revolted Negroes of Surinam...*, 2 vol. (London, 1796), II, 347. This edition differs markedly from Stedman's 1790 manuscript, recently printed under the editorship of Richard Price and Sally Price in 1988. An abridged version of this authoritative edition is *Stedman's Surinam*, ed. Price and Price, 300–1 (see the note on 339–40 for details of Kwasi's life). In 1869, the colony exported 245,622 kilos of *Quassia Amara* for medicinal purposes and to be used in English beer. See Richard Price, *First-Time: The Historical Vision of an Afro-American People* (Baltimore: Johns Hopkins University Press, 1983), 155.

13 Garden to Ellis, March 13, April 1, June 1, 1760, *Correspondence of Linnaeus* I, 474, 483, 492; Brian Vickers, ed., *Occult and Scientific Mentalities in the Renaissance* (Cambridge: Cambridge University Press, 1984), 20; Frances A. Yates, *The Occult Philosophy in the Elizabethan Age* (London: Routledge and Kegan Paul, 1979); Thomas, *Religion and the Decline of Magic*, 216, 223, 227; Lorraine Daston and Katharine Park, *Wonders and the Order of Nature, 1150–1750* (New York: Zone Books, 2001); *Theophrastus Paracelsus' Archidoxes of Magic* (1656; rpt. London, 1975); Max Weber, *General Economic History*, trans. Frank H. Knight (New York: Greenberg, 1961), 265; E. E. Evans-Pritchard, *Witchcraft, Oracles, and Magic among the Azande* (Oxford: Oxford University Press, 1937).

14 Frank J. Klingberg, *An Appraisal of the Negro in Colonial South Carolina: A Study in Americanization* (Washington, DC: The Associated Publishers, 1941), 46, 89; *South-Carolina Gazette*, October 30, 1749, or July 24, 1749, Garden to Dr. Charles Alston, January 21, 1753, Laing MSS, III, 375/42, 44, University of Edinburgh, both quoted in Philip D. Morgan, *Slave Counterpoint: Black Culture in the Eighteenth-Century Chesapeake and Lowcountry* (Chapel Hill: University of North Carolina Press, 1998), 613, 618; Wood, *Black Majority*, 187, 290; *South-Carolina Gazette*, September 20, 1735, Apr. 15, 1738; Morgan, *Slave Counterpoint*, 612–28 (quotations on 612, 613); *Pennsylvania Gazette*, December 31, 1767; *Georgia Gazette*, March 30, 1768; Winthrop D. Jordan, *White over Black: American Attitudes toward the Negro, 1550–1812* (Chapel Hill: University of North Carolina Press, 1968), 393.

15 Hilary McD. Beckles, *Natural Rebels: A Social History of Enslaved Black Women in Barbados* (New Brunswick: Rutgers University Press, 1989), 163; Long, *History of Jamaica* II, 418–19; Patterson, *Sociology of Slavery*, 265; Orlando Patterson, "Slavery and Slave Revolts," in Price, ed., *Maroon Societies*, 274; Edward A. Pearson, *Designs against Charleston: The Trial Record of the Denmark Vesey Slave Conspiracy of 1822* (Chapel Hill: University of North Carolina Press, 1999), 27; James E. McClellan, III, *Colonialism and Science: Saint Domingue in the Old Regime* (Baltimore: Johns Hopkins University Press, 1992), 55, 178, 247. See also the chapter by Regourd in this volume.

16 Graham Russell Hodges, *Root and Branch: African Americans in New York and East Jersey, 1613–1863* (Chapel Hill: University of North Carolina Press, 1999), 98, 101, 128; John Sharpe to the secretary of the SPG, June 23, 1712, *New-York Weekly Gazette*, November 30–December 7, December 13–20, 1730, and *Pennsylvania Gazette*, February 28–March 7, 1738, all quoted in Hodges, 64–8, 89, 90–1.

17 Pearson, *Designs against Charleston*, 124–5, 135, 159; Vincent Harding, "Religion and Resistance among Antebellum Slaves, 1800–1860," in Timothy E. Fulop and Albert J. Raboteau, eds, *African-American Religion: Interpretive Essays in History and Culture* (New York: Routledge, 1997), 115; Lionel H. Kennedy and Thomas Parker, *An Official Report of the Trials of Sundry Negroes, Charged with an Attempt to Raise an Insurrection in the State of South-Carolina* (Charleston, 1822).

18 Patterson, "Slavery and Slave Revolts," 189, 194; Morgan, *Slave Counterpoint*, 631; Richard Price, *First-Time: The Historical Vision of an Afro-American People* (Baltimore: Johns Hopkins University Press, 1983), 89; Pearson, *Designs against Charleston*, 25; Edwards, *Observations*, Patterson, "Slavery and Slave Revolts,"

and W. Van Wetering, "Witchcraft among the Tapanahoni Djuka," all in Price, ed., *Maroon Societies*, 239, 262, 370–88.

19 Lawrence W. Levine, *Black Culture and Black Consciousness: Afro-American Folk Thought from Slavery to Freedom* (Oxford: Oxford University Press, 1977), 65, 73.

20 Price, *First-Time*, 80; Hans Sloane, *A Voyage to the Islands of Madera, Barbados, Nieves, S. Christophers, and Jamaica, ... Wherein Is an Account of the Inhabitants, Air, Waters, Diseases, Trade, etc. of That Place...*, 2 vols (London, 1707–25), I, liv–lv; "Memoirs of the Life of Boston King, a Black Preacher," *Methodist Magazine*, 21 (1798), 105, quoted in Morgan, *Slave Counterpoint*, 619; Mark Catesby, *The Natural History of Carolina, Florida, and the Bahama Islands...*, 2 vols (London, 1731–43), in Feduccia, 102.

21 Sloane, *Voyage to the Islands*, I, cxli. Sloane continued: "But by what I could see by their practice, (which because of the great effects of the Jesuits Bark, found out by them, I look'd into as much as I could) they do not perform what they pretend, unless in the vertues of some few Simples. Their ignorance of Anatomy, Diseases, Method, etc. renders even that knowledge ... even sometimes hurtful." See also McClellan, 136; John Walduck to Petiver, September 17, 1712, Sloane MS 2302, fols. 25–6, in Stearns, "James Petiver," 319; Long, *History of Jamaica* II, 418–20; Garden to Alston, January 21, 1753, quoted in *Diary of Alexander Garden*, 32; "Inquiries Recommended to Colonel [Sir Thomas] Linch Going to Jamaica, London, December 16, 1670," in Raymond Phineas Stearns, *Science in the British Colonies of America* (Urbana: University of Illinois Press, 1970), 701.

22 Thomas Tryon, *Friendly Advice to the Gentlemen-Planters of the East and West Indies...* (London [1684]), in Thomas W. Krise, ed., *Caribbeana: An Anthology of English Literature of the West Indies, 1657–1777* (Chicago: University of Chicago Press, 1999), 66–8.

23 Sharla M. Fett, *Working Cures: Healing, Health, and Power on Southern Slave Plantations* (Chapel Hill: University of North Carolina Press, 2002), 6–34; Chireau, "The Uses of the Supernatural," in Juster and MacFarlane, eds, *A Mighty Baptism*, 177.

24 William Dawson to Robert Carter, Henry Ravenel, both quoted in Levine, 63, 64; Carter to William Berry, July 31, 1786, Robert Carter Letterbooks, quoted in Todd L. Savitt, *Medicine and Slavery: The Diseases and Health Care of Blacks in Antebellum Virginia* (Urbana: University of Illinois Press, 1978), 175–6; Walsh, *From Calabar to Carter's Grove*, 178; Mary Prince, *History of Mary Prince, a West Indian Slave...* (London, 1831), in Henry Louis Gates, Jr, ed., *The Classic Slave Narratives* (New York: New American Library, 1987), 203–4. In *The Caribbean Slave: A Biological History* (Cambridge: Cambridge University Press, 1984), Kenneth F. Kiple explains that, on large-scale Caribbean plantations, there were "sickhouse[s]" that European doctors attended, but each usually also had a slave "doctor" or "doctor women" attached to it (151–2); he remarks that "black preventive medicine was in many ways more advanced than the white variety" in the treatment of yaws and smallpox, and with regard to general herbal knowledge (154).

25 Griffith Hughes, *The Natural History of Barbados...* (London, 1687), 15–16; Long, *History of Jamaica* II, 416–20; Patterson, *Sociology of Slavery*, 186, 191.

Quoting from M. J. Field, *Religion and Medicine of the Ga People* (London: Oxford University Press, 1937), 124–5, Patterson explains that "medicine in West Africa means anything which possesses a 'power' or 'breath of life' and 'is the abode of a spiritual being or won,'" 183). See also Robin Horton, "African Traditional Thought and Western Science," *Africa* 37 (1967), 50–71; Olaudah Equiano, *The Interesting Narrative of the Life of Olaudah Equiano, or Gustavus Vassa, the African...*, 9th edn (London, 1794), in Vincent Carretta, ed., *Unchained Voices: An Anthology of Black Authors in the English-Speaking World of the Eighteenth Century* (Lexington: University of Kentucky Press, 1996), 194–5; Louis [last name unknown], "Rebel Village in French Guiana: A Captive's Description," in Price, ed., *Maroon Societies*, 314.

26   Savitt, *Medicine and Slavery*, 175; Wood, *Black Majority*, 290. Fett remarks in *Working Cures* that, in the antebellum period, "white southerners wrote slave remedies into their private recipe books even as they wrote laws curtailing the practice of enslaved doctors" (5). Summarizing this contradiction, she writes: "Fluid cross-cultural exchanges in medicine took place within a slave society characterized by a sharply defined social order" (3).

27   Carla Mulford, "New Science and the Question of Identity in Eighteenth-Century British America," in Mulford and David S. Shields, eds, *Finding Colonial Americas: Essays Honoring J. A. Leo Lemay* (Newark: University of Delaware Press, 2001), 79–103; Cotton Mather, *The Angel of Bethesda*, ed. Gordon W. Jones (Barre: American Antiquarian Society and Barre Publishers, 1972), 107; Mather, *The Negro Christianized...* (Boston, 1706), 25; *Selected Letters*, 199; Mather to John Woodward, July 12, 1716, *Selected Letters*, 214; Mather, *Small Offers towards the Service of the Tabernacle in the Wilderness...* (Boston, 1689); Mather, *Bonifacius: An Essay upon the Good, That Is to Be Devised and Designed, by Those Who Desire to Answer the Great End of Life, and to Do Good While They Live* (Boston, 1710).

28   Governor William Gooch to the secretary of state, June 29, 1729 and Gooch to Bishop Edumund Gibson, June 29, 1729, quoted in Morgan, *Slave Counterpoint*, 625; *South-Carolina Gazette*, May 9, 14, 28 1750; Wood, *Black Majority*, 289–92; Savitt, *Medicine and Slavery*, 76. Savitt argues that "new additions to the white materia medica from slave practitioners were occasionally publicized in professional journals" in the antebellum period (179); see, for example, Alexander Somervail, "Cases of Negro Poisoning," *American Journal of the Medical Sciences*, 24 (1839), 514–16. Fett found Caesar's cure copied into many plantation "scrapbooks" of medical lore and recipes in the nineteenth century; see 68–9.

29   *Maryland Gazette*, November 7, 1750.

30   "A Letter from Edward Milward, M.D. to Martin Folkes, Esq; President of the Royal Society," in RS, *Philosophical Transactions*, 42 (1742–43), 2–10, reprinted in *South-Carolina Gazette*, July 24, 1749.

CHAPTER 12

Mesmerism in Saint Domingue

Occult Knowledge and Vodou on the Eve of the Haitian Revolution

FRANÇOIS REGOURD

In August 1784, while the controversy surrounding the flamboyant physician Franz Anton Mesmer was at its height in Paris, Madame Millet, a respectable French woman settled in Saint Domingue, recorded the effects of "animal magnetism" on the inhabitants of the island in the following terms:

> A magnetizer has been in the colony for a while now, and, following Mesmer's enlightened ideas, he causes in us effects that one feels without understanding them. We faint, we suffocate, we enter into truly dangerous frenzies that cause onlookers to worry. At the second trial of the tub, a young lady, after having torn off nearly all her clothes, amorously attacked a young man on the scene. The two were so deeply intertwined that we despaired of detaching them, and she could be torn from his arms only after another dose of magnetism. You'll admit that such are ominous effects to which women should sooner not expose themselves. It produces a conflagration that consumes us, an excess of life that leads us to delirium. We will soon see a maltreated lover using it to his advantage.[1]

Mesmer's enemies could have claimed that these few words said all about mesmerism: it was nothing more than a woman's frantic disorder, leading to loose sexual conduct. However, we can say more, as historians, to unravel the tangle of facts and expose them. Mesmerism is, of course, a much more complicated phenomenon, as is the story of its rise and development in the most important French Caribbean colony of the eighteenth century.

As in Paris, mesmerism met with great popular success in Saint Domingue, confirmed by spectacular recoveries; as in Paris, it provoked contradictory and passionate reactions ranging from devotion to hatred; as in Paris, "animal magnetism" excited people's minds enough to reach the papers. Nevertheless, this chapter may be able to offer more than a mere local variant on a European phenomenon. First, the geographical distance from Paris and the colonial status of the island affected its social and intellectual status and its relation to the scientific societies in France; second, the presence at that time of more than 350,000 blacks and mulattos (including about 15,000 free colored people) beside fewer than 25,000 white colonists is a point not to be neglected, as both cultural and political tensions were involved. Saint Domingue, the wealthiest colony of the French empire, was, on the eve of the Haitian Revolution, a time bomb.[2]

The aim of this chapter is to study a fascinating episode in the history of Atlantic knowledge: how was mesmerist doctrine, and its attendant controversies, exported and appropriated in the colony? How did colonial scientists participate in both local and metropolitan debates under the watchful eye of the Royal Society of Medicine in Paris, and how did this result in a new social organization of scientific life in Saint Domingue? Beyond that point, what exactly happened with the black population of the island, when a few clues suggest that some of them may have assimilated such "magnetic" knowledge, and may have tried to use it as a weapon against their white masters? Alongside these tantalizing traces of a cross-fertilization between European and African occult knowledge and practices, this story is fundamentally about the place and the role of science, knowledge, and belief, in a context of tremendous political and social tension.

Chastenet de Puységur and the Introduction of Mesmerism in Saint Domingue

On June 8, 1784, Antoine-Hyacinthe-Anne de Chastenet de Puységur arrived in Saint Domingue, commanding his corvette *Le Vautour* (*The Vulture*). He was a bright thirty-two-year-old naval officer particularly well versed in nautical sciences: distinguished by the dignity of *chevalier de*

Saint-Louis, he had already participated in important scientific campaigns in the Atlantic, testing naval clocks under the command of Verdun de La Crenne in 1772 and learning the delicate use of the eponymous repeating circle (or *cercle de réflexion*, the best instrument at the time for precise coastal cartography) from Jean-Charles de Borda himself on the African coast in 1776. His experience of observation instruments in several military operations of the American War made Chastenet de Puységur, despite his youth, one of the most competent hydrographers of the time and fully legitimated his appointment as head of the mission to Saint Domingue in 1784.[3]

His arrival was certainly important for local naval officers and administrators, because of the expected strategic results of his hydrographic campaign. However, it had a much more important echo in the colony, because of the immediate manifestation of his "magnetic" curative powers. Six years after Mesmer's arrival in Paris and the beginning of the general craze for "animal magnetism," Saint Domingue's first "magnetic tub" was set up in the hospital of "La Providence," in Le Cap Français, the capital of the north part of Saint Domingue.[4]

Chastenet de Puységur was nothing like an ordinary charlatan: as a matter of fact, his name is very well known to historians of mesmerism, as his older brother, Armand-Marc-Jacques de Chastenet, marquis de Puységur, was one of Mesmer's most famous disciples, one who can be considered without exaggeration an enlightened precursor of medical hypnosis. For a few years, Antoine-Hyacinthe-Anne had been a fervent partisan of Mesmer, curing people himself whenever possible, and especially on the boats he was commanding: he was so convinced of the efficiency of Mesmer's curative theories that he experimented almost every day on himself and ordered the ship's surgeon not to bring any medicine aboard when he left Brest for Saint Domingue.[5]

Puységur had learnt the techniques and theory of this new medicine which was claimed to be a universal panacea (curing blindness, vomiting, paralysis, osteoarthritis, and headaches) from Mesmer himself. From him, he had learned the art of mastering the magnetic flux supposedly running through the human body to relieve pains and physical torments. Mesmerism was also called "animal magnetism" because its theory was based on the existence of magnetic fluids that flowed through animate matter (similar to but different from the familiar mineral form of magnetism). The main idea was that most illnesses were provoked by an internal obstruction impeding the regular flux of magnetic streams through the body, this fluid filling any free space between things, living beings, and even planets. Following that principle, tub séances wherein the magnetic healer magnetized the patients by means of savant passes over joints and the solar plexus were organized to provoke "crises" and "trances" which were the most efficient way to free the streams. The most visible part of the cure was the famous "tub" filled with magnetized water and metal filings. Gathered in a circle around it, patients were touching metal rods plunged in the tub, each patient holding the same rope through which the fluid was supposed to go from one to another. Some other accessories were an obsessing soft music (a fortepiano or, better, the famous glass-harmonica so precious to Mesmer) and a dark environment, to render his patients more impressionable. Information about the precise decorum that surrounded Puységur's séances in Saint Domingue is lacking, but they were probably very similar to those observed in Europe, as they were explicitly said to be organized following Mesmer's precise instructions.[6]

An interesting witness of Saint Domingue life, a certain Morange (a plantation manager who regularly wrote to France to inform the proprietor he was working for) wrote on June 11, 1784 (only three days after Puységur's arrival): "Mesmer: this physician's reputation is beginning to create a sensation here. […] M. de Puységur […] is a great master […]."[7] On June 15, he added: "M. de Puységur has established three chambers at La Providence in order to magnetize the patients."[8] A few days later, Morange continued his report: "M. and Mme Puységur have come to ask for soup. We have seen the husband mesmerize one of his officers who suffered from a bout of fever. He made him cry, then laugh, with a fortepiano. Experiments have not yet begun for want of the necessary objects. Many are coming up to be treated."[9] Last, on June 27, he noted: "M. de Puységur has begun his treatments."[10] As Moreau de Saint-Méry, a lawyer and the major chronicler of the 1780s in Saint Domingue, noted: "In the Providence Hospital you could see a tub and people with obstruction, gout or asthma assailing it."[11]

And indeed, Puységur's magnetic chambers seem to have met with quick and large popular success, attested by the *Affiches américaines*, the local newspaper, from the end of June 1784.[12] Moreover, Puységur also found disciples to accelerate the spreading of Mesmer's theories in Saint Domingue. Following Mesmer's policy (he was linked to him by a precise contract), he founded a "Society of Harmony" in Le Cap in which as "Great Master" he gathered up to three promotions of "initiates," and some "pupils," for a total of more than twenty persons, including naval officers, physicians, colonists, and even an engineer.[13] Each "initiate" had to pay a fee and sign a similar contract, including a clause punishing any disclosure of Mesmer's secrets with the payment of 50,000 *écus* (about 150,000 *livres tournois*, an enormous sum) to the master.[14] As a complement for this strategy of mesmeric expansion in the colony, he even received the support of two other disciples from the Paris

Society of Harmony: one physician and one surgeon, both recommended by Mesmer himself. This offered to Puységur the opportunity of keeping up both the teaching and practice of mesmerism in Le Cap despite long absences due to his hydrographical mission.[15]

We do not have many clues that could give us a precise account of Chastenet's and his disciples' medical activity during his first four months in the colony, but we do have a list of 131 cases, written out by the royal physician (*médecin du roi*) Arthaud, and his colleague, Côme d'Angerville, royal physician-surgeon (*médecin-chirurgien du roi*) in Le Cap Français, two of the most active opponents of Puységur's work in the colony. More than 100 patients was not a remarkably high figure for a town that counted thousands of inhabitants at that time, but those were "regular" patients, as the magnetic treatment was supposed to be continued for weeks, and often for months—and nothing guarantees, moreover, that the list was exhaustive.[16] Among them, Arthaud and his colleague noted only five recoveries and more than twenty deaths. However, persistent rumors were running in the colony, as revealed by this account from Jean Trembley, a plantation owner, perfectly aware of Parisian debates involving Mesmer:

> Two mesmeric tubs in this colony were directed by Monseigneur, the Count of Puységur, officer of the royal navy, and by other adepts. Marvelous cures that could hardly be attributed to any play of the imagination have been reported. A cripple brought from the plain to Cap François on a litter walked freely afterward. A female slave paralyzed for fourteen years was entirely cured in a short time without her realizing that she was been treated, etc.[17]

In Puységur's private papers, we can find also a certificate signed by an inhabitant of Limonade testifying to a temporary recovery.[18]

In any case, the controversy involving some of the most enlightened minds in the colony rapidly grew, not only as an echo of contemporary debates in France but as an expression of local intellectual vitality, which reveals some interesting points.

The Scientific Controversy and its Developments

As historians have shown, the frontier between such a medical and physical theory and charlatanism was not easy to define. In Saint Domingue, as in France at the same time, the general craze for the amazing powers of science was peaking: hot-air balloon flights and demonstrations of electrical machines and cures or chemical manipulations did not spare Saint Domingue.[19] In such a context, in Saint Domingue as in France, mesmerism could have been nothing more for most people than another episode in the story of men's mastery of natural powers, from Newton to Franklin, passing by the Montgolfier brothers.

Nevertheless, Mesmer was a strange figure, dangerously flirting with the usual charlatan's style, such as Cagliostro's or the Comte de Saint-Germain's, who were operating at the same time. His greed for money and his supernatural buildups also made an unfavorable impression on academic physicians and scientists of repute who led the fight against him in Paris, more or less since the beginning of the 1780s. In Paris in 1784, two commissions appointed by the government, including the main scientific authorities of the time (among them Franklin), led an investigation, and published their accounts in the fall of 1784. Both concluded that Mesmer was a charlatan and that his pretended recoveries were no more than fruits of the imagination. From that date on, Mesmer's success progressively declined in Paris under the joint accusations of caricature and scientific discredit.[20]

Thus, the Parisian controversy reached Saint Domingue at the same time as Puységur, which is not without consequence. The quick reactions of people such as Arthaud or Côme d'Angerville can be accounted for by their determination to take an active part in the Parisian debates. As soon as July 27, 1784—Puységur had by then been treating people in the colony for only a few weeks—Arthaud wrote to the Société royale de médecine; his letter was read out on September 28, 1784: "A disciple of M. Mesmer has been with us for a month. He has set up a considerable apparatus and gathered all sorts of patients. He has promised to cure them all. He has now been operating for a month; seven people have died."[21]

Such an involvement in the scientific debates of the home country is no extraordinary thing, as the intellectual elites in Saint Domingue at the end of the eighteenth century were clearly involved in the most dynamic networks of the Republic of Letters.[22] The fact is that all through the controversy that opposed physicians from Le Cap to Puységur, established physicians called the Société royale to witness, conveying information and relying on the institution to legitimate their fight. This is obviously a boon for the historian because it provides a precious source with which to approach the controversy.

In a letter to Arthaud, dated June 25, 1785, J. Fournier de Varenne, one of most fervent proponents of mesmerism in the island, brought up a novel point in the controversy, arguing that the effects of mesmerism could be different on different sides of the Atlantic Ocean:

> It is very odd and very remarkable that physicians in Paris should deny the existence of animal magnetism while physicians in the New World admit the truth of it. Are physicians in America favored

with more good faith and honesty than European ones? Or can it be that the effects of animal magnetism are more strongly felt in the equatorial area? For the physicians' sake, I go for the latter opinion, which is, besides, in conformity with our [i.e. mesmerist] doctrine.[23]

Could it be that Fournier de Varenne was invoking an American specificity as a means to disqualify the results of European scientific commissions, thus hoping to give a second chance to mesmerism on the other side of the Atlantic? As Robert Darnton underlined in his *Mesmerism and the End of the Enlightenment*, mesmerism had also reached the United States in the early 1780s, thanks to Lafayette, who went back to America in 1784 bringing magnetic theories in his luggage. An ardent proponent of Mesmer, Lafayette had planned at that time to establish a few "Societies of Harmony" in the young country; but Jefferson, who was then the U.S. ambassador in Paris, successfully undermined his efforts by sending back to his country numerous anti-mesmerist pamphlets and scientific accounts he had found in Paris—and there were very many at that time.[24] Despite Lafayette's active proselytism in Philadelphia and elsewhere, despite his visit to some Shakers in whom he had seen a manifestation of "indigenous mesmerism," it seems he did not encounter the expected success, even if a booklet called *Nouvelle découverte sur le magnétisme animal*, published circa 1800, still claimed the authority of some correspondents in America.[25]

Anyway, the fact that Fournier de Varenne claimed that magnetic streams could be different and have different effects in various points of the globe was not surprising, nor was the idea that European men and women were subject to some physiological transformation when they were transplanted under tropical and equatorial climates. The sallow-skinned Creoles and their languor or sexual disorder were commonplaces of the eighteenth-century medical discourse. Moreover, the question of magnetic variations and the various reports showing that electric fluid was less easy to produce in tropical colonies because of excess of humidity in the air had already brought to light such subtle physical differences.[26]

As we can see, the debate was spurring many questions, echoing those in France but with some "local touches," and some doubts on the quality of experts from the Old World as relevant judges for any scientific question in America. Besides, the "radical" social and political discourse that often emerged in the wake of the medical strand of mesmerism (as Robert Darnton demonstrated) was not totally absent from the debate.[27] As a matter of fact, we find a reference to this aspect of the question, once again under the quill pen of the pro-mesmerist Fournier de Varenne. He expressed the idea that Mesmer's purpose was also to perfect social institutions, and first of all medicine, ridiculing the Société royale de médecine in Paris as a "schismatic body of plotters, ... a monstrous deformed wart which has grown on the corpse of the antique Faculty, like a fungus on a decrepit tree."[28] This prompted a scathing answer from Arthaud, a royal physician (*médecin du roi*): "I suppose you should aspire to overthrow the social order and destroy it, since that would be the only means of returning to nature."[29]

Reading such passages, it is tempting to construct a sociopolitical approach to the phenomenon, which would follow Darnton's work on Mesmerism, transposing it to the colonial setting. Puységur's disciples would then be portrayed as second-rank physicians looking for some kind of social prestige and acknowledgment, trying to find their place beside the official hierarchy of science embodied in the *médecin du roi*, Arthaud, working along with the Royal Society of Medicine in Paris.[30] And indeed, for opponents of mesmerism, this episode had highlighted the lack of an authoritative scientific body in the colony, such as the Académie royale des sciences or the Société royale de médecine in Paris. As Arthaud wrote to the Société royale de médecine in September 1785, describing his campaign against Puységur and his adepts, possibly with some affectation:

> I assure you, Sir, there would be little merit in carrying out this project in any place other than a colony, but here double courage is necessary especially when one feels so distant—as we are—from the center of Government, the influence of the Graces and the encouragements which sustain the zeal and animate the undertakings of a society. Science is an exotic plant that has not yet taken root here; it is not yet known how to cultivate it, and I think it will be difficult to naturalize it here.[31]

This letter expressed much more than a simple recognition of the Paris medical society's prestige and authority, as it gives such a clear image of the idea of a scientific periphery and of the need to transplant metropolitan institutions to the colony. By that time, in fact, a small company of scientists and enlightened personalities (most of them physicians, lawyers, and planters) had already been living and developing in the colony for over a year, known as the *Cercle des Philadelphes*. And yet, this *Cercle* was born on August 15, 1784 (with its official status being approved by administrators in September 1784) in direct reaction *against* the mesmerian tidal wave that had been threatening since its beginning a month before to flood established medicine in the colony. As James McClellan noticed in his pioneering and wide-ranging book on that *Cercle*, "[N]o one disputes the role of Puységur and mesmerism in triggering the formation of the colony's scientific society."[32] The idea of Arthaud, here, as the natural leader of the nine founder fathers of the new colonial enlightened society, was clearly to embody a scientific barrier erected against Puységur and his occult powers. It also aimed at claiming support and "encouragements" from "the government" and the colonial machine as a whole.

In this sense, though it also expressed many other local and sometimes contradictory aspirations,[33] the *Cercle des Philadelphes*, built on the battlefield of an anti-mesmerist campaign under the protection of the Société royale de médecine in Paris,[34] can be seen as a transplantation of a social order of "science," claimed as universal, from one shore of the Atlantic Ocean to the other. The image of an "exotic plant" is in that way relevant, as it suggests that such a conception of science had to be acclimatized. Decked in the full attire of French academic science (with its public meetings, *jetons*, competitions, publication of approved memoirs, and so on), it stepped onto the scene as a citadel besieged by the radical proponents of supernatural and occult knowledge—who were embodied in the rival "Society of Harmony" established in Le Cap at the same time.

Thus, thanks to the royal support offered by local administrators and the navy and thanks to the scientific credit given to established medicine by the official accounts published in Paris and brought to public knowledge in Saint Domingue, mesmerism seemed to have progressively disappeared from the colony by the fall of 1785, even if the last tub séances in Le Cap cannot be dated with precision.[35] A few years later, in May 1789, the *Cercle des Philadelphes* that had proved its usefulness in a much wider range of circles than the medical ones had become the *Société royale des Sciences et Arts du Cap Français*, the first royal academy in the French colonial domain. By that time, the most enlightened people of the Island had joined the Society, including former "initiates" of the disbanded Society of Harmony.[36] The scientific acculturation of Saint Domingue's elite by the metropolitan Academic model was then completed, and the *Cercle des Philadelphes* never referred to Mesmer beyond the first year of its existence—even though the phenomenon, discredited in Paris, seemed to regain a new strength in the French provinces.[37] In 1798, Moreau de Saint-Méry could evoke the memory of this rapid fading in the following terms: "[T]he tub was deserted, and the paralytic they had brought had to be taken back home."[38]

The birth of a brand-new scientific normative institution benevolently protected by the Société royale de médecine in Paris[39] and the rise of a colonial scientific identity proudly built on the ruins of supernatural knowledge apparently sealed the destiny of mesmerism in the colony. Nevertheless, even if the official disqualification of "fanatics" and "simpletons" defending this "charlatanism"[40] had been achieved by the *Cercle des Philadelphes* by the end of 1785, the idea of mesmerism was not quite dead in Saint Domingue.

Mesmerism and Vodou: Atlantic Cultural Cross-Fertilization?

Moreau de Saint-Méry, for example, evokes the fact that mesmerism had briefly found followers in the southern part of the island, far from Le Cap, especially among slaves and people of color: "Magnetism had its disciples, its apostles, and consequently its miracles in the southern department. But it was also ridiculed, and it died." However, he adds, "[T]he miraculous was rejected by all faiths, except those that admit of the Resurrection"— meaning the island's black population.[41]

There are indeed some clues justifying the hypothesis of a spreading of mesmeric theories and practices among the black population of the colony, both in the south and north. Although no blacks had been admitted in the "Society of Harmony" in Le Cap, some seem to have been cured thanks to "animal magnetism". Arthaud and Côme d'Angerville's list of patients treated by Puységur and his local disciples during the first four months of his presence in Le Cap reveals that at least ten people of color (including at least three slaves) had been treated in Le Cap at that time, coming from various places in Saint Domingue (Ouanaminthe, Saint Marc, Port-de-Paix, and Le Dondon).[42] In all likelihood, people attending tub meetings could imitate some passes and other theatrical features to claim links with mesmerism (without having been granted (paying) permission by Puységur). Such forms of distribution are attested for example in the previously cited letter by Morange of July 23, 1784: "[T]here is a surgeon at Blin's place who magnetizes and operates faster than M. de Puységur. He sells his secret for 3300 *livres* to any newcomer. Lavaud had bought it and died of it."[43]

Moreover, the idea of using Mesmer's treatment for plantation slaves as a large-scale treatment, in particular, seems to have seduced some masters, as did the idea of inoculation some years before.[44] This is shown by a letter from Morange in June 1784, on hearing that his master, Stanislas Foache, was following such a mesmeric treatment in Le Havre, France: "We know that M. Stanislas has been initiated into the mysteries of animal magnetism and that he is making ample use of it. Should he consider the effects beneficial, he will no doubt inform us and then inform his Negroes who could profit by these things."[45] However, he continues, regretfully noting that slaves (and maybe black people as a whole) were not Puységur's major concern, "M. de Puységur who is a great master, looks extremely hesitant [to cure slaves]; he is even reluctant to touch. It is wished that he should reveal his secret for the good of humanity but he hasn't yet explained himself."[46]

However, we indeed find one more trace of diffusion among the slave population besides the three slaves and the woman paralyzed for fourteen years mentioned above. As the planter Trembley notices,

> A plantation owner on this plain [near Le Cap], made a big profit by magnetizing a consignment of cast-off slaves he bought at a low price. Restoring them to good health by means of the tub, he was able to lease them at prices paid for the best slaves. The rage for magnetism has taken hold of everyone here. Mesmeric tubs are everywhere. [But] today hardly anyone speaks of them any longer, perhaps because too much has already been said about them.[47]

Such clues are unfortunately rare and difficult to find, even if we can still hope to find some more in plantation archives or private correspondence one day.

Of great interest, then, is the mention of mesmerism in two rulings from the Conseil supérieur du Cap, in reaction to repeated nocturnal meetings of black people, just a few miles south of Le Cap, in the La Marmelade district, a recent place mainly cleared by the work of first generation slaves, most of them from Congo. In the first one, dating from May 16, 1786, the Council forbade all people of African blood to join such night meetings in which had been noticed "convulsions" and manifestations of the "false prodigies due to this would-be magnetism [...] usurped by Negroes and disguised by them under the name of *Bila*"—a term referring to Vodou practices.[48] The second one, published on November 23, 1786, condemned three slaves named Jérôme Poteau, Télémaque, and Jean, found guilty of "having held nocturnal meetings of slaves, fraught with superstition and tumult, in several houses of the La Marmelade district and other nearby places under the pretence of magnetism."[49] At the end of the century, Moreau de Saint-Méry evoked this episode in the following terms:

> One certainly does not expect to hear that La Marmelade had been the place chosen to bring to fruition the ideas of magnetism suited to the views of those who propagated them. They appeared in La Marmelade together with the hoaxes of the 'Illuminated,' the repulsive tricks of the Convulsionaries and excesses of profanation because their aim was to reap the profits of swindling.

and he then spoke about "chimerical mysteries," "superstitions," and "shameless charlatanism."[50]

However that may be, the reading of the two rulings of 1786 and of the related judiciary archives kept in the lawyer François de Neufchateau's papers reveals that what happened at that time in plantations near Le Cap, the north part of the island, was seen by many as much more than a trivial event involving unscrupulous charlatans.

The 1786 archives refer to ceremonies that systematically take place "by night in secluded places" and involve "very numerous people." During these ceremonies, the decree of May 16 continues,

> [T]he miraculous operator has the subjects who ask to submit to his power brought to him into the circle. He does not limit himself to magnetizing them in the modern sense of the word. After the magician has caused stupor or convulsions in them using both the sacred and the profane, holy water is brought to him since he pretends it is necessary to break the spell that he had previously cast on the subjects [...].[51]

The Substitute of the king's prosecutor (*substitut du procureur du roi*), in his closing speech for the prosecution, evoked in the following terms an account given by a white witness of the scene:

> M. Jacquin, M. Estève's bursar, says that in the course of last July he clearly saw through the slits in the wall of Negro Jean Lodot's cabin, Negro Jean himself [another slave judged during this trial] among a numerous assembly, kneeling before a table covered with a rug and lighted by two candles, raising a fetish at intervals; that he could not clearly distinguish the kneeling and silent Negroes during the ceremony; he adds that he has afterwards found two machetes crossed on the spot where Negro Jean had operated. A certain Dimanche, Negro slave on the Estève habitation says [...] that those assemblies were called *mayombe* or *bila*. He describes in detail the ceremonies held there: leaves taken from raspberry bushes, avocado and orange trees were put in their hands, they were asked to kneel and then given *tafia* to drink in which he mixed pepper, garlic and whiting; and once the drink had made them fall the said Negro struck them with a machete to make them stand up again. He adds that Negro Jean carried a little bag strapped on his shoulder containing a crucifix, pepper, garlic, gunpowder, nails and a small case.[52]

As we can see and as Gabriel Debien and Pierre Pluchon have shown, these accounts are clearly describing manifestations of the Vodou cult, attested at that time in similar words by several witnesses. As a matter of fact, this trial is a trial against Vodou, in which Le Cap judges saw, or feigned to see, a mere manifestation of the familiar Puységur's "animal magnetism." Even a rapid glance at the papers we have today in our hands shows clearly that what is at stake, in that story, is nothing but the place and status of black religious and magical practices, in a colony of 350,000 black people severely ruled and dominated by a minority of only 25,000 whites.

At this stage, a passage of the ruling of May 16, 1786 provides another element for analysis: "It would be extremely dangerous for this colony ... to leave in the hands of Blacks an instrument that physical science only handles with great wariness, and which lends itself so easily to excesses and conjurers' tricks, common among Negroes and venerable in their eyes."[53] First, it reveals a persistent doubt concerning the real effects of mesmeric theories and practices, and it underlines the feeling that any appropriation of a European knowledge, "usurped" and "disguised" by black people, represents a potential risk for the colony.[54] Second, it underlines that the major fear of the judges was to see any charismatic leader taking advantage of such "illusions." As a matter of fact, such a suspicion of the use by black slaves of any kind of European knowledge—knowledge supposedly giving power over men—was part of the campaigns led on the field of symbolic power by both white masters and black slaves. In that sense, the use of "holy water" in the Vodou ceremonies could be considered on the same level as the use of mesmeric "tricks," both things participating in increasing sacred and symbolic power held by Vodou officiates.[55]

At that point, indeed, older fears were revived. Behind the harsh condemnation of those occult practices loomed the fear of poisonings. In the late 1750s, that fear had taken the shape of a fugitive slave born in Africa, known as Macandal, suspected of having mounted a vast project for poisoning the entire white population of the island. His trial had brought to light his supposed knowledge of some secret toxic properties of plants but also some magical practices that had endowed him with a very special credit among black people. That was confirmed in the verdict of January 20, 1758, which found him

> guilty of crime among the Negroes, having corrupted and seduced them by tricks and encouraged them to indulge in impieties and profanations in which he was himself involved, having introduced holy objects in the composition and use of allegedly magical parcels meant for evil spells which he concocted and sold to Negroes; for having besides prepared, sold and distributed all kinds of poisons.[56]

Condemned to death and burnt alive on the public square during an execution fraught with dramatic developments, that poisoner durably and strongly impressed black people's minds. And indeed, his name is still present today in Haitian memory as it is in the Vodou cult: a *loa* or *lwa* (a major spirit participating in the cult) is called *Macandal*, and the term *macandal* still today designates a talisman made of various elements gathered into a small packet.[57]

For the colonist, that episode in which African knowledge of poisons, magical ceremonies, and phantasms of servile revolts were mixed was still looming as a threat nearly thirty years later. "And who knows"—stated the May 1786 ruling—"who could tell how far initiators or convulsionaries of the Macandal type could one day carry this fanaticism and delirium?"[58] In the minds of the colonists in 1786, magnetism was close to becoming an element of the feared black resistance to both European domination and European rationality, regularly expressed in mysterious nocturnal assemblies fraught with menace.

The idea that mesmeric elements could have been included in Vodou ceremonies is, of course, relevant, as Vodou (unlike mesmerism, which defined itself among its white European officiants as a rigid constituted knowledge) represents a body of knowledge, practices, and beliefs built on a constant cultural cross-fertilization with local and recent traditions, Catholic culture, and prestigious knowledge brought directly from different parts of Africa (especially Dahomey and Congo) by incoming slaves. However, today's Vodou does not reveal any visible or tangible connection with Mesmer: no tubs, no clear reference to Puységur, Mesmer, or any of their locally known disciples. Moreover, "mesmerism," "magnetism," and "would-be mesmerism" never appeared in judiciaries' sources of that time in Saint Domingue as anything other than European *words* used by white judges for describing various parts of Vodou rites, for which we have existing, older descriptions. Therefore, even if we should not neglect the attractive and appealing possibility of a long journey of mesmeric knowledge from Vienna to Saint Domingue's plantations, passing by Paris and Le Cap, we must be cautious, as these descriptions of black mesmerism seem to have been nothing but a smokescreen set between the rationality of French judges, and the frightening manifestation of black Vodou nocturnal ceremonies.[59] In that context, the use of words designating at that time a familiar and reassuring form of charlatanism was doubtless a way to publicly disqualify any kind of black occult knowledge and also gave words to judges to describe and condemn such a mysterious phenomenon, characterized by both impressive "trances" and hypnotic effects. In the same way, as they definitively closed the mesmeric files, the judges opportunely avoided any request for further scientific inquiries from the *Philadelphes* on such elusive, mysterious and irreducible black knowledge—deliberately excluding it from European spheres of interest.

However that may be, neither the verbal stifling of the Vodou phenomenon (which was reduced to the status of second-rate mesmerism) nor the exemplary condemnation of the nocturnal assemblies to which it gave rise in several places in the colony succeeded very far. The symbolic and social power of Vodou was actually to play a role (doubtless important, albeit difficult to evaluate) as a strong stimulation for combatants in the black Revolution which made Saint Domingue flare up in 1791.[60]

Conclusion

As we can see, the story of mesmerism in Saint Domingue had been much more than a simple episode of colonial medical life. Introduced with some success in the colony by one of its stronger supporters in 1784, it raised local debates among scientists and colonists and certainly contributed to reinforce intellectual links between colonial intellectual elites and French academicians. As a consequence, it directly led to the birth of the first French colonial academy, well known as the *Cercle des Philadelphes*, which became in 1789 the *Société royale des arts et lettres du Cap Français de Saint Domingue*. Seen as a scientific theory, mesmerism provides an interesting opportunity to observe *in vivo* the spreading of a coherent body of both scientific and occult knowledge through the Atlantic toward the European and African worlds rooted in Saint Domingue. As it provoked internal and external scientific debates under the watchful eye of royal institutions in France, its spread in the colony can be seen as a major moment in the building of a scientific identity for local enlightened minds. Used as a rebellion flag by some enlightened colonists fighting against the supposed universality of European expert knowledge and (moreover) portrayed by colonial magistrates as a threatening weapon in the hands of slaves willing to overthrow the whites' domination, mesmerism in Saint Domingue appears as a fascinating catalyst of tensions and fantasies, belonging to that mysterious part of knowledge that makes historical anthropology so interesting.[61]

The short but intense episode of Saint Domingue mesmerism is also one of the rare occasions that allow historians to glimpse a part of the secret knowledge of black slaves, usually obscured in written sources. European "mesmerism" and its "mesmeric trances," thought of by colonists, physicians, and judges as an occult power, provided them with the perfect words and tools they needed to understand and vilify such a strange (and to them incomprehensible) thing as Vodou—thus building behind a smokescreen the elusive figure of a seductive shared and changing Atlantic knowledge.

Notes

1. Gabriel Debien, "Une nantaise à Saint Domingue (1782–1786)," *Revue du Bas-Poitou et des Provinces de l'Ouest*, 83: 6 (1972), 413–36, cited and translated in James E. McClellan, III, *Colonialism and Science: Saint Domingue in the Old Regime* (Baltimore: Johns Hopkins University Press, 1992), 177.
2. For overviews of Saint Domingue and its population on the eve of the Revolution, see Stewart R. King, *Blue Coat or Powdered Wig: Free People of Color in Pre-Revolutionary Saint Domingue* (Athens, GA: University of Georgia Press, 2001); Laurent Dubois, *Avengers of the New World: The Story of the Haitian Revolution* (Cambridge, Mass.: Harvard University Press, 2004), 15–35; and John D. Garrigus, *Before Haiti: Race and Citizenship in French Saint Domingue* (New York and Basingstoke: Palgrave Macmillan, 2006).
3. Archives Nationales, Paris (hereafter AN), Marine C7 61, dossier Chastenet de Puységur. For details on his cartographical work, see François Regourd, "L'expédition hydrographique de Chastenet de Puységur à Saint-Domingue (1784–1785)," in Hubert Bonin and Silvia Marzagalli, eds, *Négoce, Ports et Océans: XVIe-XXe siècles* (Bordeaux: Presses Universitaires de Bordeaux, 2000), 247–62.
4. In 1782, the *médecin du roi* Duchemin de l'Etang had already written to the Société royale de médecine in Paris about the arguments he used against a local admirer of Mesmer, but there were at that time no tubs nor any kind of magnetic cures in the Island; Bibliothèque de l'Académie de Médecine (Paris), collection of the Société royale de médecine (hereafter BAM SRM), 167, file 1: letter from Duchemin de l'Etang, October 1, 1782.
5. "M. De Puysegur Lieutenant de Vaisseau a exigé que le Chirurgien de ce vaisseau qui mis à la voile pour l'Amérique n'embarqua dautres remedes que ces instrumens promettant bien de prévenir et guerir toutes les maladies par sa vertu magnetique": BAM, ms 10, p. 157, session of Tuesday, April 27, 1784 (Plumitif des procès-verbaux de la Société royale de médecine).
6. For a general overview of mesmerism at the end of the eighteenth century, see, among others: Robert Darnton, *Mesmerism and the End of the Enlightenment in France* (Cambridge Mass.: Harvard University Press, 1968); Charles C. Gillispie, *Science and Polity in France at the End of the Old Regime* (Princeton: Princeton University Press, 1980), ch. 4 ("Scientists and Charlatans"), esp. 257–89; Bertrand Meheust, *Somnambulisme et médiumnité, 1784–1930*, 2 vols (Le Plessis-Robinson (France): Institut Synthélabo, 1998); Jean-Pierre Peter, "Introduction," in Armand-Marc-Jacques de Chastenet de Puységur, *Un somnambulisme désordonné? Journal du traitement magnétique du jeune Hébert*, ed. Jean-Pierre Peter (Le Plessis-Robinson (France): Institut Synthélabo, 1999); Franklin Rausky, *Mesmer ou la révolution thérapeutique* (Paris: Payot, 1977); Jessica Riskin, *Science in the Age of Sensibility: The Sentimental Empiricists of the French Enlightenment* (Chicago: University of Chicago Press, 2002), 189–225.

7 "La réputation de ce médecin commence à faire sensation ici […] M. de Puységur, qui est un grand maître (…)": Maurice Begouën-Demeaux, *Mémorial d'une famille du Havre (1743–1831)*, vol. 2: *Stanislas Foache, négociant à Saint-Domingue, 1737–1806* (Paris: Société française d'Histoire d'outre-mer, 1982), 94.

8 Begouën-Demeaux, *Mémorial*, 94.

9 "M. et Mme de Puységur sont venus nous demander la soupe. Nous avons vu le mari magnétiser un de ses officiers qui avait un accès de fièvre. Il l'a d'abord fait pleurer, ensuite rire avec un forte piano. Les expériences ne sont pas encore commencées faute des objets nécessaires. Beaucoup de monde se présente pour être traité…": Begouën-Demeaux, *Mémorial*, 94 (June 19, 1784).

10 "M. de Puységur a commencé ses traitements": Begouën-Demeaux, *Mémorial*, 94.

11 "On vit à la Providence des hommes, un baquet, qu'assiégèrent des obstrués, des goûteux, des asthmatiques": Médéric-Louis-Elie Moreau de Saint-Méry, *Description topographique, physique, civile, politique et historique de la partie française de l'isle de Saint-Domingue*, 2 vols (Philadelphia: chez l'auteur, 1797–98), new edition by Blanche Maurel and Étienne Taillemite, 3 vols (Paris: Société française d'Histoire d'Outre-mer, 1984), 1: 345 (all further references are to this edition).

12 For details, see McClellan, *Colonialism and Science*, 178–9.

13 BAM, Académie royale de Chirurgie, 25, file 14. Letter from Arthaud and Côme d'Angerville, Le Cap, September 23, 1784. Correspondance Morange-Foache, cited in Begouën-Demeaux, *Mémorial*, 95. Also AN Marine 2 JJ 104, bundle 3, piece 12 (draft of a letter from Chastenet to Bellecombe, September 17, 1784).

14 AN Marine 2 JJ 104, bundle 1, piece 2 (dated February 2, 1786). Another case of a similar contract is cited in Gabriel Debien, "Assemblées nocturnes d'esclaves à Saint-Domingue (La Marmelade, 1786)," *Annales historiques de la Révolution française* 44 (1972), 273–84, at 273, note 5 (document sold by auction at the hôtel Drouot in 1967).

15 AN Marine 2 JJ 104, bundle 3, piece 12 (draft of a letter from Chastenet to Bellecombe, September 17, 1784). According to Morange, cures were interrupted for a few weeks in July 1784: "[Puységur] laisse les magnétisés dans l'embarras [et ses premiers disciples n'ont sans doute pas ses qualités…];" Begouën-Demeaux, *Mémorial*, 95.

16 One of the main criticisms made of Puységur was indeed that people were not cured, but only temporarily relieved. The list is kept in BAM, Archives de l'Académie royale de Chirurgie, box 25, file 14.

17 Gabriel Debien, "Profils de colons. I. Jean Trembley," *Revue de "La Porte Océane,"* 11: 113–14 (1955), 14–19, 8–10, translated by McClellan, *Colonialism and Science*, 177–8.

18 AN Marine 2 JJ 104, bundle 5, piece 4 (Chastenet de Puységur papers). Such certificates were common for Mesmerian practitioners, always concerned with proofs of success.

19 Charles C. Gillispie, *op. cit.*, ch. 4 ("Scientists and Charlatans"), 257 and following; Riskin, *Science in the Age of Sensibility*; Patricia Fara, *An Entertainment for Angels: Electricity in the Enlightenment* (Cambridge: Icon Books, 2002);

McClellan, *Colonialism and Science*, 168–71; François Regourd, "Sciences et colonisation sous l'Ancien Régime: le cas de la Guyane et des Antilles françaises, XVIIe–XVIIIe siècle," doctoral dissertation (Université de Bordeaux III-Michel de Montaigne, 2000).

20 For more details, see Darnton, *Mesmerism*, ch. 2 (46–81), and works cited in note 6.

21 Letter from Le Cap, July 27, 1784. BAM SRM 136, file 1, piece 24: "Nous avons icy depuis un mois un disciple de M. Mesmer. Il a établi un très grand appareil, il a rassemblé les malades de toutes les espèces. Il leur a promis à tous qu'il les guériroit. Il y a déjà un mois qu'il opère, et il y a sept personnes mortes."

22 James E. McClellan, III and François Regourd, "The colonial machine: French science and colonization in the Ancien Regime," *Osiris* 15 (2000), 31–50; François Regourd, "Lumières coloniales: les Antilles françaises dans la République des Lettres," *Dix-huitième siècle* 33 (2001), 183–99. On the general context of medical history in eighteenth-century Saint Domingue, see Isabelle Homer, "Médecins et chirurgiens à Saint-Domingue au XVIIIe siècle," thesis for the diploma of *archiviste paléographe* (École Nationale des Chartes, 1998).

23 BAM SRM, 136, dossier 1, pièce 25. Letter from Fournier de Varenne to Arthaud June, 25, 1785: "Il est très singulier et très remarquable que les médecins de Paris nient l'existence du magnétisme animal, et que les médecins du Nouveau Monde avouent sa réalité. Les médecins de l'Amérique ont-ils plus de bonne foi et de probité que ceux de l'Europe? Ou les effets du magnétisme animal sont-ils plus marqués dans le voisinage de l'Equateur? Pour l'honneur des médecins, j'adopte cette dernière opinion, qui est d'ailleurs conforme à notre doctrine."

24 Darnton, *Mesmerism*, 88–90.

25 *Nouvelle découverte sur le magnétisme animal, ou Lettre adressée à un Ami de Province, par un Partisan zélé de la Vérité* (no place, no date [circa 1800]), 61.

26 The irony of this point is that a few years later, this climatic difference between France and Saint Domingue is invoked by Moreau de Saint-Méry, as an inverse demonstration: "Soit que le climat se prêtât moins aux illusions, soit que la marche rapide des maladies qu'on y éprouve fût plus propre à montrer l'insuffisance du moyen, les faits vinrent à l'appui de la contradiction, le bacquet fut déserté et il fallut rapporter le paralytique qu'on y avait amené;" Moreau de Saint-Méry, *Description* (1984), 1: 345. See also McClellan, *Colonialism and Science*, 174, and François Regourd, "Sciences et colonisation," 351.

27 Darnton, *Mesmerism*, ch. 3 (82–105).

28 BAM SRM 136, file 1, pièce 25 (letter of June 25, 1785): "un corps d'intriguans schismatique, [...] loupe monstrueuse et difforme qui s'est élevée sur le corps de l'antique faculté, comme les agarics croissent sur les arbres tombés dans la décrépitude."

29 "Je suppose que vous devez aspirer à renverser l'ordre social et à le détruire puisque c'est le seul moien de revenir à la nature": BAM SRM 136, file 1, pièce 25 (letter of July 27, 1785).

30 The list of initiates includes, for the first promotion: "M. Laval, médecin; M. Desvarennes [i.e., Fournier de Varenne], Ch[evalier] de Saint-Louis; M. le Marquis de Cadush, habitant; et M. de Malouet d'Alibers, Commissaire de la Marine." For the second promotion: "M. Worlok [i.e., Worlock], M. de Courjolles

[i.e. Courrejolles], ingénieur; M. Laborie, avocat; M. Gulman, habitant." And for the third: "M. Joubert, habitant, ancien chirurgien; M. Rousseau, habitant; M. de Guydy, lieutenant de vaisseau; M. de Fourneau, négociant" (BAM ARC, box 25, file 14). Some of Puységur's disciples were nevertheless quite highly regarded by French scientific institutions, like Siméon Worlock, in particular, who had been the main agent of inoculation in Saint Domingue in the 1770s (which certainly explains his curiosity for innovative medical theories). He was well known and recognized by the Royal Society of Medicine, from which he received a gold medal and the official title of "correspondant" in September 1784, for an important dissertation on epizootic disease (BAM ms 10, Plumitif des procès-verbaux de la Société royale de médecine, 244, session of Tuesday, August 31, 1784; and *Journal de Physique* 25 (1784), 393).

31 BAM SRM 136 file 1, piece 26: letter from Arthaud, Le Cap, September 10, 1785: "Je vous assure, Monsieur, qu'il n'y aurois pas grand mérite à suivre ce projet dans tout autre lieu qu'une colonie, mais il faut icy un double courage, surtous lorsqu'on se voit éloigné comme nous le sommes du centre du Gouvernement et de l'influence des graces et des encouragements qui soutiennent le zèle et animent les travaux d'une société. La science est une plante exotique qui n'a point encore pris racine icy; on ne sait pas encore l'y cultiver, et je crois qu'elle s'y naturalisera difficilement."

32 McClellan, *Colonialism and Science*, 179.

33 François Regourd, "Sciences et colonisation," 396 and following. James E. McClellan, III, "L'historiographie d'une académie coloniale: le Cercle des Philadelphes (1784–1793)," *Annales historiques de la Révolution française* 320: 2 (2000), 77–88.

34 As a matter of fact, the Société royale de médecine sent to Arthaud and the Cercle some declaration showing its "satisfaction" to see such a cercle born in Saint Domingue: "J'ai transmis au cercle des phyladelphes les deux lettres dans lesquelles vous exprimés, au nom de votre compagnie, la satisfaction qu'elle a éprouvé de son établissement;" BAM SRM 136, file 1, piece 26: letter from Arthaud, Le Cap, September 10, 1785.

35 Puységur left Le Cap Français for Brest on July 31, 1785: AN Marine C6 872 ("Rôle d'équipage de la corvette du Roi *Le Vautour*, pour 1784–1786"), fol. 1. Yet in December 1784, the governor of Saint Domingue and the Minister himself strongly invited Puységur to stop the publicity around his mesmeric cures (AN Marine 2JJ 104, bundle 1, piece 4). In October 1785, one of Puységur's friends, the colonist Ladébat, confirmed in these terms the slow decline of the Lodge of Harmony in Le Cap, in a letter to Chastenet de Puységur's wife: "The Lodge keeps meeting regularly in Le Cap, every month I think. But [the lack of] M. de Chastenet's knowledge leaves an irreparable vacuum, which turns the proselytes away" ("Il y a loge au Cap régulièrement, je crois tous les mois. Mais les connaissances de M. de Chastenet font un vide irréparable qui écarte les prosélytes"); AN Marine 2 JJ 104, bundle 5, piece 8 (letter from Ladébat, October 30, 1785, to Mrs Chastenet de Puységur).

36 The engineer Courrejolles, for example, rapidly joined the Cercle (*Tableau du Cercle des Philadelphes*, AN Col F3 81, ff. 145–6), and even one of the more virulent proponents, Fournier de Varenne, appears as an official member on the 1787 list of members, as well as Worlock: *Tableau du Cercle des Philadelphes;*

établi au Cap-François avec l'approbation du Roi, le 15 août 1784 (Cap-François: de l'Imprimerie Royale, 1787), 4 pp.
37 Mesmer, in particular, made a triumphant tour all around France in 1786: Darnton, *Mesmerism*, 67.
38 Moreau de Saint-Méry, *Description* (1984), 1: 345.
39 As McClellan has shown, "seven members of the Cercle, inculding Arthaud and Warlock, became correspondants of the Paris medical society. Vicq d'Azyr, permanent secretary of the Royal Society of Medicine, pronounced the *éloges* of three Philadelphes: Lefebvre-Deshayes, Joubert de la Motte, and Cosme d'Angerville": McClellan, *Colonialism and Science*, 355 n. 24.
40 These terms are those used by Artaud in his letter to Vicq d'Azyr of September 10, 1785: "il n'y a plus que quelques fanatiques et quelques imbéciles qui deffendent cette charlatanerie," (BAM SRM 136, file 1, piece 26).
41 Moreau de Saint-Méry, *Description* (1984), 1322, cited and translated in McClellan, *Colonialism and Science*, 177. For a general overview of black medical practices and knowledge in Saint Domingue at that time, see Karol K. Weaver, "The enslaved healers of eighteenth-century Saint Domingue," *Bulletin of the History of Medicine* 76 (2002), 429–60; and her recent book *Medical Revolutionaries: the Enslaved Healers of Eighteenth-Century Saint Domingue* (Urbana and Chicago: University of Illinois Press, 2006), in which ch. 6 (98–112) deals with the mesmeric episode in Saint Domingue and its supposed influences among black healers.
42 The list includes "Le nommé Guirgny, carteron libre d'Ouanaminthe, aveugle, sorti en même état depuis trois mois"; "la nommée Rozette, mulâtresse de Madame Bailly au Cap, attaquée de douleurs de tête et migraine, sortie en même état au bout d'un mois"; "La nommée Bibiche, mulâtresse de St Marc attaquée d'une suppression au Baquet depuis un mois"; "La nommée Marianne mulâtresse de St-Marc attaquée d'une goutte sereine à l'œil au baquet depuis trois mois"; "Adam, nègre libre ancien Courier attaqué de la vue, au baquet depuis quatre mois"; "Le nommé Moreau, Mulâtre du Port de Paix attaqué de douleurs. Au baquet depuis un mois"; "La nommée Jeanne Carteronne libre fille de M. Moreau habitant au Dondon attaquée d'un Gros rhume a été magnétisée pendant un mois ce qui lui a occasionné des fièvres lentes. Son rhume a dégénéré en pulmonie, est morte en heptisie le 21 [septembre]"; "La nommée Claire, mulâtresse attaquée de la poitrine a été magnétisée pendant deux mois est devenue pulmonique et est morte le 21 [octobre]"; "Une négresse de Madame Veuve Dorée, Magnétisée par M. de Puiségur"; "Un nègre de M. de Cadusch, magnétisé par son maître"; a list to which we could even add an eleventh black person involved in the mesmeric cures as guardian of a patient: "Une Pauvre Petite Blanche âgée de dix ans dont Adam nègre libre prend soin, aveugle au baquet depuis quatre mois," BAM, fonds de l'Académie de Chirurgie, 25, file 14.
43 Letter from Morange to Foache, July 23, 1784, cited in Begouën-Demeaux, *Mémorial*, 95.
44 The years 1760–70 had seen a major inoculation wave in Saint Domingue's plantations, conducted by planters and physicians for thousands of black slaves (Moreau de Saint-Méry, *Description* (1984), 224–5, 250, 1068, and McClellan, *Colonialism and Science*, 144–5).

45 Begouën-Demeaux, *Mémorial*, 94.
46 "[… à propos des esclaves …] M. de Puységur, qui est un grand maître, paraît très réservé; il répugne même à toucher. On désire qu'il donne son secret pour le bien de l'humanité, mais il ne s'est point encore expliqué": Begouën-Demeaux, *Mémorial*, 94.
47 Gabriel Debien, "Profils de colons. I. Jean Trembley," cited and translated by McClellan, *Colonialism and Science*, 178.
48 "Faux prodiges de ce prétendu magnétisme […] usurpé par les Nègres et déguisé par eux sous le nom de Bila": AN, 27 AP 12 (papers from François de Neufchâteau), cited in P. Pluchon, *Vaudou, sorciers empoisonneurs de Saint-Domingue à Haïti* (Paris: Karthala, 1987), 66–7. "Bila" (from a Bantu word for "drawing"), according to J. Kerboull, désignates a sacrifice table (Jean Kerboull, *Le vaudou: magie ou religion* (Paris: Robert Laffont, 1973), cited in P. Pluchon, *op. cit.*, 67). On Vodou, besides Pluchon's work, see, among others: Howard Justin Sosis, "The colonial environment and religion in Haïti: an introduction to the black slave cults in eighteenth-century Saint Domingue," Ph.D. dissertation (Columbia University, 1971); Alfred Métraux, *Voodoo in Haiti* (New York: Schocken books, 1972); Laënnec Hurbon, *Les mystères du Vaudou* (Paris: Gallimard, 1993).
49 "… Avoir tenu des assemblées nocturnes superstitieuses et tumultueuses d'esclaves, dans plusieurs habitations du quartier de la Marmelade et lieux voisins, sous prétexte de magnétisme" AN, 27 AP 12 (papers from François de Neufchâteau), cited in P. Pluchon, *op. cit.*, 68. Gabriel Debien, "Assemblées nocturnes," 279 and following.
50 Moreau de Saint-Méry, *Description* (1984), 275–6: "On ne s'attend surement pas à apprendre que la Marmelade a été le lieu qu'on avait choisi pour y faire fructifier les idées du magnétisme, assorties comme en Europe, aux vues de ceux qui les propageaient. Elles ont paru à la Marmelade accompagnées des farces des Illuminés, des scènes dégoûtantes des Convulsionnaires et des abus de la profanation, parce qu'on voulait arriver aux profits de l'escroquerie." He writes later of "mystères chimériques," "superstitions" and of "charlatanisme effronté."
51 *Arrêt* of May 16, 1786 (Conseil supérieur du Cap), in AN, 27 AP 12 (papers of François de Neufchâteau), cited in Pierre Pluchon, *op. cit.*, 66: "L'opérateur miraculeux se fait représenter dans ce cercle des sujets qui demandent à subir son pouvoir. Il ne se borne pas à les magnétiser, suivant l'acception moderne de ce mot. Après que le magicien leur a causé de la stupeur ou des convulsions, mêlant le sacré au profane, il se fait apporter de l'eau bénite qu'il prétend nécessaire pour désensorceler ceux qu'il a mis en crise…"
52 AN, 27 AP 12, cited in Debien, "Assemblées nocturnes…," *op. cit.*, 279: "M. Jacquin, économe de M. Estève, dit que dans le courant du mois de juillet dernier, il vit clairement à travers le clissage de la case du nègre Jean Lodot, le nègre Jean au milieu d'une assemblée considérable, ledit nègre à genoux devant une table couverte d'un tapis et éclairée par deux chandelles, élevant à différentes distances un fétage [fétiche], qu'il n'a pas pu bien distinguer les nègres à genoux et en silence pendant cette cérémonie; il ajoute avoir trouvé deux manchettes [machettes] croisées à terre à l'endroit où opérait le nègre Jean. Le nommé Dimanche, nègre esclave de l'habitation de M. Estève, dit qu'il s'est plusieurs fois

trouvé aux assemblées que le nègre Jean tenait sur l'habitation de M. Estève, son maître; que ces assemblées se nommaient *mayombe* ou *bila*. Il ajoute le détail des cérémonies qui s'y pratiquaient, telle que de leur mettre dans les mains des feuilles de framboisier, d'avocat et d'oranger, de les faire mettre à genoux et dans cette posture de leur donner à boire du tafia dans lequel il mêlait du poivre, de l'ail, du blanc d'Espagne et que cette boisson les faisant tomber, ledit nègre les relevait à coup de manchette. Il ajoute que le nègre Jean portait sur lui en bandoulière un petit sac dans lequel était un crucifix, du poivre, de l'ail, de la poudre, des cayous, des cloux et un étui."

53 AN, 27 AP 12, cited in P. Pluchon, *op. cit.*, 66–7.
54 The expression is in the text of the ruling: "usurpé par les Nègres et déguisé par eux sous le nom de Bila," cited in P. Pluchon, *op. cit.*, 67.
55 Such a perspective can be compared to what Vincent Brown has brought to light in an article on Jamaican slave society at the same time. Studying the question of Obeah and supernatural beliefs underlying the confrontation between white masters and black slaves, he shows how much the "political potential" of any supernatural belief, whether its effects were attested or not, was of great concern for white judges: Vincent Brown, "Spiritual terror and sacred authority in Jamaican slave society," *Slavery and Abolition* 24 (2003), 24–53.
56 P. Pluchon, *op. cit.*, 171–2. See also the chapter by Parrish in this volume.
57 P. Pluchon, *op. cit.*, 218 and following. See also Laënnec Hurbon, *op. cit.*, 40 and following chapters.
58 AN, 27 AP 12, cited in P. Pluchon, *op. cit.*, 67.
59 On this point, I would be much more cautious than Weaver, *Medical Revolutionaries*, 98–112, (ch. 6, entitled "Magnetism in eighteenth-century Saint Domingue: the case of the enslaved magnetists and their fight for freedom)." Studying this possible cultural cross-fertilization, Weaver indeed asks "Was it Mesmerism?" (109), and immediately answers "We may never know," but then states, without citing convincing evidence, that "Mesmerism definitely influenced vodou" (110), that "the practice of mesmerism by slaves was a political act of revolution," and that the nocturnal vodou assemblies in La Marmelade (preceding the black insurrection in Saint Domingue) were "mesmerist meetings" (112), which is a seductive, but risky extrapolation. This point indeed fits well with her convincing thesis that enslaved healers played a major role in the Haïtian Revolution, but the fact is that we have no clear evidence, so far, that any kind of "mesmerism," strictly speaking, was a part of that story.
60 On this point, see in particular Laurent Dubois, *Avengers of the New World*, 101–2; and the articles by David Geggus collected in his *Haitian Revolutionary Studies* (Bloomington: Indiana University Press, 2002), especially "Marronage, Voodoo, and the Saint Domingue Slave Revolt of 1791" (69–80), and "The Bois Caïman Ceremony" (81–92).
61 On the links between history, memory, religion and identity in the Haitian context, see, among others: Joan Dayan, *Haiti, History and the Gods* (Berkeley: University of California Press, 1995); Michel-Rolph Trouillot, *Silencing the Past: Power and the Production of History* (Boston: Beacon Press, 1995), 31–107.

Afterword

Science, Global Capitalism, and the State

MARGARET C. JACOB

The middle phrase in my title is almost quaint, an embarrassment. Talking about capitalism has gone out of fashion. Yet, though I know that the kind of extraordinary curiosity about nature displayed by early modern explorers and travelers had many motives, to be sure none exceeded in intensity the drive for wealth and profit. It has been said of the early Spanish conquistadors that they desired not gold but more gold. Even when looking for cinnamon, they imagined gold. They were not alone. In European country after country, the search for wealth and profit fueled empire. As we try to come to terms with the massive imbalance in power that once existed between the West and the rest, one that prevailed into the 1960s and beyond, we need to wed the natural knowledge gained to the economic power that the search for profit unleashed. Motives matter. Add to the search for wealth the power of the state. At the very least, it could protect both wealth and knowledge, or at the most, in the absolutist countries, promote science actively through state-sponsored academies. The resulting knowledge, wealth, and political power laid the foundation for an unprecedented geopolitical hegemony.

Travelers through Europe's extended commercial networks seemed to know instinctively that gathering, collecting, processing, and then theorizing about what was being seen in the Atlantic world could only bring benefits for themselves. Wanting to make sense of how such motives got translated into action, our editors are surely right to lay the emphasis on networks, on local settings, on the posture of objective witnessing (or as they put it, the depersonalization of witnessing), and on the active role played by indigenous peoples and transported slaves who imparted vital knowledge to the colonizers, masters, and travelers who traversed the Atlantic.

However, awarding agency to subaltern peoples can be tricky. Agency implies some degree of autonomy, suggesting a self-determining human subject. The tension between this admirable desire to see the non-European peoples in this story as actors and the reality of their so-frequent subjugation is built into the conundrum inherent in trying to understand both humanity and hegemony.[1] Perhaps no area on the globe experienced such a merger and comingling of peoples as did the Atlantic stage: Amerindians, Africans, Portuguese, Spanish, Dutch, French, English, and more. The list is long, and while the authors of the marvelous chapters of this book aim to open windows onto the multiple forms of encounter, onto moments and episodes, they have not lost track of the fact that there were winners and losers in this process, that to take but one example, entire languages of Indian and African origin disappeared in the "new world," never to be heard on this continent again.[2] The same disappearance must have happened for much indigenous natural knowledge. Both states and churches sought to impose their languages on the newly encountered peoples.

That European participants, especially the states, knew the power stakes and valuable prizes surfaces in the opening chapter by Alison Sandman. Before any natural knowledge could be gleaned or gold mined, the cartographers and navigators had to get the Portuguese and Spanish ships there. Here we have at work the secrecy that only the state could attempt to enforce, reminding us that the gold served many purposes, mercantile to be sure but also geopolitical. Empires helped to solidify monarchical power. Keeping foreigners from access to maps and charts was essential, and the loyalty of pilots could best be assured if they came from families of property with ties to the government. However, as Sandman documents, the system continuously broke down and the monopoly of knowledge proved impossible to maintain. Here Sandman's argument resonates with what political and economic historians concerned with the rise of the West have noted. The fact that the western end of the great Eurasian land mass was broken down into competing states in the long run worked to advantage over kingdoms such as China, where something resembling a single authority could generally prevail. Try as they might, neither the

Spanish nor the Portuguese could keep the English or the Dutch off the routes that pointed toward the new wealth.

Perhaps most curiously of all, the Dutch—without a monarchy, indeed without even much of a centralized state—emerged as the leaders of the pack. We need to know more about their approach to exotic knowledge, especially given the peculiar power that the Dutch awarded their trading companies over the localities they came to control and exploit. The centrality of Dutch economic power and global imperial presence was visible for all to see by the 1640s, and competition with them became a preoccupation of both the French and the English. To this day, the sources of Dutch power have remained somewhat elusive, and this puzzle has inspired some excellent new comparative work.[3] In a superb book, Julia Adams argues that the old categories deployed by historical analysis to explain Western hegemony—state structures, class, profit-seeking, religion, and patronage—cannot fully address the complexity of that power without the addition of gender, more precisely of patrimony. In each country, she argues, "coalitions of elite patriarchs came to occupy and identify with sites of intergenerational political privilege …collectively and consciously taking in hand parts of state apparatuses and their colonial projections."[4] In each state, the result differed and depended in large measure on the ability of the central state to bond the patriarchs to its needs and interests. This sociologist has been to the archives and speaks with authority about human behavior, particularly that of the early modern Dutch male elite. Throughout, one basic assumption governs the analysis. Once the head of an early modern family acquired wealth and translated it into state offices, he fought to keep that patrimony for sons, nephews, for generations yet unborn and only imagined. Unfortunately, science is beyond the scope of Adams's interests, but her analysis of gender, state formation, and empire may go part of the way to help to explain why, as states and imperial power grew, science increasingly became a masculine domain.

As Adams notes, and as Nicholas Dew confirms, French imperial efforts were remarkably weak and underfunded throughout much of the early modern period. Dew is right to point out the benefit Newton received from French observations of pendulums near the equator and also to note how relatively little sustained enquiry made its way from the French periphery to the center. One reason why the French pendulum experiments also led to little scientific innovation may have been the absence of an agreed theoretical framework that helped to make sense out of what was being observed. The exiled Catholic king of England, James II, provides an illustration of how the French were thinking about the pendulum. Hardly a *savant*, in 1690 James turned up for his first visit to the Parisian academy.[5]

He may never have read Newton's *Principia* of 1687 (indeed few had)—with its laudatory poetry making the work appear to be an adornment to him—but he had some idea what it contained. James reported on the thinking of Newton and various others that the figure of the earth was not round, but the French academicians assured him that the different length of the swings of pendulums seen in different places in the world was solely the result of climate and "the temperament of the air." Whether scholastic or Cartesian, they were not ready to accept the effect that gravitational pull would have on the earth and its relative density.

By 1690, the French Académie des sciences also took its marching orders directly from the king's ministers. Having spent its early years occupied by alchemy and mathematics, among other pursuits, the academy eventually vexed the patience of the crown. Mechanics proved far more alluring to the absolutist state. In 1685, just after the death of Samuel Du Clos, who had led the alchemical inquiries, a representative of Louvois, the minister who now oversaw the working of the Académie, came before it with this angry indictment of alchemy (often called the Great Work), and with a directive for future research:

> I understand by useful research that which could relate to the service of the King and the State;—not the Great Work which also includes the extraction of Mercuries from all sorts of metals, their transmutation or multiplication, which Monseigneur de Louvois does not wish to hear spoken of—or else the investigation and examination of the mines and open-cast workings of France, as well as of all the sulphur-containing compounds used in war, or those able to desalinate sea water and make it fit to drink.[6]

The search for the philosopher's stone, distillation, and the transmutation of metals had become far less relevant to a belligerent state that had enemies it perceived at home and abroad. In one of the first historical examples of the state's intervention in the agenda of science, the king's representative complained about the time wasted on distillations and begged the company to turn its attention toward material, rather than organic, substances: "the other research more suited to this Company and which would be more to the taste of Monseigneur de Louvois concerns everything that could explain physics and serve Medicine, these two things being almost inseparable because Medicine draws on the results and profits from new discoveries of the physical."[7] It is not clear in 1700 what value, if any, the French crown put in overseas scientific explorations. But had high value been awarded such activity, I suspect much more of it would have been done in the French colonies. Yet, as Neil Safier demonstrates, the pace of

French scientific exploration quickened in the course of the eighteenth century, partly I suspect under the pressure that Newtonian science put on French assumptions about the world but also by the intervention of other institutions, such as the Jardin du Roi. By the mid-eighteenth century, ironically just as the French were losing imperial ground to the British, the pace of their inquiries accelerated. Joseph de Jussieu is offered by Safier's chapter as an exemplar of what could be accomplished. In Jussieu we have someone actually interested in cinnamon and being inspired by the myths of gold associated with its region. His difficulties in communicating with the center are poignant reminders of just how hard such interests were to pursue. He may have counted himself a failure but we can admire his dogged determination and inquisitiveness.

I can think of no French (or for that matter Dutch) colonial of the eighteenth century, however, who can match Benjamin Franklin in sheer curiosity, and ultimately, eclectic brilliance. Joyce Chaplin describes his pioneering efforts to understand the Gulf Stream, to make public the knowledge that had once been the private domain of craftsmen such as mariners. As market forces played havoc with the ability of the guilds to protect and control wages and prices, many just disappeared or—in the case of the stonemasons—transformed themselves into voluntary associations to which Franklin himself belonged.[8] The Royal Society had been one of the first organized bodies to see the importance of craft secrets and attempted during the Restoration to write histories of all the crafts, a project it never completed. By the eighteenth century, as Chaplin emphasizes, knowledge had come to be seen—and this insight goes back to the Spanish and Portuguese—as nationalist in purpose, part of the vast network of competitions for power and domination. Motivated by the passion for wealth and profit, Franklin—like the nations to and from which his allegiance would shift—sought to make money through overseas trade and exploration. Had he ever doubted it, there was now no mistaking the importance of mapping the seas, and Franklin the investor turned to the craftsmen-mariners for assistance, particularly with understanding the Gulf Stream. As he migrated from Englishman to revolutionary, Franklin's unique understanding and systematic study of matters oceanic was put in service of the war and American independence.

Though Franklin's scientific style became very much his own and in addition aptly associated with nascent American virtues, I would like to emphasize its British roots. Through Isaac Greenwood, Franklin would have known much about the experimental practices and the scientific ideology of Jean Desaguliers.[9] Greenwood lived and worked with Desaguliers during the many months that Franklin knew Isaac during their stay in London. There was no one more skilled at drawing attention to the role that scientific knowledge could play in advancing commercial and—in Desaguliers's vision of machinery—entrepreneurial and industrial wealth. In his vast enterprise of itinerant lecturing, Desaguliers laid out the relationship between machines, labor saving, and profit. Franklin returned to the colonies familiar with such ideas, and with the faith that all natural knowledge had the potential not only to enlighten but to profit its possessors.

Newtonian philosophers such as Desaguliers invented civil engineering and laid one of the cultural foundations of the first Industrial Revolution.[10] Thus we associate their physico-theology with apologies for power, industry, order, and empire. Ralph Bauer, however, wants to remind us of an earlier natural philosophical tradition, one that laid emphasis on the search for the elixir of life and that was open to magical, secretive, and occult practices. Ingeniously, he argues that "occult Philosophy functioned ideologically and rhetorically in the early historiography about the New World by synthesizing…mercantile values…with the aristocratic values of the courts, upon whose sponsorship or favors early transoceanic expansionism critically depended." Could it be that the alchemists were more open to the foreign and the exotic than the mechanists? Could these men who also handled business affairs for their princes have been more at ease with beliefs and values that a mere hundred years later would simply become superstitions, pagan and savage?

There is considerable evidence that the pronouncements and practices of the alchemists certainly favored a cosmopolitan approach to all learning. We think of alchemy as secretive, magical, hence irrational. It could be all of those things while at the same time, paradoxically, being cosmopolitan. Devotees meant it when they took the code name "Cosmopolitiae" as a pseudonym, or made sure that *cosmopolite* appeared in their title.[11] They saw themselves as wanting to publish in every European language because they aimed "at nothing, but the undeceiving of the world." They claimed to want to search every country in the quest to find the philosopher's stone, the key to the elixir of life.[12] Their search to benefit all of mankind promised life itself, through an elixir that would "perfectly regenerate and promote… an endless life."[13]

Alchemy flourished at the courts of Europe and could be found in every country, equally among the Latinate and highly learned and among the semiliterate. Indeed, one of the most surprising conclusions presented by the study of the early modern languages of science concerns alchemy. For a time it gave a specifically cosmopolitan and universalist set of values to the scientific enterprise. Of the three major ways of describing nature available in early modern Europe—scholastic, mechanical, and alchemical—only the alchemical provided an avowedly idealist message of renewal, rejuvenation, the crossing of borders and class barriers, in the search for the secrets of nature or the key to longevity. The diffusion and transmission of alchemy depended on adepts who wrote, visited, and traded recipes across national borders, seemingly at will. It also relied on a vast storehouse of practices employed by pharmacists and artisans, people who made a living by using their hands.

In contrast to the scholastic and the mechanical, alchemy also offered the most universalist and idealist message, emphasizing reform and salvation.[14] Janus-faced, it could do the bidding of highborn patrons and kings while at the same time promising wisdom only to the diligent and the pious. When under the influence of alchemy, Robert Boyle, the young student of nature and genteel Protestant visionary, built his *"invisible...philosophical college,"* as he called it, to "put narrow-mindedness out of countenance, by the practice of so extensive a charity, that it reaches unto every thing called man, and nothing less than an universal good-will can content it...[and] take the whole body of mankind for their care."[15] Boyle even believed that Europeans had new things to learn from the peoples of the world, however different, exotic, or in the terms of the age, savage.

However, the alchemical, or more generally the occult, did not guarantee catholicity in the face of the different or the exotic. Cosmopolitan communication enjoyed a central place in the Godly reformation that so moved Boyle, but access to it was qualified. Participants were assumed to be Protestant and pious. In 1651, as the new English republic took shape, Boyle feared that the Jews who were to be offered toleration as part of the millenarian expectations of the Puritan state "may seduce many of those numerous Unprincipled (& consequently) Unstable Souls, who hav[e] never been solidly or settledly grounded in the Truth... ."[17] Just as the participation of Jews worried him, undoubtedly Boyle would have had deep reservations about even educated Catholics whom he would have associated with a propensity for royal absolutism, let alone peoples whose descent was non-European.

The biased treatment of Catholics in Ireland, where Boyle was born as the youngest son of the Earl of Cork, should make clear the limitations that bound the Protestant version of the cosmopolitan. If Boyle had trouble in coming to terms with the Irish or the Jews in his midst, we can more easily understand the mind-set, the effortless sense of superiority that all Europeans could bring to the imperial enterprise. They may have been open for a time to the magic witnessed in the colonies, as Júnia Ferreira Furtado demonstrates in her discussion of the Portuguese in Brazil, they could take up folk remedies of the indigenous, "but their systems of knowledge as a unit were discarded as mere superstitions." Whether occult or mechanical, Western theories of nature, found either among Catholics or Protestants, were packaged in a binding that was deeply indebted to Christianity, although gradually many of its central assumptions eroded to be replaced by a relativistic approach to the non-Western.[10] We may justly ask whether any peoples of the world in the early modern period possessed the imaginative capacity to adopt the other's worldview or culture. Would Europeans have fared much better at the hands of the peoples of Africa or the Americas, had the prerogative of technological superiority been on the other side? Yet, as Furtado's chapter indicates, when it came to the treatment of diseases that neither natives nor colonizers actually understood, a common humanity could sometimes surface, with some Portuguese doctors seeing that their system of enslavement lay at the root of the suffering. A similar dedication to the Hippocratic oath stands out in Safier's description of Jussieu's work as a doctor among the peoples he encountered throughout Ibero-America. In most cases, however, the local and indigenous curers came to be seen as practicing a sometimes unsavory form of backwardness.

The power of culture to shape and mold, indeed to determine, was probably believed even more fervently in the early modern period than it is today. As Jan Golinski shows, American settlers in the new world actually believed that their civilizing mission could "mold nature to their wishes." Jefferson believed human beings would shape their climate. With the same self-confidence he could edit the Bible, taking out the bits and pieces that did not conform to his rationalizing Unitarianism.[18] There is a sensibility here toward nature that we might want to characterize as simply naive, but a more authentic description should see it as naive but epistemologically anchored as realism. These were enquirers who privileged what they saw and their own ability to make sense out of it. Without such confidence, I believe much of the rampant curiosity we see among colonial travelers, investigators, and settlers might never have found expression. Not least, it enabled Americans such as Franklin and Jefferson to enlist climate "as an asset for the newborn nation." Sometimes naiveté re-enforced idealism.

Naiveté combined with the lust for profit also turned explorers into empiricists. That is the message contained in Antonio Barrera's chapter on the Spanish in the new world. He tells us that "natural things and instruments together with the artisans, royal officials, and scholars who manipulated them produced institutions and practices that encouraged empirical approaches to nature." The Spanish and the Portuguese led the way, as he persuasively argues. Even in decline, as Daniela Bleichmar aptly demonstrates, the Spanish state commissioned scientific investigations and through them hoped to reverse its fortunes. Perhaps the turn toward the quantitative and empirical—so distinctive of Western science by the middle of the seventeenth century—can best be explained by reference to the challenges thrown up by overseas exploration and trade. If Barrera's contention is right, then the mystery only deepens as to why the Iberian peninsula, once so advantaged, could ultimately fall behind in both science and empire. However explained, even in decline the Spanish fostered a network of naturalists who in turn sought their own independence as investigators. All saw in nature's bounty the prospect of profit.

As Bleichmar demonstrates, the so-called peripheries in trade and exploration gradually came to resemble new centers. James Delbourgo shows how British-Americans became experimenters in their own right, employing as his example electrical machines used after the 1740s by Franklin and many others. Quickly, the colonists were making their own machines and transforming them from play-things into philosophical apparatus. In the 1750s, Franklin's achievement looked like another example of British scientific prowess but, as we have seen, by the 1770s these machines had become exemplars of a national American genius. In the words of Delbourgo, "electricity could unite the Atlantic world, and it could sunder it." Science, with electricity as a symbol of power through shock and awe, could serve a new revolutionary and romantic sensibility that challenged empire.

The Atlantic world gave bountifully and materially to the enterprise of modern science: plants, animals, bones, trees from the new world—as Susan Scott Parrish puts it, all "were a major material source, from 1492 onward, for the development of botany, pharmacology, zoology, paleontology, geology, and ethnology, among other sciences." Europeans had come to depend on the locals in the colonies, not just as sources of wealth but now as knowledge resources. In this traffic, she sees one of the roots of the Enlightenment and its "laborious reckoning of world-wide matter." This hunger to know, she argues, extended even to those with the least agency among the many actors assembled in this splendid volume, to the African slaves and their "secret" natural knowledge. In the darkest side of this Western quest for usable natural knowledge, their bodies could become testing grounds. Some healing may have occurred, but we can be sure that much pain was inflicted in the process of experimenting on the bodies of African slaves. Yet the slaves were not bereft of their own knowledge and ingenious imitation of the collecting techniques taught them by their masters. John Stedman in his famous *Narrative* (1796) even tells us about a former slave in Surinam who had come to be called by his admirers a professor of herbology. His case offers cold comfort in a story that is about exploitation, curiosity and greed.

Once forged out of global experiences, after the mid-eighteenth century, European knowledge systems became global. Even mesmerism, shunned by the scientific establishment in Paris but for a time embraced by thousands of Europeans (and eventually to become one of the sources for the practice of hypnotism), had its uses. In the colony of Saint Domingue, as François Regourd explores its shadowy history, mesmerism may have come to be seen by local Europeans as ultimately useless, but the black people of the colony may have received it differently, even been empowered by it. It could have worked in Haiti as it worked in places in France. There, larger numbers of medical doctors and male and female seekers of various sorts, often with reformist agendas, had gravitated toward the mesmerist societies during the 1780s. They wrote about "the magnetic fluid…[that] is able to produce a great revolution…for honest and sensitive souls."[19] Perhaps among the African slaves recently brought to the colony, mesmerism mixed with voodoo also offered sensitive souls rejuvenation, conceivably of a revolutionary nature. As we know a few years later, revolution in Haiti would acquire a very specific meaning, just as it did in France. The upheavals around mesmerism in the 1780s may have signaled that establishments of all sorts, not just scientific, were increasingly seen to be corrupt and removable. The states that had built the empires had not reckoned on how colonials and slaves, as well as their own people—at least in America, France, the Low Countries and Haiti—had grown tired of their monopolies.

With empire came hegemony and vast abuses. The peoples of the Atlantic world had also come to know much more about nature. The Atlantic knowledges described here were many and varied, as simple as specimens collected, as complex as electrical devices exported and exploited, as magical as native folk practices and mesmerism. All opened the human imagination in unprecedented ways. All promised profit: knowledge turned into economic and political power for states, traders, explorers, or just the curious gone off to see the world. Though only a few may have profited hugely, and even a lesser number may have embraced the ethic that Weber said became the defining ascetic of capitalism, all—regardless of their power or powerlessness—served to put in place a global capitalist system of markets and communications the importance of which we are just beginning to grasp and attempt to understand.

Notes

1. For wise thoughts on the problem, see Jon E. Wilson, "Subjects and Agents in the History of Imperialism and Resistance," in David Scott and Charles Hirschkind, eds, *Powers of the Secular Modern: Talal Asad and His Interlocutors* (Stanford: Stanford University Press, 2006), 180–205.
2. A point made repeatedly in Peter Burke, *Languages and Communities in Early Modern Europe* (Cambridge and New York: Cambridge University Press, 2004), 81–2, 166.
3. Julia Adams, *The Familial State: Ruling Families and Merchant Capitalism in Early Modern Europe* (Ithaca: Cornell University Press, 2005).
4. Adams, *The Familial State*, 165.
5. Douglas McKie, "James, Duke of York, F.R.S.," *Notes and Records of the Royal Society of London* 13 (1958), 6–18.
6. Paris, Académie des Sciences, Procès-verbaux, Reg.11, fol.157r–158r and cited and translated in G. G. Meynell, *The French Academy of Sciences, 1666–91: A Reassessment of the French Académie Royale des Sciences under Colbert (1666–83) and Louvois (1683–91)*, to be found at http://www.haven.u-net.com/6text_7B2.htm#Utilite. This and other themes are discussed in Margaret C. Jacob, *Strangers Nowhere in the World: The Rise of Cosmopolitanism in Early Modern Europe* (Philadelphia: University of Pennsylvania Press, 2006).
7. Louvois quoted in Meynell, *The French Academy of Sciences*.
8. See Margaret C. Jacob, *The Origins of Freemasonry: Facts and Fictions* (Philadelphia: University of Pennsylvania Press, 2005). Controversy exists as to whether Franklin joined the freemasons in London for the first time or in Philadelphia after his return from London.
9. J. A. Leo Lemay, *The Life of Benjamin Franklin, Volume 2: Printer and Publisher, 1730–1747* (Philadelphia: University of Pennsylvania Press, 2006), 83, 488.
10. For the development of this argument see Margaret C. Jacob, *Scientific Culture and the Making of the Industrial West* (New York and Oxford: Oxford University Press, 1997).
11. A. Everaerts used "cosmopolitiae"; Michael Sendivogius (1566–1636) had his works collected under the title, *Les Oeuvres du Cosmopolite*—see the catalogue of the Bakken Library, Minneapolis. The National Medical Library in Washington, DC lists a work by Sendivogius as *Cosmopolite; ou Nouvelle Lumière Chimyque* (Paris, 1669). For one of many alchemical explications of "the universal spirit of Nature" see [Thomas Vaughn] Eugenius Philalethes, *Anima Magica Abscondita; Or Discourse of the Universal Spirit of Nature* (London, 1650). The tract begins by attacking the scholastics and then proceeds to describe the universal spirit. For a clear statement of the principles, see J. Malbec de Tresfel, *Abrege de la Théorie et des veritables Principes de l'Art appellé Chymie qui est la troisième Partie ou Columne de la vraye Medecine Hermetiques...dedié a M. Valot, Conseiller du Roy* (Paris, 1671), 15–28 on the principles; "la Chymie a la connoissance des elemens inferieurs, du Sel, du Soulfre, & du Mercure... ." at 40; and 43, where the goal is "great and admirable remedies that conserve health, reestablish lost energy (les forces perdues), ... and prolong life." Paracelsus is also invoked (51).

12 For one such fanciful journey see [Anonymous,] *Les Avantures du Philosophe Inconnu, en la Recherche & en l'Invention de la Pierre Philosophale* (Paris, 1646), esp. 46–7.
13 *Otto Tachenius his Hippocrates Chymicus, Discovering the Ancient Foundations of the Viperine Salt*, trans. J. W., unpaginated introduction; [John Frederick Houppreght,] *Aurifontina Chymica; or a Collection of Fourteen small Treatises Concerning the First Matter of Philosophers, For the discovery of their Mercury… for the benefit of Mankind in general* (London, 1680).
14 For good illustrations of these points, see the tales told in William R. Newman, *Gehennical Fire: The Lives of George Starkey, An American Alchemist in the Scientific Revolution* (Cambridge, Mass.: Harvard University Press, 1994), and J. T. Young, *Faith, Medical Alchemy and Natural Philosophy: Johann Moriaen, Reformed Intelligencer, and the Hartlib Circle* (Brookfield: Ashgate, 1998).
15 Robert Boyle to Francis Tallents, February 20, 1647 (italics in the original) in Michael Hunter, Antonio Clericuzio and Lawrence M. Principe, eds, *The Correspondence of Robert Boyle*, 6 vols (London: Pickering and Chatto, 2001), 1: 46.
17 Boyle to John Mallet, *Correspondence of Robert Boyle*, 1: 104. See Anthony Pagden, *The Fall of Natural Man: The American Indian and the Origins of Comparative Ethnology* (Cambridge: Cambridge University Press, 1984); for a new relativism see Bernard Picart, *Cérémonies et Coutumes Religieuses* (Amsterdam: J. F. Bernard, 1723–1743).
18 Thomas Jefferson, *The Jefferson Bible: The Life and Morals of Jesus of Nazareth* (Boston: Beacon Press, 1989).
19 University Library, Strasbourg, MS 1432, f. 70, discussed in Margaret C. Jacob, *Living the Enlightenment: Freemasonry and Politics in Eighteenth-Century Europe* (New York and Oxford: Oxford University Press, 1991), 199–200.

Contributors

Antonio Barrera-Osorio is Associate Professor of History at Colgate University, where he teaches the history of science in the sixteenth and seventeenth centuries, Atlantic World history (1500–1800), and early modern Spanish history. He is the author of *Experiencing Nature: The Spanish American Empire and the Early Scientific Revolution* (University of Texas Press, 2006); "Local Herbs, Global Medicines: Commerce, Knowledge, and Commodities in Spanish America," in Pamela Smith and Paula Findlen (eds), *Merchants and Marvels: Commerce, Science, and Art in Early Modern Europe* (New York: Routledge, 2001); and "Things from the New World: Commodities and Reports" in *Colonial Latin America Review* (2006). He is currently the coordinator of the Latin American Studies Program at Colgate University.

Ralph Bauer is an Associate Professor of English at the University of Maryland, College Park. He is the author of *The Cultural Geography of Colonial American Literatures: Empire, Travel, Modernity* (Cambridge University Press, 2003); *An Inca Account of the Conquest of Peru, by Titu Cusi Yupanqui* (Colorado University Press, 2006); and of numerous articles on colonial British and Spanish American literature, as well as the (co)editor of various collections of essays.

Daniela Bleichmar is Assistant Professor in the Departments of Art History and Spanish and Portuguese at the University of Southern California. Her research focuses on the visual culture of natural history in the Spanish empire, the history of collecting and display, and the history of print. She is the author of multiple articles on these topics, among them "Painting as Exploration: Visualizing Nature in Eighteenth-Century Colonial Science," *Colonial Latin American Review* (2006); "Training the Naturalist's Eye in the Eighteenth Century: Perfect Global Visions and Local Blind Spots," in Cristina Grasseni (ed.), *Skilled Visions: Between Apprenticeship and Standards* (Berghahn, 2006); and "Books, Bodies, and Fields: Sixteenth-Century Transatlantic Encounters with New World Materia Medica," in Londa Schiebinger and Claudia Swan (eds), *Colonial Botany: Science, Commerce, and Politics* (University of Pennsylvania Press, 2004). She is currently at work on a book provisionally entitled *Painting as Exploration: Visual Culture and Colonial Botany in the Eighteenth-Century Spanish World*.

Joyce E. Chaplin is James Duncan Phillips Professor of Early American History at Harvard University and author, most recently, of *The First Scientific American: Benjamin Franklin and the Pursuit of Genius* (Basic Books, 2006). She is currently working on a history of circumnavigation.

James Delbourgo is Assistant Professor of History and Chair of History and Philosophy of Science at McGill University. He is the author of *A Most Amazing Scene of Wonders: Electricity and Enlightenment in Early America* (Harvard University Press, 2006). His work explores the relationship between science and empire in the British Atlantic world, with a current focus on Hans Sloane and Jamaica, and underwater exploration. Most recently, he published a commissioned essay on Sloane's engagement with slavery for the British Museum.

Nicholas Dew is Assistant Professor in the History department at McGill University, in Montréal, Québec, where he teaches early modern French history and history of science. He is the author of *Orientalism in Louis XIV's France* (Oxford University Press, 2008), and is currently working on scientific expeditions in the French Atlantic World, *c*. 1670–1760.

Júnia Ferreira Furtado is Full Professor of History at the Universidade Federal de Minas Gerais, Brazil, and was a Visiting Professor in the History Department at Princeton University in spring 2001. With a Ph.D. in social history from the Universidade de São Paulo, she has published numerous books and articles on Colonial Brazil, including *Chica da Silva e o contratador dos diamantes: o outro lado do mito* (Companhia das Letras, 2003) (of which a translation is forthcoming from Harvard University Press), and *Cartografia da conquista das minas*

(Kappa, 2004). She is also a contributor to the *History of Cartography*, volume 4, edited by Matthew Edney and Mary Pedley.

Jan Golinski is Professor of History and Humanities at the University of New Hampshire, where he currently serves as Chair of the Department of History. He is the author of *Science as Public Culture: Chemistry and Enlightenment in Britain, 1760–1820* (Cambridge University Press, 1992); *Making Natural Knowledge: Constructivism and the History of Science* (new edition, University of Chicago Press, 2005); and *British Weather and the Climate of Enlightenment* (University of Chicago Press, 2007). With William Clark and Simon Schaffer, he co-edited *The Sciences in Enlightened Europe* (University of Chicago Press, 1999).

Margaret C. Jacob is Distinguished Professor of History at UCLA. She has written extensively on science and industrialization, the Enlightenment, and most recently on early modern cosmopolitanism (University of Pennsylvania Press). Currently she has embarked upon a study of Bernard Picart and his multivolume attempt to depict all the religions of the world (1723–43).

Susan Scott Parrish is an Associate Professor in the Department of English Language and Literature and the Program in the Environment at the University of Michigan; she is also a Fellow at the Graham Environmental Sustainability Institute (UM). Her recent book *American Curiosity: Cultures of Natural History in the Colonial British Atlantic World* (University of North Carolina Press, 2006) has been awarded the Jamestown Prize and the Ralph Waldo Emerson Prize; the Emerson prize is given by the Phi Beta Kappa Society to one book each year for its contribution to understanding "the intellectual and cultural condition of humanity." Her current projects include work on slavery and portraiture in the eighteenth-century Atlantic world, a new edition of Robert Beverley's 1705 *History and Present State of Virginia*, and a book-length study of the ecological imagination of the U.S. South in the first half of the twentieth century.

François Regourd is Maître de Conférences in the History Department at the University of Paris X, Nanterre, where he has taught since 2001. For ten years, he has been working on French colonial science, particularly in French Guyana and the French Antilles in the seventeenth and eighteenth centuries. He has published more than fifteen articles, both in French and English, and has co-edited *Connaissances et pouvoirs: les espaces impériaux (XVIe-XVIIIe siècles): France, Espagne, Portugal* (Bordeaux, 2005).

Neil Safier is Assistant Professor of History at the University of British Columbia in Vancouver. He received his Ph.D. from the Johns Hopkins University in 2004 and has subsequently held visiting appointments at the University of Pennsylvania and the University of Michigan. A specialist in the geographical and natural historical exploration of South America, he is the author of *Measuring the New World: Enlightenment Science and South America*, forthcoming from the University of Chicago Press.

Alison Sandman is an Assistant Professor of History and Interdisciplinary Liberal Studies at James Madison University. She is currently at work on a book on early modern Spanish navigation and cartography.

Index

A

Abreu, Aleixo de, 130
Absolutism, 11, 333, 336, 339
Académie des Sciences (Paris), 16, 55, 57–62, 64–65, 207, 318, 336
Academies: *see* Scientific academies
Acosta, José de, 108
Actor–network theory: *see* Metrology; Latour, Bruno; Measurement
Adams, John, 268
Adams, Julia, 337
Adanson, Michel, 269
Admiralty (British), 78, 262
Aepinus, Franz, 259
African and Afro-American knowledge, 5, 11–12, 19–21, 128, 210, 219, 263, 281–305, *292*, 312, 320–326, 334, 341–342 (*see also* Afro-American religions; Slavery and slaves; Free blacks; Obeah; Occult knowledge; Poisons; Vodou)
Africans and Afro-Americans: *see* Slavery and slaves
Afro-American religions, 20, 263, 283, 295–298, 300–301, 322–324 (*see also* Obeah; Vodou)
Agriculture, 9, 17–18, 153–156, 160–164, 166–170, 181–182, 184, 210, 230, 232, 236, 238, 256, 260, 305 (*see also* Improvement; Plantations)
Agrippa von Nettesheim, Heinrich Cornelius, 100, 109
Air quality, 62–63, 135, 140, 154–155, 158–160, 163–165, 169–170, 317, 336 (*see also* Climate; Heat; Humidity; Meteorology)
Alchemy, 6, 17, 103, 106–108, 110, 112–115, 117, 119, 178, 336, 338–339
Aldea Urries, Victoriano de, 236
Alexander VI (Pope), 99
Allamand, Jean, 269–270
Alzate y Ramírez, José Antonio de, 265
Amazon River, 118, 204, 212, 214–215
Amazonia, 217
American Philosophical Society, 90, 92, 161, 164
American Revolution, 16, 19, 81, 156, 162, 258, 261–262, 265–270, 295, 341
Amerindian knowledge, 11–12, 17, 19, 100–101, 104, 108–113, 116–119, 128–129, *134*, 135–138, 140–142, 146–147, 161, 177, 183–184, 206, 219, 226, 272, *273*, 274–275, 281–283, 285, 288, 297–298, 304, 334, 340, 342 (*see also* Amerindians)
Amerindians, 4–5, 8–9, 11–12, 17, 19–21, 54, 82, 84, 100–101, 103–104,

350 • Index

109–113, 115–119, 128–129, 132, *134*, 135–138, 140–142, 146–147, 155–156, 160–161, 166–169, 177, 183–184, 206, 215, 219, 226, 270, 272, *273*, 274–275, 281–283, 285, 288, 297–298, 304, 334, 340, 342 (*see also* Amerindian knowledge; Race)
Amsterdam, 54
Anatomy, 269
Andagoya, Pascual de, 182
Andes, 18, 204–207, 213–215, 220
Animal magnetism: *see* Mesmerism
Anomalies, 61–62, 153, 158, 272
Anson, George (Commodore), 262
Antigua, 263
Antilles: *see* Caribbean Islands
Aquinas, Thomas, 105
Arawak Indians, 63, 274 (*see also* Kasum; Wapisiana)
Aristocracy, 102, 105–106, 262, 267, 338
Aristotle, 105, 107
Aristotelianism, 100, 105, 107, 119, 130, 178, 299 (*see also* Scholasticism)
Armero, Francisco, 240
Arthaud (French physician), 315–316, 318–320
Artisans, 5, 18, 31–33, 64, 74–75, 86, 101, 106, 108, 119, 124 n32, 178–183, 186, 188, 257, 259–260, 267, 336–337, 339–340 (*see also* Guilds; Mariners)
Astrolabes, 36, 39–40, 44, 190, 192–193
Astrology, 99–100, 107, 114, 145, 178, 261, 293, 300
Astronomy, 5, 7, 11, 13, 15, 42, 57–60, 64–66, 78, 105, 135, 188, 193, 257, 265, 282 (*see also* Transits of Venus; Telescopes)
Atlantic history, 6–7, 11, 20–21
Atlantic Ocean and Atlantic World, 7–9, 14, 16, 20–21, 53–54, 56–57, 59, 65–66, 67 n10, 72 n43, 73–92, *80*, *88*, *91*, 182, 188, 190, 205–206, 212–213, 218–221, 257–258, 261, 333–342
Atlases (nautical), 77, 88–89
Audiencia of Quito, 210, 240, 242–243
Australia, 239
Auzout, Adrien, 58
Avicenna, 107
Ayala, Marco de, 183–184

Aztec, 161

B

Bacon, Francis, 1–4, 203, 217, 261, 282, 287–288
Bacon, Roger, 107, 109
Baconianism, 1–4, *3*, 16, 100, 101, 104, 119, 157, 203, 217, 261, 282, 287–288, 299, 304
Badianus, Juannes, 109
Bahia, 132, 143–144
Bancroft, Edward, 4–5, 270–271
Banks, Joseph, 11, 235, 238–239
Barbados, 157, 294, 298, 300
Barleo, Gaspar, 133
Barometers, 62, 90, 157
Becher, Johann Joachim, 106
Behn, Aphra, 269
Belknap, Jeremy, 165–166, 169
Beltrán, Gonzalo Aguirre, 107
Benalcázar, Sebastián de, 117–118, 229
Benin, 301
Berrio, Antonio de, 116
Berry, William, 300
Black Legend (*Leyenda Negra*), 304
Blake, William, 78
Bloomfield, William, 116
Bogotá, Santa Fé de, 19, 49n35, 117, 228, 231, 235, 239, 242, 244
Bollstädt, Albert graf von, 105
Bologna, 259–260
Bonnet, Charles, 14–15
Bonpland, Aimé, 4, 271
Borda, Jean-Charles de, 313
Bordenave, Juan de, 219
Boston, 81, 83, 87, 157, 260, 285, 302–303
Botanical drugs, 2, 17–19, 109–111, 113, 128, 131–133, *139*, 140–142, 144, 177, 207, 226, 236, 240, 242–244, 290–291, 294, 296, 298–299, 301–305, 323 (*see also* Medicine and healing; Botany; Botanical gardens; Natural history; Cinchona)
Botanical gardens, 60, 133, 135–136, 204, 207–208, 210, 225, 228–229, 231–233, 235–240, 244, 337 (*see also* Botany; Botanical drugs; Medicine and healing; Jardin du Roi; Kew Gardens; Royal Botanical Garden of Madrid)

Botany, 5–7, 11–13, 15, 18–20, 57, 60, 109–111, 113, 129–133, 135–138, 141–142, 177, 180–182, 203–221, 225–245, 256, 271, 274, 289–290, 293–298, 301–305, 337, 341–342 (*see also* Botanical drugs; Botanical gardens; Medicine and healing)
Bougainville, Louis Antoine de, 5
Bourbon reforms, 18–19, 226
Bourdaz (French agent), 213
Bourgeoisie, 106, 113, 266–267
Boyle, Robert, 13, 76, 339
Braudel, Fernand, 8
Brazil, 6, 9, 17, 121–133, 135–142, *141*, 144, 146–147, 210, 339–340; Dutch invasion of (1620s–1640s), 131–138; *141*
Brest, 313
Bristol, 37
Buarque de Holanda, Sérgio, 141–142
Buccaneers: *see* Piracy and pirates
Buckley, Francis, 85
Buenos Aires, 238
Buffon, George Louis Leclerc, comte de, 14, 160, 162, 165–168, 259
Burgess (Caribbean physician), 304
Bute, Lord, 268, *268*
Bynkershoek, Cornelius van, 1, 10
Byrd, William, II, 289

C

Caballero y Góngora, Antonio, 231, 240, 242
Cabbala, 100
Cabinets of curiosities, 7, 9, 21, 135–136 (*see also* Curiosity)
Cabot, Sebastian, 4–5, 36–37, 48n, 190–193
Cadiz, 212, 237
Calicut, 104–105
Cambridge, 54, 56
Canada, 57, 85
Canary Islands, 9, 44, 59, 238
Canelos, or Land of Cinnamon, 118, 204, 213–217 (*see also* Cinnamon)
Cape Verde Islands, 99
Capitalism, 333–342
Caracas, 235, 242

Cardosa de Miranda, João, 140, 142, 144–147
Caribbean Islands, 2, 6, 12, 20, 55–57, 59, 103–105, 132–133, 154, 160, 165, 181–182, 188, 207–208, 210, 238, 260, 262–263, 265, 282–283, 289, 293–295, 297–299, 301, 304–305, 312
Carijós Indians (Brazil), 142
Caro, Juan, 192, 201n61
Cartagena, 238
Carter, Robert, 300
Cartesianism, 62, 336
Cartography, 6, 8, 9, 12, 16, 31–46, 57–58, 65–66, 73–92, *88*, 100, 128, 177, 184–185, *187*, 188–193, 219, 238, 344 (*see also* Cosmography; Hydrography)
Casa de la Contratación, 11, 16, 18, 34–36, 40–41, 60, 177–180, 183–191, *185*, 201n61
Cassini, Gian-Domenico, 56–61, 64–65
Castillo, Juan del, 236
Catalina de Ayanbex, Taino, 183
Catesby, Mark, 290, 298
Catholicism, 8–9, 13, 115–116, 146, 304, 324, 339–340 (*see also* Christianity; Missionaries; Protestantism; Inquisition)
Cavanilles, Antonio José, 243–244
Cayenne, 16, 53, 55, 57, 61, 62, 63, 64, 65, 70n27, 28
Center vs. periphery, 6, 10–15, 19, 60–66, 73–74, 78, 84, 89, 128, 137, 157, 203–204, 217–221, 226–229, 237–245, 249n57, 257–259, 261–263, 270–271, 281–282, 302, 304, 312, 318, 335, 337, 341 (*see also* Science and empire; Empire; Local knowledge; Networks; Provinces and provincialism)
Cercle des Philadelphes, 20, 318–319, 325
Céspedes, Andrés García de, 44–45
Cervantes, Vicente, 240
Ceylon, 230–231
Chalmers, Lionel, 157–160, 164
Chaplin, Joyce, 118–119
Charlatanry, 205, 211, 265, 313, 315–316, 319, 321–322, 324 (*see also* Enthusiasm; Imagination)

352 • Index

Charles I (Spain), 181, 194
Charles III, 225
Charles V, 3, 4, 40, 108, 177, 182, 183, 188, 192
Charleston, 157–160, 164, 289, 295, 298
Chaves, Alonso de, 35, 192–193
Chaves, Jerónimo de, 191, 193–194
Chemistry, 107, 226, 231, 315
Cheyne, George, 159
Chile, 38, 181, 189, 225, 231
Chimborazo (volcano), 204
China, 2, 9, 57, 115, 163, 230, 260, 265, 334
Christianity, 2, 8–9, 12–13, 16, 99–100, 110, 181, 256, 293, 295, 297, 300, 302–303, 340 (*see also* Catholicism; Protestantism; Missionaries; Millenarianism)
Chronometers, 91–92
Cialli, Antonio, 146
Cinchona (quinine), 18, 211–212, 226, 235, 242–244
Cinnamon, 18, 118, 142, 204, 214–217, 223n29, 226, 229–231, 235–236, 240, 333, 337 (*see also* Canelos, or Land of Cinnamon)
Circulation of knowledge, 56–66, 73–92, 131, 133, 144, 147, 182, 205–206, 211–213, 218–221, 228, 230–231, 234–237, 240–241, 337–339 (*see also* Secrecy vs. circulation of knowledge; Center vs. periphery; Science and empire)
Civilization, 154–170, 263, 265, 281, 285, 340 (*see also* Improvement)
Claggett, William, 260
Classification, 7, 62, 63, 127, 129–131, 133, 135, 137–138, 216, 230–231, 241–244 (*see also* Taxonomy; Botany; Natural history)
Clayton, John, 156–157
Clifton, Chief Justice (Florida), 290
Climate, 14, 16–17, 61–63, 70n29, 130, 135, 138, 140, 153–170, 213–214, 216, 261, 271, 285, 317, 336, 340 (*see also* Meteorology; Air Quality; Heat; Humidity; Record-keeping)
Clusius, Carolus, 133, 137, 186, 208
Coffee, 159, 230, 238, 242, 260

Coffins family, 82
Coimbra, 129
Colbert, Jean-Baptiste, 57
Colden, Caldwallader, 82–83
Colegio de Santa Cruz, 109
Collecting, 6, 15, 18–20, 57, 109, 127, 136, 179–180, 184–186, 188, 192–193, 208, 212–216, 225–229, 234–235, 237, 244, 282–285, *284*, 287, 289–291, 293, 296, 305, 333 (*see also* Botany; Natural history; Classification; Taxonomy; Cabinets of curiosities)
College of New Jersey, 167
College of Philadelphia, 257
Collinson, Peter, 262, 291, 305n1
Colombia, 182, 226
Columbian Exchange, 73, 181, 281–282
Columbus, Christopher, 2, 6, 7, 99, 105–106, 180–181
Columbus, Hernando, 192–193
Commerce, 7, 11, 20–21, 81, 105–106, 112, 132, 144, 156, 178, 183, 185–186, 189–190, 212, 219, 225–226, 228–237, 240–245, 258–260, 262–263, 265, 281, 291, 333, 335, 337, 341–342 (*see also* Monopolies and monopoly trading companies; Travel: Commercial; Triangular trade; Networks)
Compass, 41–42, 191, 193
Conchillos, Lope de, 189
Condorcet, marquis de, 219
Congo, 321, 324
Conjuring: *see* Magic; Occult knowledge; Amerindian knowledge; Obeah; Vodou; African-American religions; Witchcraft
Conquistadors, 101, 166, 215, 229, 333
Conspiracy, 267–268, *268*, 295, 321, 323–325
Cook, James, 4, 5, 15, 56, 77, 87–88, 226, 263
Copernicanism, 64
Cortés, Hernán, 106, 108, 117
Cosa, Juan de la, 189
Cosmography, 16, 32–46, 48n23, 103–104, 177, 186–194 (*see also* Cartography; Hydrography)
Cosmology, 109–111, 258, 261, 291 (*see also* Natural theology)

Cosmopolitanism, 15, 261–262, 267, 269, 338–340
Costa da Mina, 144
Cotes, Roger, 53–55
Cotes, William, 293–294
Cotopaxi (volcano), 205, 210, 219
Council of the Indies, 18, 34–35, 37, 39–40, 43, 178–180, 186, *187*, 190–192, 194, 196n10
Credibility: *see* Trust
Creole Americans, 4–6, 14–16, 19–21, 60, 72n43, 101, 154–170, 206, 209, 219, 225–226, 234, 239, 261–267, 270–271, 274, 317, 341 (*see also* Creole-American knowledge; Local knowledge; Provinces and provincialism)
Creole-American knowledge, 101, 154–170, 206, 239, 265, 270–271, 341 (*see also* Creole Americans; Local knowledge; Center vs. periphery)
Cruz, Martinus de la, 109
Cuba, 184, 226
Cuenca, 211
Cuon, Alberto, 182–183
Curiosity, 13, 204, 214–216, 266, 287, 290, 333, 337, 340, 342 (*see also* Cabinets of curiosities)
Currie, William, 163, 169
Curvo Semedo, João, 147
Cuzco, 219

D

D'Ahlbergh (Swedish naturalist), 290
d'Angerville, Côme, 315–316, 320
d'Orta, Garcia, 133, 137
Dampier, William, 78–79, *80*, 84
Darnton, Robert, 317–318
Darwin, Charles, 74
Darwin, Erasmus, 260
Dávila, Pedro Franco, 234
Dawson, William, 300
de Brahm, William Gerard, 86
de Pauw, Cornelius, 166, 168
Dear, Peter, 180
Debien, Gabriel, 322–323
Dee, John, 114–116
Defoe, Daniel, 79
Deforestation, 161–165

Degeneracy, 14, 166–168, 261, 267, 317 (*see also* Dispute of the New World)
Denmark, 237, 295
Derham, William, 157
Desaguliers, Jean, 339–340
Descartes, René, 6
Deshayes, Jean, 59, 61–62, 64
Díaz de Solís, Juan, 189, 201n58
Díaz del Castillo, Bernal, 108
Diet, 143–145, 211, 233
Dioscorides, 107, 130
Diplomacy, 32–33, 40–41, 45, 48n21
Discipline (in science), 13–14, 19, 33, 64, 257–261, 269–270, 275 (*see also* Trust; Travel questionnaires; Scientific instruments)
Discoverie of the Large, Rich, and Bewtiful Empyre of Guiana (Ralegh, 1596), 17, 103, 110–120
Discovery narratives, 5, 100, 102, 112, 115–117, 130, 217 (*see also* Expeditions; Travel narratives)
Disease 2, 4, 73, 107, 110, 127, 130–132, 135, 137, 139–146, 151n44, 155–156, 159–161, 165, 169, 205, 209–212, 226, 287, 293, 298–299, 301–303, 313, 340; Malaria, 226; Scurvy, 131, 144–145, 151n44, 159; Smallpox, 155, 205, 209, 212, 287, 293, 302–303; Yellow Fever 131, 165 (*see also* Medicine and healing; Botanical drugs; Physicians; Poisons)
Dispute of the New World, 14, 55, 67n7, 160, 165–168 (*see also* Degeneracy; Hemispheric difference)
Doctors: *see* Physicians
Dombey, Joseph, 231
Domjen, Samuel, 260
Douglass, Frederick, 287–288
Douglass, William, 303
Du Close, Samuel, 338
Dudley, Paul, 157
Dunbar, James, 162, 170
Drake, Francis, 37–38, 40
Drayton, Richard, 107–108, 110
Dyes, 18, 178, 182–184, 194, 236, 242, 260

E

Eamon, William, 101, 107

Earthquakes, 63, 158
East India Company (Dutch), 60, 230
East Indies, 45, 262
Ecuador, 55, 226
Eden, Richard, 103–105, 107
Edwards, Bryan, 296
Egypt, 4, 255, 271
El Dorado, 17, 100–101, 103, 110–112, 114–118, 126n56, 214–215 (*see also* Manoa del Dorado)
El Salvador, 144
Electric battery, 269–270, 272
Electric eels, 19, 258, 269–275, *273* (*see also* Kasum)
Electric machines, 19, 255–275, *256*, *268*, 315, 341–342
Electricity, 19, 74, 84, 158, 255–275, *256*, *264*, *268*, *273*, 315, 317, 341–342 (*see also* Lightning; Kasum)
Electrotherapy, 260
Elizabeth I, 112, 115
Ellis, John, 287, 290–291, 293
Empire, 5, 8–10, 14–15, 20–21, 73–74, 79–81, 83–85, 87, 89–90, 101, 104, 114–116, 118–120, 154, 178, 186, 189, 218–221, 226–229, 235, 257–258, 261–263, 265, 275, 283, 333–342 (*see also* Science and empire; Center vs. periphery; Provinces and provincialism; Race)
Empiricism, 6, 17–18, 21, 101–108, 119, 127–147, 148n4, 177–195, 281, 283, 285, 288, 299, 340–341 (*see also* Experience)
Encyclopédie (Diderot and d'Alembert, 1751–72), 75
Engineering, 338
Enlightenment, 15, 19, 53, 58, 166, 164, 169–170, 234, 257–258, 261–265, 275, 281–282, 291, 295, 305, 325, 341
Enthusiasm, 261, 267–269, 317, 321, 324 (*see also* Imagination; Charlatanry)
Enrique, Micer, 182–183
Enríquez, Martin, 184
Equator, 55, 60, 62, 64, 75, 105, 317
Equiano, Olaudah, 301
Error, 61, 63, 69n26; 105, 190, 219, 263
Espinosa, Mariano, 236

Ethiopic Sea: *see* Atlantic Ocean and Atlantic World
Ethnography, 12, 60, 100, 110–112, 133, *134*, 135, 282, 286, 341 (*see also* Amerindians)
Eurocentrism, 7, 16, 78, 161, 281–283, *284*, 339
Evans, Lewis, 84
Expeditions: general, 36–37, 99, 104–105, 108, 110–120, 154, 177, 188, 215, 218, 262, 333, 337, 341; scientific, 4–5, 10–11, 42–43, 53, 57–66, 78, 136, 179, 203–221, 225–226, 271–274, 290, 313, 336–337 (*see also* Travel: commercial; Travel narratives; Discovery narratives; Science and empire)
Experience, 13, 16, 18, 32–33, 41, 45, 48n23, 92, 108, 130, 133, 142, 144, 146, 163, 178–181, 184, 195, 203–205, 257–260, 268–272, 275, 281–282, 298, 304 (*see also* Experiment; Empiricism)
Experiment(s), 6, 11, 16, 18–19, 53–66, 74, 76, 84, 92, 101, 107, 114, 129–131, 136, 163–164, 181–183, 194, 237, 243, 255–275, 286, 287, 295, 313, 337–338, 341 (*see also* Experimental demonstrations; Experience; Empiricism; Natural philosophy; Pendulum experiments)
Experimental demonstrations, 2, 255–257, 259–263, 265–266, 315–316, 338 (*see also* Experiment)
Ezpeleta, José de, 236

F

Fabian, Johannes, 218
Facts, 13, 65, 69n26, 100, 157, 257, 281–282, 304 (*see also* Empiricism)
Fagon, Guy-Crescent, 207
Farabee, William Curtis, 274–275
Ferdinand the Catholic (Ferdinand V of Castile), 188–189
Ferreira da Rosa, João, 131
Fetishism, 263, 322
Feuillée, Louis, 207–208
Fishermen, 38, 88, 90, 272–273, 290
Florida, 38, 115, 181, 290
Fludd, Robert, 114
Foache, Stanislas, 320

Folger, Timothy, 84, 88, *88*, 90
Folger family, 82, 87–88
Folk knowledge, 108, 210, 297, 340, 342
Folkes, Martin, 304
Forster, Johann, 263
Fournier de Varenne, Jean-Jacques-Julien, 316–318
Fragonard, Jean-Honoré, 266
Franklin, Benjamin, 16, 19, 73–75, 81–92, *88, 91,* 158, 168, 258–262, *264,* 266–267, 270, 316, 337–338, 340–341 (*see also* Electricity; Gulf Stream)
Franklin, Josiah, 81
Free blacks, 295, 312, 320
Free trade, 232, 242–243
Freile, Andrés, 34–35
French Revolution, 168

G

Galen, 107, 130, 299
Galvani, Luigi, 260
Gálvez, José de, 229, 232
Gandía, Enrique, 118
García de Torreño, Nuño, 190
Garden, Alexander, 287, 290, 293–294, 298
Geodesy, 65–66
Georgia, 158
Gerbi, Antonello, 168
Gilbert, Humphrey, 114
Glos, Guillaume de, 59
Godin, Louis, 214, 224n14
Gold, 12, 106–108, 113, 116–118, 139–140, 143–144, 215, 333–334, 337
Gomes Ferreira, Luís, 140, 142–147, *143*
Gómez Ortega, Casimiro, 229, 231–241, 243–244
Gorée, Island of, 55, 62, 64, 70n27
Graman Quacy: *see* Kwasi
Grateo, Pedro, 36
Gravesande, Laurens van 's, 269–270
Gravesande, Willem van 's, 269–270
Gravity, 7, 53–66, 267, 336
Greenwood, Isaac, 157, 337–338
Grotius, Hugo, 9
Grove, Richard, 165
Gruzinski, Serge, 101
Guadeloupe, 207
Guaiqueri Indians (New Granada), 272
Guatemala, 178, 233
Guayaquil, 214
Guiana (Dutch), 19, 269–271, *273,* 274 (*see also* Surinam)
Guiana (English), 103, 112–113, 115–116, 119
Guiana (French), 16, 17, 55, 57, 60–61, 301
Guilds, 41, 75, 129, 183, 337 (*see also* Artisans; Secrecy vs. circulation of knowledge)
Guinea, 11–12, 296, 303
Gulf of Mexico, 82–83
Gulf Stream, 16, 73–74, 82–83, 86–92, *88, 91,* 337
Gullah Jack (slave), 295–296
Gumilla, José, 230
Guthrie, John, 294–295
Gutiérrez, Diego de, 190–191, 193
Gutiérrez, Sancho, 34–35, 19

H

Hague, The, 291
Haig, Harry, 296
Haiti: *see* Saint Domingue
Haitian Revolution, 9, 312, 325, 332n59, 342 (*see also* Saint Domingue)
Hakluyt, Richard, 104, 187–188
Hall, David, 84–86
Halley, Edmund, 57–59, 76–79, 82
Haring, Clarence, 188
Harris, George, 289
Harris, Steven, 7
Harvard College, 157
Havana, 235–236, 238, 260
Hawthorne, Nathaniel, 256
Healers: *see* Medicine and healing; Hospitals; Physicians; Surgeons; Amerindian knowledge; African and Afro-American knowledge
Heat (atmospheric), 62, 70n29, 78, 114, 153, 156, 158–159, 161–166, 271, 285 (*see also* Climate; Humidity; Meteorology)
Hemispheric Difference, 14, 16, 60–62, 153, 158, 160, 162, 165–168, 316–317 (*see also* Dispute of the New World; Laws of nature; Universalism)

Hermetic tradition, 99, 103, 114, 179 (*see also* Magic; Occult knowledge)
Hernáiz, Ramón, 236
Hernández, Francisco, 109, 196n10, 232, 234
Herzo y Mendigaña, Lucas, 240
Hester, John, 114
Hillary, William, 157
Hippocrates, 107, 138
Hispaniola, 177–178, 181 (*see also* Saint Domingue; Santo Domingo)
Historia General (Sahagún, c.1570), 12
Historia Medicinal (Monardes, 1574), 109
Historia Natural (Acosta, 1590), 125n45
History of America (Robertson, 1777), 160
History of Jamaica (Long, 1774), 294
Holland, Samuel, 87–88
Holy Roman Empire, 184
Hooke, Robert, 8
Hospital of La Providence (Saint Domingue), 313–314
Hospital Real de Todos os Santos (Lisbon), 129
Hospitals, 85, 129, 186, 228, 243, 301, 313–314 (*see also* Medicine and healing; Physicians)
House of Trade: *see* Casa de la Contratación
Hudson's Bay, 85, 158
Hughes, Griffith, 300
Huguenots, 59
Humanism, 103, 105, 119, 183
Humboldt, Alexander von, 4–5, 65, 74, 219, 243–244, 260, 271–274, *273*
Hume, David, 162, 286
Humidity, 62–63, 70n29, 157–159, 164, 213, 216, 265, 271, 317 (*see also* Climate; Heat)
Humoralism, 14, 261, 285
Hunter, John, 160
Hunter, William, 265–266
Huygens, Christiaan, 58, 61, 63, 69n24
Hydrography, 12, 18, 31–46, 57, 73–92, *88*, 186–193, 313, 337 (*see also* Cartography)
Hypothesis, 54, 64

I

Imagination (Power of), 272, 315–316 (*see also* Enthusiasm; Charlantry)
Impressment, 75, 77
Improvement, 12, 17, 19, 159–166, 168–170, 241, 245, 260–261, 265, 271, 286, 340 (*see also* Agriculture)
Inca, 115, 117–118, 161
India, 154, 192, 260, 265, 296
Industrialization, 256–257, 276n2, 338
Innocent VIII (Pope), 99
Inoculation, 302–303, 320 (*see also* Medicine and healing; Physicians)
Inquisition, 99, 131
Instauratio Magna (Bacon, 1620), 2, 3, 287
Invention(s), 2, 6, 263–266
Islamic science, 107
Itinerancy, 204–206, 259–260, 262, 338

J

Jacobs, Harriet, 287–289
Jamaica, 156, 233, 263, 260, 265, 294–296, 298, 300–301
James II, 337–338
Jardin du Roi (Paris), 204, 207–208, 210, 231, 235, 238, 337
Jefferson, Thomas, 14, 162–163, 165–168, 170, 256, 269, 286, 300, 317, 340
Jefferys, Thomas, 77
Jesuits, 8, 13, 57, 60, 206, 214, 230, 265
Jonson, Ben, 114
Josselyn, John, 82
Juan, Jaime, 42, 196n10
Junta of Badajoz, 177, 192
Jurin, James, 157
Jussieu, Antoine-Laurent de, 204, 208, 231
Jussieu, Bernard de, 208, 219–220
Jussieu, Joseph de, 4, 18, 203–221, 222n12, 223n32, 337, 340

K

Kaási (Lángu leader), 297–298
Kalm, Pehr, 83, 161, 164
Kasum, 274–275
Kayashuta (Seneca chief), 263
Kelly, Edward, 114
Kew Gardens, 238

King, Boston, 298
King, Silvia, 297
Kinnersley, Ebenezer, 260–263
Kwasi (slave), 20, 290–291, *292,* 296

L

La Condamine, Charles-Marie de, 58, 212, 219, 230, 269, 271
La Coruña, 177
La Crenne, Verdun de, 313
La Hire, Philippe de, 61–63, 70n27
La Pérouse, Jean François de Galaup de, 10–11, 58, 238–239
La Rochelle, 39, 206, 208
Laboratories, 205, 258–259, 270
Laet, Johannes de, 133, *134,* 135
Lafayette, marquis de, 317
Lamo y Zúñiga, Joaquin de, 219
Lángu (Saramaka), 297
Latitude, 16, 32, 40–42, 44–45, 153, 191 (*see also* Longitude)
Latour, Bruno, 10–11, 228, 238–239, 257
Laws of nature, 7, 54, 158 (*see also* Universalism; Hemispheric difference; Natural philosophy)
Laws of navigation, 34–35, 39–40 (*see also* Navigation)
Le Cap Français (Saint Domingue), 313–316, 319–324
Le Havre, 320
Le Jau, Francis, 294
Le Rouge, George Louis, 89–90
Ledesma, Pedro de, 184
Leibniz, Gottfried Wilhelm, 1
León, Cieza de, 181
Léry, Jean de, 269
Letter Concerning Toleration (Locke, 1689), 54
Library Company of Philadelphia, 258
Libya, 78
Lightning, 83, 153, 158, 259, 262–263, 266, 274–275 (*see also* Climate; Meteorology; Lightning rods; Electricity; Kasum)
Lightning rods, 259, 261–266, *264* (*see also* Lightning; Electricity; Invention)
Ligon, Richard, 13
Lima, 19, 206, 209, 213–214, 217, 220, 228, 239, 242, 244–245

Lind, James, 159–160
Lining, John, 157
Linnaeus, Carolus, 12, 20, 137, 161, 290–291, 296, 298
Lisbon, 34, 38
Loaisa, Joffrey de, 177
Local knowledge, 7, 10, 12, 15–17, 19, 21, 32, 37, 40–41, 45, 53–56, 60–66, 74, 84, 89, 101–103, 108, 110–113, 116–120, 127, 131, 136, 145, 154, 165–166, 203–206, 209–211, 228–229, 238–240, 242, 244–245, 270–271, 275, 281–282, 324, 334, 341 (*see also* Center vs. periphery; Creole-American knowledge; Amerindian knowledge; African and Afro-American knowledge; Science; Science and empire; Metrology; Networks)
Locke, John, 54–56, 61, 65
Lodot, Jean, 322
Loja, 211–212, 242–243
London, 77, 84, 153, 157, 259–260, 262, 270–271, 283, 338
Long, Edward, 294, 298, 300–301
Longitude, 16, 32, 40–45, 57–58, 81 (*see also* Latitude)
López Ruiz, Sebastián, 243, 244
Louis XIII, 207
Louis XIV, 53, 57, 65 207
Louisbourg, 86
Louisiana, 235
Louvois, 338
Lozières, Baudry des, 260
Ludlam, Richard, 293
Lunar eclipses, 42–44, 58
Lusitano, Abraão Zacuto, 130
Luxury, 159, 168 (*see also* Civilization)
Lynch, Thomas, 298

M

Macandal (slave), 20, 295, 323–324
McCartney Embassy, 265
McClellan, James, 234, 318–319
Macclesfield, Lord, 262
Machine in the Garden (Marx, 1964), 256–257
Madrid, 19, 228–229, 231, 237–245
Magellan, Ferdinand, 36, 105, 190

358 • Index

Magic 5, 17, 20, 99–120, 137, 145–146, 178–179, 255, 257, 275, 283, 285–286, 293, 296–297, 299–301, 304–305, 322–324, 338–339, 342 (*see also* Occult knowledge; African and Afro-American knowledge; Amerindian knowledge; Witchcraft)
Magic lanterns, 263
Magnetic tubs, 311, 313–314, 319–321, 324 (*see also* Mesmerism)
Magnetism, 78, 191, 193
Magnus, Albertus, 100, 105, 107
Makandal, François: *see* Macandal
Malaspina, Alejandro, 4
Manila, 228, 260
Manoa del Dorado (Golden King), 112, 115 (*see also* El Dorado)
Mar del Norte: *see* Atlantic Ocean and Atlantic World
Mare Liberum Doctrine, 9
Marggraf, Georg, 133, *134,* 135, 137–138, 208
Marine (French Navy), 214, 313, 315, 319
Mariners, 1–2, 5, 7, 16, 31–46, 47n16, 48n23, 59, 73–92, 101, 128, 180–181, 184–194, 207, 295, 334, 337 (*see also* Navigation; Nautical instruments; Artisans; Secrecy vs. circulation)
Maroons, 283, 294–296, 301 (*see also* Rebellions; Slavery and slaves; African and Afro-American knowledge)
Martínez, Juan, 116
Martinique, 207–209, 230, 238
Marx, Leo, 256–257
Maryland, 153, 157, 260
Massachusetts, 157, 267
Matawais (Surinam maroons), 296
Mathematics, 6–7, 42, 76, 86, 180, 191–192, 267, 336
Mather, Cotton, 84, 302–303
Mathison, Gilbert, 286–287
Maurepas, Jean-Frédéric Phélypeau, comte de, 214
Mauricius, Jan, 291
Maurits, Johan, count of Nassau, 132–133, 135–136
Measurement, 4–5, 12, 16, 43–44, 55–56, 60–66, 69n22, 69n26, 91–92, 154, 157–158, 164, 214, 238, 271–272 (*see also* Metrology; Latour, Bruno)
Mechanical philosophy, 118, 338, 340
Medicine and healing, 6, 12, 18, 20, 104, 107–111, 113, 127–147, 156, 159, 177, 183, 207–212, 217–218, 225, 233, 236, 240, 242–244, 282, 294–305, 311–326, 336, 340–342 (*see also* Physicians; Botany; Botanical drugs; Natural history; Disease; Hospitals; Amerindian knowledge; African and Afro-American knowledge; Mesmerism; Surgeons)
Medina, Pedro de, 190–191, 193
Mendes, José Antônio, 140, 145–147
Merchants, 9, 11, 37, 59, 82, 85, 101, 106, 128, 179, 181, 189–190, 194, 207, 233, 260 (*see also* Commerce; Monopolies and monopoly trading companies)
Meridian (Prime), 9, 77, 84, 89
Mesmer, Franz Anton, 311–319, 324
Mesmerism, 6, 20, 295, 311–326, 332n59, 342
Mesmerism and the End of the Enlightenment (Darnton, 1968), 317–318
Mestizaje and mestizos, 5, 9, 101, 136, 140, 219, 226
Metamorphoses (Ovid), 117
Meteorology, 62, 81, 153–170, 205, 219 (*see also* Climate; Barometers; Measurement; Record-keeping)
Metrology (actor–network theory), 10–13, 16, 56, 60–66, 238–239, 249n57, 257–259, 266, 270–271, 275 (*see also* Discipline; Measurement; Networks; Latour, Bruno)
Mexico, 106, 117, 179, 183, 189, 226, 233, 260, 265 (*see also* New Spain)
Mexico City, 19, 228, 239, 242, 244
Microscopes, 8, 204
Midwives, 129, *139* (*see also* Women)
Millenarianism, 288, 339
Miller, Charles A., 167
Miller, Shannon, 112
Millet, Madame, 311
Milward, Edward, 304
Minas Gerais (Brazil), 17, 140–143, *141,* 146

Mineralogy and minerals, 18, 100, 104–105, 108–111, 113–115, 117–118, 125n45, 127, 140, 142–146, *143*, 210, 226, 234, 284, 293–294, 299, 313 (*see also* Mining; Gold)
Mining, 142, 184, 186 (*see also* Mineralogy and minerals; Gold)
Ministry of the Indies (Spain), 233
Missionaries, 8, 13, 109–110, 207, 214, 230, 265, 293, 295 (*see also* Jesuits)
Mociño, José, 239
Modernity, 6, 8, 10, 100, 274–275
Monardes, Nicolás, 12, 109–110, 116, 133, 137–138
Monceau, Duhamel de, 234
Monopolies and monopoly trading companies, 4, 34, 38, 59–60, 74, 85, 132, 183, 226, 230–231, 235, 242–243, 335, 342 (*see also* Commerce)
Montagu, Lady Mary Wortley, 302
Montesquieu, Charles-Louis de Secondat (Baron), 162
Montpellier, 207
Morange (Haitian planter), 314, 320
Moreau de Saint-Méry, 314, 319–321
Mourão, Simão Pinheiro, 130–131
Muisca Indians (Colombia), 117
Mungo (slave), 263
Münster, Sebastian, 104–105
Mutis, José Celestino, 231, 236, 239–244
Mysticism, 17, 78, 103, 106–107, 113–116, 119, 268–269

N

"Nanny" (maroon), 296
Nantes, 207
Nantucket, 82, 84, 87, 90
Napoleonic Wars, 77, 81
Narrative of a Five Years' Expedition (Stedman, 1796), 291, 341
Nassau: *see* Maurits, Johan, count of
Natural history, 60, 63, 100, 108, 128–135, *134*, 137–138, 154, 156, 158, 167, 186, 203, 220–221, 225, 232, 234–235, 269–271, 281–282, *284*, 289 (*see also* Botany; Collecting; Classification; Taxonomy; Cabinets of curiosities; Royal Natural History Cabinet of Madrid)
Natural philosophy, 4, 7, 12, 53–66, 76, 84–86, 106–108, 119, 157, 180, 203–204, 258, 263, 267, 335–336, 338, 341 (*see also* Alchemy; Astronomy; Chemistry; Gravity; Electricity; Experiment; Laws of nature; Scientific instruments; Universalism; *Naturphilosophie*)
Natural rights, 55
Natural theology, 257–258, 261, 293, 338 (*see also* Cosmology)
Naturphilosophie, 4, 271
Nautical Instruments, 15, 34, 36, 39, 58, 64, 188–193 (*see also* Navigation; Mariners; Hydrography; Scientific instruments; Astrolabes; Chronometers; Compass)
Navigation, 31–46, 76–92, 180, 187–194, 334 (*see also* Mariners; Nautical instruments; Hydrography; Latitude; Longitude)
Navigators: *see* Mariners
Neo-Platonism, 100, 103, 108, 178–179 (*see also* Occult knowledge)
Nerlich, Michael, 105–106
Netherlands (United Provinces), 9, 11, 23n13, 59–60, 69n24, 132, 229–232, 269–270, 335 (*see also* East India Company; West India Company; Brazil; Guiana; Surinam)
Networks, 6–13, 15–16, 18–19, 21, 56–60, 65–66, 81, 157, 163, 186, 188, 205, 218–220, 226–229, 234–237, 241–242, 257–258, 260–263, 281–283, 333–334, 341 (*see also* Science; Science and empire; Center vs. periphery; Commerce; Travel: commercial; Metrology; Local knowledge)
Neufchateau, François de, 322
New Atlantis (Bacon, 1627), 1–4, 6
New England, 82, 118, 153, 158, 165, 256, 260, 270, 288–289
New France, 9
New Granada, 19, 116, 196n10, 226, 231, 236, 239, 240, 242–244, 260, 270 (*see also* Colombia; Venezuela; Ecuador)
New Hampshire, 165, 169
New Historicism, 100, 102
New Jersey, 84, 161, 295

New Spain, 6, 42, 44, 109, 182–184, 196n10, 226, 229, 239, 243, 259 (*see also* Mexico)
New York (city), 87, 295
New York (province), 84, 260, 287
Newfoundland, 38, 87–88, 153
Newton, Isaac, 6–7, 14, 53–57, 63, 65, 71n37, 207, 257–258, 267, 316, 335–336
Newtonianism, 7, 14–15, 53–57, 257, 337–338
Nicholl, Charles, 117
Niderburg, Sigismund, 260
Nollet, Jean-Antoine, Abbé, 259, 262, 270
Northern Sea: *see* Atlantic Ocean and Atlantic World
Northwest Passage, 85–86, 115
Notes on the State of Virginia (Jefferson, 1787), 162

O

Obeah, 263, 265, 296–298, 300–301 (*see also* Vodou; Afro-American religions; African and Afro-American knowledge)
Observatoire (Paris), 58, 62
Occult knowledge, 99–120, 178–179, 293, 295, 311–326, 338–340 (*see also* Alchemy; Astrology; Hermetic tradition; Magic; Neo-Platonism; Obeah; Vodou; Witchcraft; Supernatural)
Oceanography, 92
Old March (slave doctor), 300
Oldenburg, Henry, 64, 298
Oliver, Peter, 267
Ondériz Ambrosio, Pedro, 44–45
Onesimus, 302
Ordás, Diego de, 117
Orellana, Franciso de, 118, 215–216
Orinoco River, 113, 116–117, 224–225
Orreries, 257
Oviedo, Gonzalo Fernández de, 108

P

Pacific Ocean, 4–5, 10, 58, 77, *81,* 85, 87, 154, 182, 226
Pagel, Walter, 107–108
Paine, Tom, 168

Palatine Meteorological Society, 164
Palau, Antonio, 236
Panama, 181–182, 206, 208–209
Papaw, James, 303
Paracelsus, 107–108, 114
Paradise, 17, 100, 106, 130, 132, 135, 138
Paris, 16, 18, 20, 57, 59, 64, 90, 165, 167, 204, 207, 213, 220, 238, 259, 271, 311–313, 316–319, 324, 342
Parliament (British), 11, 81, 85, 263
Pascual, Juan, 37–38
Pastoralism, 256
Pavón, José, 225, 231, 236–237, 239, 243–244
Peking, 265
Pendulum experiments, 16, 55–57, 60–66, 335–336
Penn, Thomas, 258
Pennsylvania, 85, 157, 161, 257–258
Pepper, 18, 177, 182, 211, 226, 232–235, 242
Pérez de Oliva, Juan, 180–181
Pernambuco, 132, 135
Persia, 260
Peru, 2, 48n21, 115–116, 118, 181–182, 212, 225, 229–231, 235, 242–243
"Peter the Doctor" (free black), 295
Petiver, James, 289, 291, 298
Philadelphia, 82, 84–85, 87, 161, 163–164, *256,* 258, 260, 262, 263, 317
Philip I, 189
Philip II, 8, 32, 109
Philippine Islands, 226, 229–230, 242
Physicians, 5, 108, 109, 128–130, 132–133, 140, 147, 157, 159–161, 163–164, 170, 209, 211, 228, 243, 270, 294–295, 298–301, 304, 311, 314–318, 325, 340 (*see also* Medicine and healing; Surgeons; Hospitals)
Physico-theology: *see* Natural theology
Picard, Jean, 58, 61, 63–64, 71n39
Pico della Mirandola, Giovanni, 99
Pillars of Hercules, 2, *3,* 4, 288
Piracy and pirates, 9, 33, 37–39, 59, 79, 81, 218, 304
Piso, Willem, 133–137, *134,* 208, 210, 298
"Pite": *see* Silkgrass
Pizarro, Gonzalo, 117–118, 215–216, 229

Pizarro, José García de Léon, 240
Planispheres, 40, 68n12
Plantations, 12, 20, 59, 230, 241, 262, 269–270, 283, 285–287, 289, 294, 296–299, 301, 303–304, 320–322, 324 (*see also* Slavery and slaves; Agriculture)
Pluchon, Pierre, 322–323
Plumier, Charles, 207, 208, 216
Poisons, 19–20, 105, 111, 135, 137, 283, 285–286, 293–298, 301, 303–305, 323–324 (*see also* African and Afro-American knowledge; Slavery and slaves; Botanical drugs)
Poor Richard's Almanack (Franklin), 82, 262
Postal services: British, 87–89; general, 213; Spanish, 241
Postcolonialism, 100
Pozo, Carlos del, 260
Preternatural, 102, 158 (*see also* Supernatural)
Priestley, Joseph, 164–165, 259, 263, 266
Prince, Mary, 300
Principia Mathematica (Newton, 1687, 1713, 1726), 53–57, 63, 336
Pringle, John, 159
Printing press, 239
Privateers: *see* Piracy and pirates
Protestantism, 9, 13–14, 54, 156, 158, 256, 263, 266, 304, 339–340 (*see also* Puritanism; Christianity; Catholicism)
Provinces and provincialism, 15, 19, 257–265, 267, 269 (*see also* Center vs. periphery; Empire; Creole-Americans)
Psychological distress (travelers'), 18, 111, 205, 215, 218, 220–221
Pueblo Indians, 9
Puerto Rico, 44, 233, 235–236
Puritanism, 158, 256, 339 (*see also* Protestantism)
Puységur, Antoine-Hyacinthe-Anne de Chastenet de, 312–316, 318–319, 321, 323–324
Puységur, Armand-Marc-Jacques de Chastenet, marquis de, 313

Q
Québec (Province), 157, 161
Quiteños, 219
Quito, 117, 205–207, 209–210, 212–214, 216–217, 231
Quixos, 214, 216

R
Race, 14–15, 166–167, 258, 261, 263, 283, 285–286, 203, 304 (*see also* Slavery and slaves; Amerindians)
Rackstrow, Benjamin, 263
Ralegh, Walter, 17, 103, 106, 110–120
Ramos, José Eusebio, 231
Ramsay, David, 164–165
Rathbun, Valentine, 261
Rationalism, 100, 108, 128, 137–138, 340
Ravenel, Harry, 300
Real Sociedad Económica de Madrid, 229
Réaumur, René-Antoine de, 270–271
Rebellions, 9, 19–21, 118, 211, 263, 265, 267, 283, 285, 288, 294–296, 324–325 (*see also* Revolution; American Revolution; Haitian Revolution; Maroons)
Record-keeping (meteorology), 154–157, 159, 163–164 (*see also* Climate; Meteorology; Measurement; Metrology)
Regourd, François, 234
Relativism, 340
Religious toleration, 54
Republic of Letters, 57, 316
Republicanism, 16–19, 168–169, 256–257, 261, 266–269
Revolution, 4, 15–16, 19, 21, 81, 258, 261, 265–269, 274–275, 288, 312, 317–318, 325, 341–342 (*see also* Rebellions; American Revolution; Haitian Revolution)
Ribault, Jean, 38
Ribeiro, Diego, 177–178, 194
Rice, 12
Richelieu, Armand Jean du Plessis de, Cardinal de, 207
Richer, Jean, 16, 57, 59–65
Rio, Martín del, 107
Rio de Janeiro, *141*, 143
Rio Grande do Norte, 137–138

362 • Index

Rittenhouse, David, 257
Robertson, William, 160–162, 166, 168, 170
Robie, Thomas, 157, 158
Robinson Crusoe (Defoe, 1719), 8, 79
Rodríguez de Campomanes, Pedro, 229
Rodríguez Fonseca, Juan, 189
Roemer, Ole, 61
Rome, 13
Rosicrucianism, 115
Royal Botanical Garden (Madrid), 228–229, 231–233, 236–239
Royal College of Physicians (London), 156
Royal Natural History Cabinet (Madrid), 225, 228, 234–235, 237
Royal Navy (British), 11, 75–76, 78–79, 82, 165
Royal Pharmacy (Madrid), 228, 231, 243
Royal Society (London), 13, 61, 78–79, 81, 84, 156–159, 180, 188, 261–262, 282, 287–288, 302–304, 312, 337
Royal Society of Medicine (Paris), 312, 316, 318–320
Ruiz, Hipólito, 225, 231, 236–237, 239, 243–244
Rush, Benjamin, 164–165

S

Sacrobosco, 193
Sahagún, Bernardino de, 12, 109
Sailors: *see* Mariners
Saint Domingue, 9, 20, 60, 206, 208–210, 230, 260, 262, 295, 311–326, 328n26, 329n30, 329n35, 332n59, 342 (*see also* Haitian Revolution)
St. John's (Antigua), 263
St. Petersburg, 259
Sainte Croix de la Montagne, 220
Sakhalin (Siberia), 238–239
Salem: *see* Witchcraft
San Juan Island, 181
San Marcos University, 214
Santiesteban, Miguel de, 242
Santo Domingo, 177, 179, 183, 235
São Francisco River, 136, *141*
São Paulo, *141,* 142
Saramaka River, 296
Sarmiento de Gamboa, Pedro, 116

Schaffer, Simon, 11–12
Schiebinger, Londa, 205
Scholasticism, 102, 105, 107–108, 119, 179, 288, 336, 339 (*see also* Aristotelianism)
Science, 6–15, 20–21, 53–56, 65–66, 100, 108, 179, 205, 245, 257, 275; Colonial, 60, 73–74; Historiography, 6–7, 10–15, 20–21, 65–66, 100–101, 119–120; Institutionalization of, 178, 186, 188–189, 194, 318–319, 333, 339, 342 (*see also* Science and empire; Science and the state; Local knowledge; Metrology; Networks; Scientific instruments; Scientific academies)
Science and empire, 10–15, 20–21, 73–74, 77–78, 83, 101, 114–115, 119, 128, 135–137, 147, 178–180, 185–186, 194, 205–207, 218–221, 225–230, 235, 237–245, 257–258, 261–265, 271, 275, 318–319, 333–342 (*see also* Science; Science and the state; Local knowledge; Metrology; Networks; Race)
Science and the state, 2–6, *3,* 8–11, 15–16, 18–19, 32–34, 57–60, 65–66, 77–78, 131, 181–188, 192, 194, 206–207, 226–230, 235, 242, 318–319, 333–342 (*see also* Science; Scientific Academies; Science and Empire; Networks; Empire)
Science in Action (Latour, 1987), 10–11
Scientia, 104, 107–108
Scientific academies, 11, 16, 20, 55–62, 64–65, 102, 107–108, 179, 188, 207–208, 221, 228, 235, 245, 318–319, 325, 333, 335–336 (*see also* Science; Science and the state)
Scientific instruments, 4–5, 13, 15–16, 36, 42, 56–66, 92, 154, 157, 177–178, 191, 193–194, 214, 313, 340–341 (*see also* Science; Discipline; Nautical instruments; Telescopes; Microscopes; Barometers; Thermometers; Electric machines)
Scientific revolution, 194
Seconds pendulum: *see* Pendulum experiments

Secrecy vs. circulation of knowledge, 15–17, 31–46, 48n21, 75–81, 314, 320–321, 334, 337 (*see also* Circulation of knowledge; Secrets of nature; Occult knowledge)
Secrets of nature, 99–120, 132, 136–137, 147, 283, 286, 297, 303, 305, 338–339 (*see also* Secrecy vs. circulation of knowledge; Occult knowledge)
Sebastian, King of Portugal, 132
Segovia, , 183
Selkirk, Alexander, 79
Senegal, 59, 62, 269
Senegal Company (French), 59
Senegambia, 62
Seniergues, 209, 211–212, 217–218
Sessé, Martín de, 239
Seven Years War, 81, 86, 89, 261
Seville, 11, 16, 33–37, 39–40, 44, 153, 177–179, 183, 189, 190–193
Shakespeare, William, 104
Shamanism: *see* Amerindian knowledge; African and Afro-American knowledge; African-American religions; Occult knowledge
Shape of the Earth debate, 55, 56–66, 207, 207, 221, 336 (*see also* Pendulum experiments; Astronomy)
Sharpe, John, 295
Sherard, William, 290
Shipwrecks, 38, 83, 104, 187, 217, 237, 290
Sigerist, Henry, 109
Silk, 63, 182
Silkgrass, 63
Silva, Nuño da, 37
Skepticism, 108
Slavery and Slaves, 4–5, 8–9, 12–14, 17, 19–21, 59, 128, 135, 140, 142–145, 166, 183, 206, 208–210, 219, 233, 235, 262–263, 265, 270, 281–305, *292*, 311–312, 320–326, 334, 340–342 (*see also* African and Afro-American knowledge; Afro-American religions; Free blacks; Maroons; Rebellions; Race; Plantations)
Sloane, Hans, 156, 287, 294, 297–298, 304, 309n21
Smith, Pamela, 106

Smith, Samuel Stanhope, 167, 170
Smith, Theophus, 285, 306n3
Smith, William, 257, 263–265
Société Royale des Sciences et Arts (Saint Domingue), 319, 325
Societies of Harmony, 314–315, 317, 319–320 (*see also* Mesmerism)
Society of Jesus: *see* Jesuits
Society for the Propagation of the Gospel (SPG), 293–295
Sonnini, Charles, 255, 257
South Carolina, 157, 159, 164, 287, 289, 293–294, 298, 300–301, 303
Specimens (natural), 10, 12, 18, 109, 136, 165, 204, 207–208, 212–217, 219–220, 225–228, 231, 234–237, 239, 241–245, 261, 282, *284*, 288, 290–291, 298, 305, 341–342 (*see also* Botany; Natural history)
Spice trade, 7, 230 (*see also* Cinnamon; Pepper)
Sprat, Thomas, 282, 306n2
Stamp Act (British, 1765), 86
Starbuck family, 82
Starrenburgh, John, 289
Stedman, John Gabriel, 291, 307n12, 341–342
Strait of Magellan, 37, 40
Sublime, 273–274, *273*
Sugar, 13, 59, 132, 136, 139, 262
Supernatural, 158, 263, 289, 295–296, 316, 319, 332n55 (*see also* Preternatural; Occult knowledge; Magic)
Superstition, 101, 110, 118, 137, 275, 296, 321, 338, 340
Surgeons, 17, 108, 129–130, 132, 138–147, *139*, *143*, 159, 209, 314–315, 320 (*see also* Medicine and healing; Physicians)
Surinam, 20, 269, 290–291, 296–298, 341
Surveying, 42–43, 77, 84, 87, 154 (*see also* Astronomy; Navigation; Latitude; Longitude)
Swaine, Charles, 85
Sydenham, Thomas, 302

T

Tacky's Rebellion (Jamaica), 263
Tafalla, Juan José, 237, 244

Tamanac Indians (New Granada), 272
Taxonomy, 7, 12–13, 69n26, 130, 216–217, 231, 243–244 (*see also* Classification; Botany; Natural history)
Tea, 18, 159, 226, 242, 260
Telescopes, 58, 64, 71n38
Temperate zone, 161–162–285 (*see also* Climate; Air quality)
Tempest, The (Shakespeare, c. 1611), 104
Tenerife, 59
Tenochtitlan, 117
Terrall, Mary, 217
Tetzel, John, 184
Thames, River (London), 262
Theoretical knowledge, 33, 56, 61–63, 105, 130, 138, 178–180, 186, 189–191, 193–194, 262, 313, 315, 334–335, 340
Thermometers, 62, 90–92, 157–158
Thevet, André, 208
Thouin, André, 235, 238–239
Timonius, Emanuel, 302
Tlatelolco, 109
Tobacco, 131
Todd, Anthony, 87, 90
Toderini, Giambattista, 261
Torrid zone, 62, 105, 130, 285 (*see also* Climate; Air quality; Heat; Humidity; Tropics)
Torture, 263, 265, 295
Toulouse, 183
Townshend duties (British, 1767), 86
Trade: *see* Commerce
Trade winds, 78, *80* (*see also* Hydrography)
Traders: *see* Merchants
Trading companies: *see* Monopolies and monopoly trading companies
Transits of Venus, 58, 72n43, 78
Travel: commercial, 2, 5, 16, 20, 58–59, 65, 82, 87, 105–106 (*see also* Expeditions; Travel narratives)
Travel narratives, 2, 4, 13, 19, 79, 85, 103, 105–106, 99–120, 154, 215–217, 229 (*see also* Expeditions; Discovery narratives; Travel: commercial; Travel questionnaires)
Travel questionnaires, 13, 42, 185, 196n10, 234 (*see also* Travel: commercial; Travel narratives; Discipline)
Treaty of Tordesillas, 9, 31, 34, 177
Trembley, Jean, 315, 321
Triangular trade, 8, 16
Trinidad, 116
Tropics, 55–56, 61–62, 113, 127–147, 159–160, 165, 270, 317 (*see also* Climate; Caribbean Islands; Heat; Humidity; Torrid zone)
Trust (epistemological), 13–14, 34, 45, 61, 64–65, 109–110, 128, 164, 283, 303–304, 319, 323 (*see also* Discipline)
Tryon, Thomas, 299
Turkey, 260, 302

U

Ulloa, Antonio de, 219, 230
Ulloa, Jorge Juan de, 230
United States, 156, 162–163, 168, 317
Universalism, 7–8, 11, 16, 53–56, 64, 115, 257, 275, 281, 291, 316–317, 319, 335, 338–339 (*see also* Laws of nature; Hemispheric difference; Natural philosophy)
Universidad de Mareantes, 41, 50n39
Universities, 41, 50n39, 133, 164, 178, 183, 188, 193, 207
University of Leyden, 133
University of Pennsylvania, 164
Uppsala, 283, 291
Ursúa, Pedro de, 118
Utility, 2, 12–13, 16, 41–42, 86, 136, 138, 146–147, 178–179, 186, 204, 208, 232, 234, 237, 245, 257, 259, 275, 336, 338, 341

V

Valdecazana, marquis de, 243
Valladolid (Yucatán), 183
Valle, Marqués del, 184
Varin, le Sieur, 59, 64
Velasco, Juan Lopez de, 43–44, *187*
Velasco, Luis de, 183–184
Venezuela, 182, 226, 265, 271
Vesey, Denmark, 295
Vespucci, Amerigo, 36, 105, 185, 189–190
Vespucci, Juan (Giovanni), 36, 201n58

Villaluenga, Juan José de, 241–242
Villarroel, Domingo de, 35, 37, 39, 40,
Virginia, 115, 153, 156–157, 162–163, 167, 260, 269, 294, 298, 300–301, 303
Virginia Company, 4
Vodou, 20, 295, 321–325, 332n59, 342 (*see also* African and Afro-American knowledge; Afro-American religions; Slavery and slaves; Mesmerism)
Volcanoes, 204–205, 210, 219
Volney, Constantin-François de, 157, 163, 168–169
Volta, Alessandro, 269–270, 272
Voltaire, François-Marie Arouet, 53, 65
Voyage to the Islands (Sloane, 1707–25), 287, 304

W

Walduck, John, 298
Walsh, John, 270
Wapisiana Indians (Arawak), 274
War of Jenkins's Ear (1739–44), 83
War of the Austrian Succession (1744–48), 83
Warfare, 81, 111, 159–160, 205, 245, 305, 313, 337
Waterspouts, 84, 86
Watson, William, 262–263
Weather: *see* Climate; Meteorology
Weather journals: *see* Record-keeping
Wedderburn, Alexander, 267
West, Benjamin, 266
West India Company (Dutch), 132

West Indies: *see* Caribbean Islands
Western Ocean: *see* Atlantic Ocean and Atlantic World
Whalers, 82, 84, 87
Whirlwinds, 84
Whitehead, Neil, 111–112
Willem V, Prince of Orange, 291
Williams, Hannah English, 289–290
Williams, Samuel, 163–165, 167–169
Williamson, Hugh, 161–162, 164
Winthrop, John, 157, 158
Witchcraft (Salem), 283, 288 (*see also* Occult knowledge; Magic; Obeah; Vodou; Afro-American religions)
Witnessing (scientific), 13, 109, 204–205, 262, 303, 334 (*see also* Discipline)
Women, 5, 75, 77, 285, 287–289, 294, 296, 298, 300, 304, 311–312, 330n42, 342 (*see also* Midwives)
Wonder(s), 12–13, 137, 158, 177, 263, 265, 341
Woodward, John, 302

Y

Yáñez Pinzón, Vincente, 189
Yucatán, 122n9, 235

Z

Zamorano, Rodrigo, 36, 39, 188
Zaruma, 215
Zauzo, Alonso de, 177, 181, 194
Zea, Francisco Antonio, 243–244
Zilsel, Edgar, 179